Creating Healthy Workplaces

Psychological and Behavioral Aspects of Risk Series

Series Editors: Professor Cary L. Cooper and Professor Ronald J. Burke

Risk management is an ongoing concern for modern organizations in terms of their finance, their people, their assets, their projects and their reputation. The majority of the processes and systems adopted are very financially oriented or fundamentally mechanistic; often better suited to codifying and recording risk, rather than understanding and working with it. Risk is fundamentally a human construct; how we perceive and manage it is dictated by our attitude, behavior and the environment or culture within which we work. Organizations that seek to mitigate, manage, transfer or exploit risk need to understand the psychological factors that dictate the response and behaviors of their employees, their high-flyers, their customers and their stakeholders.

This series, edited by two of the most influential writers and researchers on organizational behavior and human psychology explores the psychological and behavioral aspects of risk; the factors that:

- define our attitudes and response to risk;
- are important in understanding and managing "risk managers"; and
- dictate risky behavior in individuals at all levels.

Titles Currently in the Series Include:

Human Frailties
Wrong Choices on the Drive to Success
Edited by Ronald J. Burke, Suzy Fox and Cary L. Cooper

The Fulfilling Workplace
The Organization's Role in Achieving Individual and Organizational Health
Edited by Ronald J. Burke and Cary L. Cooper

Occupational Health and Safety
Edited by Ronald J. Burke, Sharon Clarke and Cary L. Cooper

Corporate Reputation
Managing Opportunities and Threats
Edited by Ronald J. Burke, Graeme Martin and Cary L. Cooper

New Directions in Organizational Psychology and Behavioral Medicine
Edited by Alexander-Stamatios Antoniou and Cary Cooper

Creating Healthy Workplaces

Stress Reduction, Improved Well-being, and Organizational Effectiveness

Edited by

CAROLINE BIRON
Laval University, Canada

RONALD J. BURKE
York University, Canada

and

CARY L. COOPER
Lancaster University, UK

GOWER

Gower Applied Business Research
Our programme provides leaders, practitioners, scholars and researchers with thought provoking, cutting edge books that combine conceptual insights, interdisciplinary rigour and practical relevance in key areas of business and management.

Published by
Gower Publishing Limited
Wey Court East
Union Road
Farnham
Surrey
GU9 7PT
England

Gower Publishing Company
110 Cherry Street
Suite 3-1
Burlington
VT 05401-3818
USA

www.gowerpublishing.com

Reprinted 2014

British Library Cataloguing in Publication Data
A catalogue record for this book is available from the British Library.

The Library of Congress has cataloged the printed edition as follows:
Biron, Caroline.
  Creating healthy workplaces : stress reduction, improved well-being, and organizational effectiveness / by Caroline Biron, Ronald J. Burke and Cary L. Cooper Gower.
     pages cm. -- (Psychological and behavioural aspects of risk)
  Includes bibliographical references and index.
  ISBN 978-1-4094-4310-0 (hbk) -- ISBN 978-1-4094-4311-7 (ebk) -- ISBN 978-1-4724-0240-0 (epub) 1. Work environment--Psychological aspects. 2. Job stress. 3. Industrial safety. 4. Organizational effectiveness. I. Title.
  HD7261.B57 2014
  658.3'12--dc23

                                                                                      2013026261

ISBN 978 1 4094 4310 0 (hbk)
ISBN 978 1 4094 4311 7 (ebk – PDF)
ISBN 978 1 4724 0240 0 (ebk – ePUB)

Printed in the United Kingdom by Henry Ling Limited, at the Dorset Press, Dorchester, DT1 1HD

Contents

List of Figures   *ix*
List of Tables   *xi*
About the Editors   *xiii*
About the Contributors   *xv*
Preface   *xxiii*
Acknowledgements   *xxv*

Chapter 1   Improving Individual and Organizational Health:
               Implementing and Learning from Interventions   1
               Ronald J. Burke

**PART 1   CREATING HEALTHY WORKPLACES: MODELS AND
               APPROACHES**

Chapter 2   The WHO Global Approach to Protecting and Promoting
               Health at Work   23
               Evelyn Kortum

Chapter 3   Work and the Dynamics of Development: An Integrated Model   37
               Antonella Delle Fave and Marta Bassi

Chapter 4   ACT: A Third Wave Behavioural-Cognitive Approach to
               Creating Healthy Workplaces   51
               Julie Ménard and Brent Beresford

**PART 2   WORK–LIFE BALANCE AND PHYSICAL HEALTH
               INTERVENTIONS**

Chapter 5   Improving Employee Health and Work–Life Balance:
               Developing and Validating a Coaching-based ABLE (Achieving
               Balance in Life and Employment) Programme   67
               *Arla Day, Lori Francis, Sonya Stevens, Joseph J. Hurrell, Jr. and
               Patrick McGrath*

Chapter 6   A Case Study in the Design, Implementation and Evaluation
               of a Worksite Wellness Programme for Reducing Cardiovascular Risks   91
               Douglas W. Roblin, Brandi E. Robinson and Stacey A. Benjamin

Chapter 7   Effects of Workplace-based Physical Exercise Interventions on
               Cost Associated with Sickness Absence and on Productivity   109
               Ulrica von Thiele Schwarz, Henna Hasson
               and Petra Lindfors

Chapter 8   Services Vouchers: Tools to Prevent Stress and Enhance Well-being   129
               Nathalie Renaudin

PART 3   CIVILITY, ENGAGEMENT AND PARTICIPATION

Chapter 9   Civility, Respect and Engagement in the Workplace (CREW):
               Creating Organizational Environments that Work for All   147
               Katerine Osatuke, Maureen Cash, Linda W. Belton and Sue R. Dyrenforth

Chapter 10   CREW as a Work Engagement Intervention   169
               Michael P. Leiter

Chapter 11   Participative Climate as a Key for Creating Healthy Workplaces   183
               Keiko Sakakibara Seki, Hirono Ishikawa and Yoshihiko Yamazaki

PART 4   LEADERSHIP INTERVENTIONS

Chapter 12   Transformational Leadership Training for Managers: Effects on
               Employee Well-being   205
               Meghan Donohoe and E. Kevin Kelloway

Chapter 13   How Positive Psychology and Appreciative Inquiry Can Help
               Leaders Create Healthy Workplaces   223
               Sarah Lewis

PART 5   IMPLEMENTING INTERVENTIONS

Chapter 14   Interventions to Prevent Mental Health Problems at Work:
               Facilitating and Hindering Factors   239
               Nathalie Jauvin, Renée Bourbonnais, Michel Vézina, Chantal
               Brisson and Sandrine Hegg-Deloye

Chapter 15   Implementation of an Organizational Intervention on Quality
               of Life at Work: Key Elements and Reflections   261
               Caroline Biron, France St-Hilaire and Jean-Pierre Brun

Chapter 16   Merging Occupational Health, Safety and Health Promotion
               with Lean: An Integrated Systems Approach (the LeanHealth
               Project)   281
               Terese Stenfors-Hayes, Henna Hasson, Hanna Augustsson, Helena
               Hvitfeldt Forsberg and Ulrica von Thiele Schwarz

Chapter 17   eHealth Interventions for Organizations: Potential Benefits
              and Implementation Challenges   299
              Henna Hasson, Ulrica von Thiele Schwarz, Karin Villaume
              and Dan Hasson

Conclusion: Positive vs. Stress Interventions – Does it Really Matter?   321
Caroline Biron

Index   *327*

List of Figures

2.1	The four avenues of influence	26
2.2	Leadership engagement	27
2.3	The WHO Healthy Workplace model and framework	28
2.4	The link between poverty and sub-standard working conditions in developing countries	31
2.5	Influences on health according to the WHO 'Healthy Workplace' model (Macdonald, 2011)	32
3.1	An integrated model of development	45
4.1	The Hexagon Model of Psychological Flexibility (Bond et al. 2006)	55
5.1	Impact of ABLE intervention on life satisfaction	82
5.2	Impact of ABLE on perceived stress	82
5.3	Impact of ABLE on negative mood	83
5.4	Impact of ABLE on hassles	83
6.1	Common sense model of health behaviour and health outcomes for a worksite wellness programme to reduce cardiovascular risks	97
8.1	Vouchers scheme principle	130
8.2	Food vouchers compared to other social policies (Bulgaria)	134
8.3	The cycle of poor nutrition and low national productivity	134
8.4	Annual cost of absenteeism to small businesses (per employee)	136
8.5	Work–life balance and childcare: The vicious circle	138
9.1	Decreased rates of informal and formal complaints to the VA Office of Resolution Management regarding Equal Employment Opportunities at a VA medical centre since it started participating in CREW	157
9.2	Illustration of connection between psychological safety and civility	160
9.3	Relationship of civility and psychological safety to an important organizational outcome: Costs incurred from sick leave usage	161
9.4	The thank you note from the second example	165
10.1	Profile of civility scores before and after CREW	177
15.1	Steps of an organizational stress intervention (Brun et al. 2009)	268
16.1	A Kaizen note, used at department level as part of the Kaizen work at the hospital in Enköping to document the improvement work process	286
17.1	Mean values on self-rated health for employees at four departments during 18 months	306
17.2	Employees' and leaders' ratings of expectations concerning the tool at baseline	310
17.3	Employees' and leaders' ratings of their leader's attitude and information sharing at baseline	311
17.4	Employees' and leaders' ratings of opportunities for participation in the implementation at baseline	311

List of Tables

2.1 Global financial and mental health impact of work-related stress 30

5.1 Study 1: Means, standard deviations and intercorrelations among the study variables 79

5.2 Study 1: means, standard deviations and intercorrelations among study variables 80

5.3 Study 1: Repeated measures means, standard deviations, and t-tests for outcomes 81

5.4 Study 2: Summary of the well-being means and standard errors for the ABLE treatment participants and the control group at Time 1 and Time 2 82

6.1 Worksite wellness programme content topics 96

6.2 Sample size and effect estimates for achieving statistical significance (alpha of 0.05, 80 per cent power) 100

6.3 Components of a successful worksite programme and design and implementation issues from the pilot cardiovascular risk reduction programme 102

6.4 Estimated worksite wellness programme costs per participant (amortized on 60 baseline participants) 103

7.1 A conceptual model delineating how physical exercise during work hours can affect productivity and sickness absence at the employee and workplace levels 118

9.1 Civility scale 153

9.2 CREW testimonials from employee participants (also displayed on CREW webpage, NCOD Internet site) 155

10.1 Correlations of engagement with social constructs 179

10.2 Multiple regression on engagement 2009 179

11.1 Demographic characteristics of participants (N=625) 190

11.2 Means and SDs of work demand and work control items 191

11.3 Range, number of items, means, SDs and means per items of the subscale of participative climate (N=625) 191

11.4 Range, means, SDs and correlations of the variables used in this study (N=625) 192

11.5 Standardized regression coefficients of work factors on fatigue, depression and work motivation (N=625) 192

12.1 Types of organizational leadership styles 208

14.1 Brief description of the projects: Common methods 242

14.2 Pertinent sources of information and main results of the intervention effects 244

14.3 Factors which facilitated or hindered the intervention process 246

15.1 Four critical factors for the success of interventions and concrete indicators characterizing these factors 273

15.2   Comments from participants 12 months after the start of the intervention   274
17.1   The content of the web-based tool   302
17.2   A programme theory of the web-based tool   308

About the Editors

Caroline Biron

Caroline Biron is an assistant professor in Occupational Health and Safety Management, Faculty of Business Administration, Laval University, Québec, Canada. Recipient of the Best Intervention Award at the 2011 Work, Stress, and Health Conference (American Psychological Association, National Institute for Occupational Safety & Health, Society of Occupational Health Psychology), her work was recognized as an outstanding evaluation of interventions in which researchers partner with industry and labour to prevent occupational injuries and illnesses and promote workplace safety and health. Her current research includes organizational stress interventions, the evaluation of process and context issues during interventions, the roles of line managers in implementing healthy management practices, and workplace presenteeism. She has published several journal articles and book chapters, and has presented her work in numerous international conferences.

Email: caroline.biron@fsa.ulaval.ca
Webpage: http://www4.fsa.ulaval.ca/cms/site/fsa/page26618.html

Ronald J. Burke

Ronald Burke is Emeritus Professor of Organizational Studies, Schulich School of Business, York University. His current research interests include voice in the workplace, women in management, work and well-being, work and family, the dark side of leadership, and interventions to improve individual and organizational health. He was the Founding Editor of the *Canadian Journal of Administrative Sciences* and currently serves on the editorial boards of a dozen journals. He was also Associate Dean – Research, head of both the Research and PhD committees and department head of Organizational Studies at Schulich.

Cary L. Cooper

Cary L. Cooper is a distinguished professor of Organizational Psychology and Health, Lancaster University Management School (UK). He is also Chair of the Academy of Social Sciences, Editor of the journal *Stress and Health* and was honoured by the Queen with Commander of the British Empire for his contribution to occupational health.

About the Contributors

Hanna Augustsson has a master's degree in public health and is a PhD student at the Medical Management Centre (MMC), Karolinska Institutet, Sweden. Her research concerns the implementation of complex interventions in health care organizations and how the implementation process affects the outcomes of such interventions.

Marta Bassi is Assistant Professor in General Psychology at the Medical School of the University of Milan, Italy. She has conducted research on the quality of experience in daily life, and on its implications for well-being promotion at the individual and community levels. She is founding member of the Italian Society of Positive Psychology, co-editor of its trimonthly newsletter, and member of the International Positive Psychology Association (IPPA).

Linda W. Belton, FACHE, joined the Veterans Health Administration National Center for Organization Development (NCOD) as Director in Organizational Health in 2008. Her primary interests are Civility, Respect and Engagement in the Workplace (CREW), patient-centred care, servant leadership and organizational transformation. Prior to her current position, she served as Director, Veterans Integrated Service Network (VISN) 11, Administrator of the Division of Care and Treatment Facilities, Department of Health and Social Services, State of Wisconsin, and Vice-president, Mercy Medical Center Oshkosh, Wisconsin. Ms. Belton holds a master's degree in administration and completed the programme for senior executives in government, Harvard University.

Stacey A. Benjamin, MA, MCHES, CWPC, is currently an independent wellness consultant and life coach. Ms. Benjamin previously was employed for six years as the Manager of Wellness and Health Promotion for Kaiser Permanente Georgia, the state's largest non-profit health plan. Prior to working at Kaiser Permanente, she worked as the Director of the Cancer Education and Risk Reduction Program for the State of Georgia. In her native New York City she served on the board of the Public Health Association of New York City (PHANYC) and was an adjunct instructor at Herbert H. Lehman College in the Health Services Department for eight years.

Brent Beresford is a doctoral candidate in industrial/organizational psychology at Université du Québec à Montréal (UQÀM) under the supervision of Dr Julie Ménard. He completed a Bachelor of Science in occupational therapy at McGill in 2002. Subsequently, while pursuing clinical work as an occupational therapist in various fields, he undertook part-time studies at Concordia University in order to obtain his Bachelor of Psychology. Brent practices mindfulness meditation and incorporates these techniques into his clinical treatments. His research interests include: work rehabilitation, musculoskeletal injuries, cognitive-behavioural therapies, stress reduction, mindfulness, multidisciplinary approaches in rehabilitation, and acceptance and commitment therapy (ACT).

Renée Bourbonnais has a PhD in epidemiology from University Paris V and a post-doc in occupational epidemiology. She is associate professor at the faculty of medicine, Laval University, Québec, Canada. Since 1983 she has been a member of an interdisciplinary research group that developed an original approach to the study of psychosocial work environment and health. Since 1992, she is also member of a research group on personal, organizational and social interrelations at work. She has studied the impact of restructuring in the health care system and realized evaluative intervention research aimed at reducing adverse psychosocial work factors.

Chantal Brisson is full professor at the Department of Social and Preventive Medicine of Laval University, researcher at the Population Health research unit of Laval University and director of a multidisciplinary research team on work and health since 1992. Her research interests include psychosocial stressors at work and health inequalities. Dr Brisson is internationally renowned for her work in occupational epidemiology and is specialized in the psychosocial determinants of cardiovascular diseases and mental health problems.

Jean-Pierre Brun is a full professor in the Management Department of the Faculty of Business Administration at Université Laval in Québec City. He has a PhD in Ergonomics and a master's in Industrial Sociology. He is the Director of the Chair in Occupational Health and Safety Management. His interests include several complementary questions, in particular management systems for occupational prevention, OHS in small- and medium-sized firms, stress and violence at work. He is also the author of numerous articles published in scientific and professional reviews. Jean-Pierre Brun likewise acts as a guest speaker and a consultant in OHS management and occupational mental health.

Maureen Cash, PhD, is a supervisory programme analyst for the Veterans Health Administration National Center for Organization Development (NCOD). She received her doctorate in clinical psychology from Michigan State University in June 1993. Dr Cash worked full-time at the Cincinnati VAMC as staff psychologist in the addiction treatment programme. She later became Director of Psychology Training, and took on the additional role of Chief of Psychology for the medical centre in March 2007. Dr Cash began organizational health work with NCOD in May 2010. She is currently providing national oversight for the VHA's Civility, Respect, and Engagement in the Workplace (CREW) initiative.

Arla Day is Canada Research Chair and Full Professor in Industrial/Organizational Psychology at Saint Mary's University in Halifax, Canada. Arla is a founding member of the CN Centre for Occupational Health and Safety, and she chairs the Nova Scotia Psychological Healthy Workplace Program. She is on the steering committee of APA's Psychology & Work Network, which oversees the Psychologically Healthy Workplace Program across the United States and Canada. Arla is an associate editor for the *Journal of Occupational Health Psychology*. She has received funding and authored articles and book chapters pertaining to healthy workplaces, work stress, employee well-being, and work–life balance.

Antonella Delle Fave, MD specialized in clinical psychology, is full professor of psychology at the Medical School, Università degli Studi di Milano, Italy. Her research interests concern eudaimonic well-being, optimal experience in daily life, and psychological selection, that is the long-term differential replication and cultivation of activities and competencies. Her studies have produced the largest international data bank available on these topics. She is involved in intervention projects aiming at resource implementation in conditions of disability and social maladjustment. She is past President of the International Positive Psychology Association (IPPA) and Editor in Chief of the *Journal of Happiness Studies*.

Meghan Donohoe is a PhD candidate at Saint Mary's University, in Halifax, Nova Scotia. She currently works as an internal organizational effectiveness practitioner at an international oil and gas company in Calgary, Alberta. Meghan is responsible for the succession planning and development of future senior leaders, and works to support managers in their team building and mentoring initiatives. She has served on numerous committees, including the Nova Scotia Psychologically Healthy Workplace Program, and the professional development committee for the Human Resources Association of Calgary.

Sue R. Dyrenforth, PhD, began her VA career in 1989 as a clinical psychologist in mental health. She quickly progressed to the Director in Substance Abuse Rehabilitation, then Assistant Chief of Psychology, and finally Director, Veterans Health Administration National Center for Organization Development (NCOD). She is active in workplace enrichment programmes: the Civility, Respect, Engagement in the Workplace (CREW) initiative, the All Employee Survey administration, analysis, and action planning. She provides executive coaching to VA leaders and maintains a private practice (psychotherapy, executive coaching, organizational consultation). She has been involved in psychological research, practice and education for over 30 years and consulted to organizations for more than 20 years.

Lori Francis has a PhD in industrial/organizational psychology from the University of Guelph. Lori is an associate professor in the Department of Psychology at Saint Mary's University in Halifax, Nova Scotia. She has broad research interests in occupational health psychology including work stress and aggression. She is a member of the CN Centre for Occupational Health and Safety and has served on the Board of Directors for the Nova Scotia Health Research Foundation.

Dan Hasson is an associate professor in public health. He is currently working at Karolinska Institutet and at the Stress Research Institute at Stockholm University, Stockholm, Sweden. His research mainly concerns web-based interventions for individual and organizational health promotion, stress management and interventions for systematic improvements of the psychosocial work environment.

Henna Hasson is an associate professor at the Medical Management Centre, Karolinska Institutet, Stockholm, Sweden. Her research concerns theoretical and measurement issues related to implementation of organization level interventions. Currently, she is working with integration of process and outcome data on workplace health-promotion interventions to investigate the significance of various implementation components for the intervention effects on employees' health and well-being.

Sandrine Hegg-Deloye is a PhD candidate (kinesiology, Laval University). Her research interests are directed towards occupational stress and its effects on physiological and mental health.

France St-Hilaire, PhD, is an assistant professor in human resources management, Faculty of Administrative Sciences, Sherbrooke University, Québec, Canada. During her doctorate, she worked on managers' skills and workers' skills linked to stress at Laval University in Canada. Her main current funded research aims to first design a model and then an intervention to develop managers' and workers' skills and behaviours to improve well-being and reduce stress. She also works on psychological risks specific to managers.

Joseph J. Hurrell, Jr. is an adjunct professor of psychology at St. Mary's University in Halifax, Nova Scotia and the current editor of the *Journal of Occupational Health Psychology*. He holds a BA and PhD in psychology from Miami University and an MA in clinical psychology from Xavier University. He was a researcher, for over 30 years, at the US National Institute for Occupational Safety and Health (NIOSH), has authored over 100 scientific publications and is internationally recognized for his work in the area of occupational stress. He is also a founding member of the Society for Occupational Health Psychology.

Helena Hvitfeldt Forsberg has a PhD in medical management with a focus on clinical informatics. Her thesis concerns informatics tools in the context of chronic care improvement focusing on patient participation and process simulation. In 2011 Helena was certified in coaching health care improvement teams, attending the Dartmouth Institute educational programme. She works at Karolinska Institutet, Sweden.

Hirono Ishikawa, PhD, is Associate Professor of Health Communication at School of Public Health, University of Tokyo, where she teaches and does research on interpersonal communication in healthcare. She received her PhD in social and behavioural science from Johns Hopkins University Bloomberg School of Public Health and the University of Tokyo Graduate School of Medicine. Her primary research interest is on physician–patient relationship and communication, health literacy, and communication skills training in medical education.

Nathalie Jauvin has a PhD in applied human sciences (Montréal University) and a post-doc at the Research Chair on Occupational Integration and the Psychosocial Environment Work (Laval University). She is researcher at CSSS de la Vieille Capitale and associate professor at the faculty of medicine (Laval University). Since 1998, she is member of RIPOST (Research Group on Personal, Organizational and Social Interrelations at Work). She is interested in mental health at work and in preventive interventions aimed at reducing adverse psychosocial work factors. She also developed an expertise on interpersonal violence and psychological harassment at work.

E. Kevin Kelloway is the Canada Research Chair in Occupational Health Psychology and Professor of Psychology at Saint Mary's University. A prolific researcher, he is a fellow of the Association for Psychological Science, the Canadian Psychological Association and the Society for Industrial/Organizational Psychology. His research focuses on topics such as leadership, occupational health psychology and workplace violence. He holds several editorial positions and actively consults to public and private sector organizations in these areas.

Evelyn Kortum holds a MSc from London University, Birkbeck College and a PhD in applied psychology from the University of Nottingham. Since 1989, she has been working for the World Health Organization across several programmes ranging from Human Resources, the Global Programme on AIDS, to Water Sanitation and Health, and since 2000 as a technical officer in the Occupational Health Programme. Areas of responsibility and scientific interest include work organization and psychosocial risks at work; developing and promoting the WHO healthy workplace approach as the global coordinator of the initiative; managing and coordinating a large international Network of Collaborating Centres for Occupational Health as the WHO Focal Point; maintaining other expert networks; and issuing occupational health publications and advocacy materials. Evelyn is a member of several professional societies including ICOH, the APA and the BPS. She has published papers about the WHO Global Plan of Action on Workers' Health, and in the area of healthy workplaces, psychosocial risks at work in developed and developing countries, and the work of the Global Network of Collaborating Centres in Occupational Health.

Michael P. Leiter is Professor of Psychology at Acadia University in Canada and Director of the Center for Organizational Research & Development that applies high quality research methods to human resource issues confronting organizations. He holds the Canada Research Chair in Occupational Health and Wellbeing at Acadia University. He is a registered psychologist in Nova Scotia, Canada. Dr Leiter has received ongoing research funding for 30 years from the Social Sciences and Humanities Research Council of Canada as well as from international foundations for his work on job burnout and work engagement. He is internationally renowned for his work on job burnout and work engagement.

Sarah Lewis is an associated fellow of the British Psychological Society and a principal member of the Association of Business Psychologists. Sarah is a consultant facilitator and managing director of Appreciating Change, a boutique change consultancy. All her work with organizations or individuals is informed by a social constructionist, and appreciative, view of organizations. She also has particular expertise in appreciative inquiry, systemic consultation, open space, world café, future search and positive psychology. She can be contacted at sarahlewis@appreciatingchange.co.uk and there is more information about services and workshops at www.appreciatingchange.co.uk.

Petra Lindfors is Associate Professor of Psychology, Senior Lecturer of Work and Organizational Psychology and Research Fellow at Stockholm University, Department of Psychology and board member of the International Society of Behavioral Medicine and the Swedish Society of Behavioral Medicine. She is also on the steering committee of Stockholm Stress Center. Her research focuses on how different factors in occupational and educational settings relate to psychobiological mechanisms, health and well-being in adults, children and adolescents, with a particular emphasis on gender and social position.

Patrick McGrath is a clinical psychologist who has extensive experience as an applied scientist and as a decision-maker. His research has focused on two areas, pain in children and the delivery of health care using technology. He is the editor in chief of the *Oxford Textbook of Paediatric Pain*. His Strongest Families programme is now delivered through the not-for-profit Strongest Families Institute. He is currently the Integrated Vice President Research and Innovation at the Capital District Health Authority and the IWK Health Centre. His academic appointment is as Professor of Psychology, Pediatrics and Psychiatry and Canada Research Chair at Dalhousie University.

Julie Menard is Professor of Psychology at Université du Quebec in Montreal, Canada. Her main current research interests are recovery from work and mindfulness-based stress management interventions. During her doctorate, she worked on authenticity and well-being at the University of Montreal in Canada. She also worked as a postdoctoral fellow at City University in London, United Kingdom.

Katerine Osatuke, PhD, is a supervisory health scientist and research director at the Veterans Health Administration National Center for Organization Development (NCOD). She provides data analytic support to management initiatives, conducts research studies, and participates in designing nationwide organizational intervention programmes. She has conceptual and empirical background and research interests in models of psychological change, including how change is defined, empirically measured and tracked through time. She is a licensed clinical psychologist experienced at working with diverse individuals and groups. She has co-authored 30 publications and over 100 national and international conference presentations on aspects of clinical and organizational change.

Nathalie Renaudin graduated in European legal and social studies (Master of European law and Master of Occidental Societies). While preparing a PhD, she was associate tutor at Paris VII University. She is now Public Affairs Director within Edenred which operates in 39 countries worldwide. Her 10-year work experience within Accor Services involved a technical knowledge of workplace health promotion where she designed and managed various projects and partnerships with governmental, non-governmental and international organizations, as well as with the private sector. She also contributed to a guide led by CSR Europe with best practices and tips for implementing a successful well-being strategy at work.

Brandi E. Robinson, MPH, is a Research Project Manager with the Center for Health Research/Southeast at Kaiser Permanente Georgia. Her research interests include health literacy, chronic disease epidemiology, racial/ethnic health disparities, adolescent health and community-based participatory research. Ms Robinson received her MPH with a concentration in epidemiology from the University of Alabama at Birmingham.

Douglas W. Roblin, PhD, is a senior research scientist with the Center for Health Research/Southeast at Kaiser Permanente Georgia and an adjunct assistant professor in health policy and management with the Rollins School of Public Health at Emory University. For nearly 30 years, he has been employed in various positions related to health services research, quality improvement and health care operations management. His research interests include health care organization and financing, social epidemiology and chronic disease management. Dr Roblin has an MA from the University of Chicago and a PhD from the University of Michigan, both in social anthropology.

Keiko Seki Sakakibara, MPH, is a PhD candidate of the University of Tokyo, Graduate School of Medicine. She received MPH from the University of Tokyo, School of Public Health. Prior to entering graduate school, she was with Fleishman-Hillard Japan, a global communication consulting firm, where she worked as a consultant of the change management communication team. Her primary research interest is on organizational communication and climate, mentoring at work and career stress.

Terese Stenfors-Hayes has a PhD in medical education and is assistant professor at Karolinska Institutet, Sweden. She is currently a postdoctoral fellow at the Centre for Health Education Scholarship at the University of British Columbia, Canada.

Sonya Stevens is a consultant with Knightsbridge Human Capital Solutions in Toronto, Ontario. She works in the Leadership Solutions practice where she contributes to the creation of healthy and productive organizations by enhancing the capacity of leaders. She has served on numerous boards and committees including the executive committee for the Nova Scotia Psychologically Healthy Workplace Program. Sonya has authored articles and delivered presentations on a variety of organizational psychology topics for both academic and industry audiences. She received her MSc and PhD in Industrial/Organizational Psychology from Saint Mary's University, in Halifax, Nova Scotia.

Ulrica von Thiele Schwarz is a psychologist and an associate professor in psychology. She is currently working at the Department of Psychology, Stockholm University and Medical Management Centre, Karolinska Institutet. Her main research theme is applying a behavioural perspective to interventions and continuous improvement within the work setting. She has studied interventions such as physical exercise, team work, integration of health promotion and Lean and leadership development, and outcomes such as psychosocial work environment, health, sickness absence costs, productivity and quality of care. Overall, the research is concerned with furthering the understanding of effective implementation and links between work conditions, health and productivity.

Michel Vézina has a specialty in community health and has been a tenured professor at Laval University in the social and preventive medicine department since 1983, and a consultant in workplace health at the Institut national de santé publique du Québec since 2000. He holds a master's degree in public health from Harvard University. He has released numerous scientific publications on the effects of the organization of work on mental and cardiovascular health, and on psychological harassment at work. His expertise mainly concerns the social and psychological impacts of work and strategies that can be implemented to prevent them.

Karin Villaume has a background in public health (Karolinska Institutet) and in psychology (Stockholm University). She has been involved in a large-scale organizational health promotion intervention project. The project involves a web-based system for individual and organizational health promotion and stress management. She will pursue her PhD at Karolinska Institutet in her project focusing on personality in relation to health promotion interventions and adherence.

Yoshihiko Yamazaki, PhD, is Professor, Faculty of Social Welfare, Nihon Fukushi University. He is former Chair of Department of Health Sociology, Graduate School of Medicine, the University of Tokyo. He received his PhD in health science from the University of Tokyo Graduate School of Medicine. His primary research interest is on salutogenesis, sense of coherence (SOC), and their application to health promotion in workplace, school and community.

Preface

This book aims to bring together a number of researchers and practitioners from several countries to contrast and compare interventions for stress and well-being. During the last decade, several types of interventions have emerged. Indeed, the recent developments in positive organizational scholarship (Cameron and Spreitzer 2012) raise questions regarding the "traditional" stress interventions which were either focused on reducing exposure to psychosocial risks (i.e. job demands, job control, social support, rewards) or on improving individuals' capacities to cope with the demands of the workplace (Cooper and Cartwright 1997, Giga et al. 2003). Although this book does not aim to resolve the fundamental differences that might exist between these two domains (Biron et al. 2012; Fineman 2006; Hackman 2009; Luthans and Avolio 2009), researchers and practitioners describe their strategies for reducing stress, and for improving individual and organizational health and well-being. We hope that their contributions will:

1. set concrete examples of best practices to create healthy workplaces;
2. promote theoretical developments in both the fields of stress prevention and positive organizational scholarship, by emphasizing the differences, similarities and intervention practices in each of these fields;
3. help practitioners develop, implement and evaluate these complex interventions.

References

Biron, C., Cooper, C.L. and Gibbs, P. 2012. Stress interventions vs positive interventions: Apples and oranges? In *Oxford Handbook of Positive Organizational Scholarship*, edited by K.S. Cameron and G.M. Spreitzer. New York: Oxford University Press, 938–52.

Cameron, K.S. and Spreitzer, G.M. 2012. *Oxford Handbook of Positive Organizational Scholarship*. New York: Oxford University Press.

Cooper, C.L. and Cartwright, S. 1997. An intervention strategy for workplace stress. *Journal of Psychosomatic Research*, 43(1), 7–16.

Fineman, S. 2006. On being positive: Concerns and counterpoints. *Academy of Management Review*, 31(2), 270–291.

Giga, S.I., Cooper, C.L. and Faragher, B. 2003. The development of a framework for a comprehensive approach to stress management interventions at work. *International Journal of Stress Management*, 10(4), 280–296.

Hackman, J.R. 2009. The perils of positivity. *Journal of Organizational Behavior*, 30, 309–19.

Luthans, F. and Avolio, B.J. 2009. Inquiry unplugged: Building on Hackman's potential perils of POB. *Journal of Organizational Behavior*, 30, 323–8.

Acknowledgements

We would first like to thank all the contributors who kindly agreed and to share their work and research experiences in this volume. We are very grateful for their hard work and the wonderful contributions they provided. We would also like to thank Maude Villeneuve, PhD candidate at Laval University, for her valuable contribution in editing this volume. We would also like to thank Marie-Esther Paradis, research associate, and Marie-Eve Caouette, Master's degree student for their precious help with the production process.

1

Improving Individual and Organizational Health: Implementing and Learning from Interventions

RONALD J. BURKE

Abstract

This introduction sets the stage for the chapters that follow. Factors that diminish or enhance levels of individual and organizational health are identified from the literature. A case is then made for the benefits of individual- and organizational-level interventions, although the evidence is mixed. Here we present a number of different interventions, efforts undertaken by researchers and organizations together, to improve the mental and physical health of employees through a number of different change initiatives. Some interventions target individuals and their attitudes and behaviours, others target workplace relationships, still others target work units and wider organizational features. Outcomes include various individual health and well-being outcomes such as levels of smoking, obesity, depression, elevated levels of blood pressure, accidents and workplace injuries, and absence and turnover. We identify factors associated with the success of these interventions (e.g. clear goals, top management commitment, employee participation) and indicate how interested individuals and organizations might proceed to develop worksite interventions on their own. These themes, and others, are addressed in more detail in the following contributions.

Introduction

This volume builds upon the already published *Risky Business* (Burke and Cooper 2010), *The Fulfilling Workplace: The Organization's Role in Achieving Individual and Organizational Health* (Burke and Cooper 2012), *New Directions in Organizational Psychology and Behavioral Medicine* (Antoniou and Cooper 2011), and *Improving Organizational Interventions for Stress and Well-being: Addressing Process and Context* (Biron et al. 2012). In *Risky Business*, we examined the effects of human frailties on individuals, their families and workplaces, and the effects of toxic organizations on employees and organizational performance. In *The Fulfilling Workplace: The Organization's Role in Achieving Individual and Organizational Health*, we focused on organizational initiatives that address the incidence and effects of human frailties and organizational toxicity. In *New Directions in Organizational Psychology and Behavioral Medicine*, the authors considered antecedents and consequences of workplace stress and ways to reduce potentially adverse effects. And, in *Improving Organizational Interventions for Stress and Well-being: Addressing Process and Context*, we made the case for more organizational-level intervention efforts and how these might be undertaken and evaluated.

Consider the following conclusions based on research studies about men and women in a variety of jobs:

- increasing violence against hospital-based nursing staff, primarily but not exclusively in emergency and psychiatric units;
- increasing levels of workplace incivility more generally (Leiter et al. 2011);
- recent mine disasters in the United States, Chile and China resulting in injuries and deaths;
- increases in obesity and being overweight among adults and children in many countries and their associated health-related risks;
- increasing incidence of workplace violence where disgruntled employees take out their complaints with attacks and shootings of those they hold responsible (Lester 2011);
- abusive supervision with corrosive effects on employees (Tepper 2000);
- abusive customers with negative effects on employees in the retail sector;
- doctors (usually physician interns) falling asleep during surgeries, patient interviews, and in a few cases while driving home from work (Burke 2010);
- Post-Traumatic Stress Disorder (PTSD) reported by first responders, police officers, fire fighters, and soldiers (Burke 2011);
- increasing levels of job insecurity among public sector employees as governments try to reduce their budget deficits;
- increases in stress associated with organizational restructurings, mergers and downsizings (Barling et al. 2005, Schabracq et al. 2003);
- the stresses associated with international travel, and the stresses experienced by expatriates working in new countries;
- about half the occupants of managerial and professional jobs are falling short in their performance (Hogan and Hogan 2001);
- a growing number of surveys of mangers and professionals have found that their levels of work stress has increased, and this has increased levels of psychological distress and diminished physical health (Cooper et al. 2009, Antoniou and Cooper 2005);
- more managers and professionals have reported working more hours in more intense jobs and taking less than their full allotment of vacation time (Hewlett and Luce 2006, Burke and Cooper 2008); this results in employees being fatigued, lacking in energy, indicating diminished well-being and prone to mistakes (Loehr and Schwartz 2003, Schwartz 2007, Seldman and Seldman 2008);
- increasing numbers of managers and professionals report work–family and family–work concerns (Drago 2007, Friedman and Greenhaus 2000);

- concerns have been raised about the character (greed, materialism, corruption) of managers and professionals given the increasing number of corporate scandals worldwide (Burke et al. 2011);
- increasing numbers of women who obtain business education, start their careers in large organizations, and then "opt out" (Maineiro and Sullivan 2006). Women continue to make little progress in career advancement in organizations worldwide (Davidson and Burke 2011, Wittenberg-Cox and Maitland 2008);
- most people are less ethical than they think they are. We have considerable knowledge as to the rationalizations they use to justify their unethical choices and why they are blind to their own and others' unethical choices and behaviours;
- toxic organizational environments that diminish employee well-being and limit organizational performance (Kusy and Holloway 2009, Lubit 2004).

Our Rationale for This Collection

Several factors have come together for us in the development of this collection. First, there is a body of evidence suggesting that both individual factors and workplace experiences are associated with individual mental and physical health concerns and organizational well-being. There is a vital need for more intervention work that addresses these concerns. Second, this distress spills over into families and society as a whole in terms of emotional, physical and financial tolls. Third, we know a lot about the factors associated with these adverse outcomes. Fourth, we are gaining an increased understanding of the actual benefits to individuals and their workplaces following interventions designed to address particular health and well-being outcomes, and how best to undertake intervention projects in the workplace (Bambra et al. 2009, Caulfield et al. 2004). We have come to know a lot about implementing successful change, but it is not easy or guaranteed (Briner and Reynolds 1999). Successful change is complex. It is time consuming and requires that some resources be made available to carry out the project successfully.

Fifth, there is a need to showcase interventions that have worked, what was successful, what challenges remain, and what lessons were learned from these applied interventions. Learning from men and women working in actual organizations has been shown to be more useful than getting the same information from academics. We see this volume as performing a critical "sharing" function. Sixth, interventions that have been successful in one setting cannot be transferred directly into other workplaces but need to be tailored to fit particular work settings. "Off the shelf" or "one size fits all" thinking is likely to produce actions that fall short (Biron et al. 2009). But these case studies provide examples of success, how this success was achieved, and what workplaces need to think about before undertaking similar – or different – interventions. Seventh, interventions, when undertaken, need to be monitored, maintained, invigorated to prevent "fade out," and evaluated. These processes indicate aspects of success and failure, and contribute to lessons learned for future efforts.

Interventions can be at different levels: individual, work unit, and total organization. Interventions can have different targets, including aspects of individual well-being such as satisfaction, work–family integration, depression, blood pressure, obesity, mental and physical heath, workplace relationships and interpersonal processes such as increasing workplace civility and reducing incidences of workplace bullying and abusive supervision, and organizational outcomes such as accidents and injuries, attendance and turnover, and violence against employees (e.g. nurses). Intervention content includes the following: education and training, data collection and feedback for employee problem solving and action planning, individual coaching and counseling, team building, identifying and changing workplace norms and rewards in ways that improve well-being and successful performance. There is a long history of organizational intervention covering at least the past 50 years. In the 1960s, organizational development (OD) interventions were undertaken to improve organizational performance (see Sashkin and Burke 1987, for a review, Argyris 1970), for example changes in the design of jobs (Davis and Cherns 1975, Hackman and Oldham 1976, 1980) and changes to organizational culture (Schein 1978). Unfortunately there has been relatively little effort made to conduct research studies on the intervention process and the success of organizational interventions since then. This situation is now slowly changing (see Biron et al. 2012).

In this collection, we assemble current thinking and research evidence relevant to enhancing the satisfaction and mental and physical health individuals achieve from all their life roles – the whole person – with particular attention to workplace experiences and their association with a wide range of health and well-being outcomes. Thus it is applied and practical. Managers and policy makers have indicated that they learn more from interventions that have been undertaken in actual workplaces and from the experiences of individuals working in these organizations.

WHAT DO WE MEAN BY WELL-BEING AND BY HEALTHY ORGANIZATIONS?

Most writers and researchers on individual well-being typically see well-being as including physical and psychological health. And health is not just the absence of illness but also a state of optimal functioning: individual and organizational health have both positive and negative aspects (Csikszentmihalyi 1990, 2003, Fredrickson 2003, Seligman 2002, Tetrick et al. 2012, Hoffman and Tetrick 2003). This also means that the development and implementation of organizational interventions must consider both positive and negative aspects of the work environment. Positive and negative aspects of the work environment often co-exist (Cameron 2007). The contributors in this volume share their experiences of various types of interventions at both the individual and organizational levels, with targets ranging from stress reduction to improved well-being. As Biron discusses in the concluding chapter of this book, we posit that creating healthy workplaces entails the implementation of sustainable interventions at the individual and organizational levels. Whether these interventions aim to improve well-being or to reduce stress, they all aim to create healthier workplaces.

Academics and practitioners have addressed the question of what competitive advantages make one organization more effective than another (Likert 1961, 1967, Peters and Waterman 1982, Lawler 2003, Katzenbach 2000, Sisodia et al. 2007). Lencioni (2012) offers organizational health as the key competitive advantage. Successful organizations are healthy organizations. For Lencioni, an organization is healthy when it is whole, consistent and complete, having a unified management, operations and culture. Healthy organizations support the development of satisfied, engaged and healthy employees. He identifies four steps to organizational health: building a cohesive top leadership team, creating a clear vision and strategy, clearly over-communicating this vision and strategy, and continually reinforcing this clarity.

An example of healthy organization is provided by Galt (2012) who describes an organizational-level intervention undertaken by Union Gas in Canada to turn around what senior executives concluded was a dysfunctional top team, following prodding by its parent company Houston Energy. The top team at Union Gas exhibited a very low level of trust. Working with an external consultant, top team members began by indicating the behaviours they thought each person should exhibit, continue exhibiting or stop. This process then moved down the ranks, helping managers clarify their roles and responsibilities, and how they could support others. Four priorities were emphasized following this exercise: strategy, financial performance, employee engagement and organizational culture. Since then Union Gas has met all performance targets and has been cited during 2010 and 2011 as one of Canada's top 100 employers. Unfortunately there are more unsuccessful efforts to bring about positive organization change than successful ones (By 2005, Kotter 1995). Balogun and Hope Hailey (2004) conclude that about 30 per cent of organizational change efforts are successful. Jauvin et al. (Chapter 14, this volume) provide answers as to why some of these change efforts fail to produce the intended results.

'TH AND FOR THIS VOLUME

There has been increased attention devoted to the field of occupational health psychology over the past decade both by professional associations and by the creation of academic and practitioner journals including intervention research projects (Quick and Tetrick 2011). The former would include the US based Society for Occupational Health Psychology and the American Psychological Association/National Institute for Occupational Safety and Health (NIOSH) partnership which has hosted a biannual work, stress and health conference. In addition North American business school conferences (Academy of Management, Administrative Sciences Association of Canada) and European psychology and management conferences (European Academy of Occupational Health Psychology, European Academy of Work and Organizational Psychology, British Academy of Management) have increasingly included symposia on work and health associations, with some attention to intervention projects.

Europe, in general, and Scandinavia in particular, is at the forefront in terms of intervention activities. An important by-product of this collection would be to make this work available to researchers and practitioners in countries that have not advanced as far in the application of current knowledge to address work and health issues. This interest is also reflected in the development of academic journals focusing on work and health, and on the role played by interventions. These include: the *Journal of Occupational Health Psychology*, *Work and Stress*, *Health Psychology*, the *Journal of Applied Behavioral Science*, and the *European Journal of Work and Organizational Psychology*, to name a few.

We see the content of this collection as being of interest to several audiences. These include government agencies such as NIOSH, currently active in supporting intervention work, government policy-makers tasked with understanding and addressing work and health issues, health-care providers and insurers working to reduce work and health difficulties and their associated costs, line managers interested in factors influencing the attitudes and performance of their employees, medical officers and human resource managers responsible for employee health and well-being and health and safety issues, and academic writers and researchers interested in applying their work to improve the work life of employees and their organizations.

Various levels of government (national and local) have become increasingly concerned about the escalating costs of health care, and given budgetary constraints are eager to examine potentially important avenues for increasing levels of employee health and well-being. This volume addresses important avenues for preventing employee ill health, thus lessening the burden on the health-care system. The content combines research evidence, self-assessment tools where relevant and practical implications.

A CHANGING ORGANIZATIONAL ENVIRONMENT

Organizations today are facing a number of new demands in order to remain competitive. These include increasing competition from developing countries (Brazil, India, China and Russia), increasing globalization, new technologies, more demanding consumers, changes in expectations of the workforce, the need to control costs, increasing rate and scope of change, increasing speed in market change, increasing importance of knowledge capital, changing demographics of the workforce, and a global search for talent.

Organizations will need to increase their effectiveness in order to meet these changes. This will increase the need to manage the health and performance of their workforces. Some countries (e.g. the UK, the Netherlands, several Scandinavian countries) have already introduced legislation to support such efforts. In other countries, organizations will have to undertake these efforts on their own initiative to remain successful.

A large and increasing body of research evidence has identified workplace stress as a major factor in employee well-being and health (Barling et al. 2005, Cartwright and Cooper 2009, Langan-Fox et al. 2007). There are massive financial costs associated with work-related illnesses and injuries (Hurrell 2005). We now have a reasonably good understanding of job, work and organizational factors that diminish employee well-being. These include role stressors, work schedules, poor leadership, harassment and discrimination, and the physical work environment, among others.

There is increasing realization that individual health contributes to organizational health and organizational health contributes to individual health (Burke and Cooper 2012, Quick et al. 1997, Tetrick et al. 2012, Lowe 2011, Lencioni 2012). Coming to grips with and addressing workplace demands increases both individual health and organizational effectiveness. Organizations are now more interested in managing psychosocial risks to employee health (e.g. job demands, abusive supervisors, job dissatisfactions) by planning and implementing interventions that enhance employee well-being (Biron et al. 2012). Employee health and well-being is a performance issue.

Lowe (2011) makes the case for the emerging integration of interest in employee health and health promotion, workplace changes to reduce stress and improve individual health and well-being, and organizational performance and success; areas that historically have been considered separately. This integration strengthens the link between people and organizational performance, between healthy people and healthy organizations, and between healthy organizations and benefits to all societal stakeholders. Lowe proposes looking at organizations through a health lens. Healthy people and healthy organizations have the same qualities (e.g. vigorous, engaged, resilient). Lowe identifies four building blocks of a healthy organization: a positive culture, inclusive leadership, a vibrant workplace, and inspired employees. Lowe then develops a healthy organization value chain in which culture and leadership lead to effective people practices, which in turn lead to a vibrant workplace, which in turn leads to inspired employees, which in turn lead to sustainable success in the form of value created for employees, customers, shareholders and communities. Thus organizational leaders need to think in more holistic/integrative terms and with a long-term horizon.

CORPORATE WELLNESS INITIATIVES

This thinking by corporate leaders has recently been reflected in the emergence of interest in corporate wellness initiatives (Report on Business 2012). A Harvard University study reported reduced absenteeism of 1.7 days per employee per year and a saving of $274.00 per employee. European studies report reduced absenteeism of 1.5 days per employee and benefits of $251.00 per employee. US human resources firm Towers Watson reported that firms having wellness programmes reported 11 per cent higher revenues per employee than industry peers, with less absenteeism, fewer disability claims, turnover and medical costs. These firms also indicated 28 per cent higher returns to shareholders as well.
~~ducted in 2010 involving a survey of 4,000 respondents~~

most Canadians are falling short here. This is due to willpower (61 per cent), a lack of time (46 per cent) and a lack of money (39 per cent). Organizations are therefore creating a culture supporting attitudinal and behavioural change.

A survey of 677 organizations in 2011, the Buffett National Wellness Survey found that 72 per cent of organizations offered at least one employee wellness programme such as employee assistance for family or legal matters, flu shots or nutrition counselling. Wellness programmes also make organizations employers of choice. In addition, wellness programmes are relevant to keeping an ageing workforce committed and productive. Wellness programmes are particularly relevant to the overall health of employees indicating potential risk factors.

The following factors have been found to be associated with effective wellness programmes (Isaac and Ratzan 2013):

1. Setting goals and measuring progress. These include: weight, blood pressure, cholesterol, overall healthy lifestyle habits, health satisfaction, weight loss, and hours spent exercising.
2. Taking a strategic, multi-pronged approach. Targeting attitudes and behaviours associated with key health outcomes.
3. Using incentives to encourage participants, offering rewards for taking part.
4. Senior management involvement. Senior managers need to promote, support and take part in the wellness offerings.
5. Creating a strong communications initiative to challenge participation and commitment. This typically involves the following: email reminders, an easy-to-access website, newsletters, banners, bulletin boards, information on goals and progress, and ongoing communication from senior management.

In the same vein, Dollard (2012) reports on "best practice principles" embraced in an intervention project with the Australian government. These practices included:

1. Support of top management through involvement and commitment.
2. Human resource management practices include leadership and management capability and accountability.
3. A positive work group climate – colleague and manager support.
4. Involvement of all layers of the organization.
5. Integration into the day-to-day operation of the organization.
6. Risk identification, risk assessment and implementation of risk controls at organizational and work unit levels.
7. Participation and consultation with employees, unions and health and safety representative – a bottom-up approach.

Interventions: Definitions and Types

Cartwright and Cooper (2005) view organizational interventions as actions undertaken by organizations aimed at managing exposure to work stressors and their negative consequences. Organizational interventions can be undertaken at the individual level as well as at the organization level.

Many writers on intervention draw a distinction between primary, secondary and tertiary levels of intervention. Primary interventions are focused on changing, reducing or eliminating environmental sources of stress in the workplace that result in the experience of stress. Examples would include changes in the design of jobs, improving communication, enhancing leadership skills, and changes to the organizational structure itself. Secondary interventions focus on increasing the personal resources of employees so they are better able to cope with the workplace stressors they experience. Examples include stress management programmes. Tertiary interventions focus on treating employees who have developed serious stress-related emotional and/or physical health problems. Examples would include counselling, employee assistance programmes. This dimension focuses on when to intervene, a time dimension. Individual-level interventions would typically be secondary and tertiary; organizational-level interventions would typically be primary and tertiary.

Barling et al. (2005) also distinguish three levels of population inclusiveness in a taxonomy of work stress interventions. Universal interventions target all employees, selective interventions target vulnerable subgroups, and indicated interventions target individuals who are experiencing adverse outcomes. This dimension identifies who will be targeted for intervention – all employees, or a subset of those at higher risk or already in distress. Cartwright and Cooper (2005) identify the following individual-level interventions: health promotion activities, stress management training programmes, relaxation and meditation techniques, exercise, cognitive behavioural therapy, and counselling. Hurrell (2005) places primary organization-level interventions into two categories: psychosocial and socio-technical. Psychosocial interventions include participatory action research, job redesign, and changing the behaviours of managers and supervisors. Socio-technical interventions reviewed include: workload interventions, work scheduling interventions, and work process and procedure interventions. He includes, as tertiary prevention organizational level interventions, medical care, psychological counselling and therapy, and post-traumatic stress interventions. Passmore and Anagnos

(2009) show how organizational coaching and mentoring can reduce levels of stress of focal individuals and others around them.

There are more individual-level interventions than organization-level interventions since the former are easier to carry out and require fewer resources (Giga et al. 2003). Organizational-level interventions are difficult to conduct, require considerable cooperation from organizations, need considerable resources – time, financial and skill – and organizational needs may be in conflict with researcher needs. Yet, a review by Lamontagne et al. (2007) suggests that positive effects, such as stress reduction, are more likely when interventions integrate primary, secondary and tertiary levels. As mentioned, organizational interventions are rather difficult to conduct and to evaluate. As Biron et al. (2012) highlight, the lack of attention to the process and context of the intervention can explain why so many interventions have failed to produce the intended results. In the next section, we consider some of these process and contextual issues influencing the outcome of interventions.

behaviour and organizational psychology (Anderson 2007). It is not surprising that it exists in the undertaking of workplace interventions to improve individual and organizational health and performance managers, and the consultants that often advise them, are less interested and capable of undertaking sound evaluation studies than are some academics and researchers. Interventions are often implemented without evaluation and without public discussions of their merits. Some published evaluation studies of organizational interventions are limited because of research design flaws. Flaws include the absence of a theoretical model underpinning the intervention, lack of a control group, no a priori evaluation of risks, lack of a prospective design, a too short follow-up period after the implementation of the intervention, and failure to provide qualitative information on the context and implementation of the intervention (Semmer 2006). One of the flaws that has been flagged quite frequently is the lack of attention paid to process and contextual factors influencing the intervention (Cox et al. 2007, Griffiths 1999, Nielsen et al. 2010). In this section, we describe recent developments on this topic of intervention process and context.

PROCESS EVALUATION MODEL

Randall and Nielsen (2012) propose a model which includes three components to evaluate the process and context of interventions:

1. context;
2. intervention design and implementation; and
3. participants' mental models.

Context refers to both the organizational context (e.g. pre-intervention healthiness, fit between organizational culture and the intervention, ceiling effects that might prevent further improvements) and the discrete context (e.g. the context surrounding the intervention that could influence it, such as organizational change or conflicting priorities). The second and third components of their model refer to process factors. These include how the intervention was carried out, who did what, where and why and to what effect (Cox et al. 2007: 353). Process factors apply to all phases of an intervention. The second component of their model is the intervention design and implementation strategy, which refers to issues such as who initiated the intervention and for what motives, whether the intervention activities targeted the problems of the workplace, what were the roles of stakeholders such as middle managers, senior management, participants, external consultants, what exactly was the substance and nature of changes, whether interventions reached their target and were communicated properly. Finally, Randall and Nielsen suggest evaluating a third component, participants' mental models which refers to individuals' perceptions and appraisal of the intervention. Ideally, interventions need to be assessed at their development, their implementation, and in terms of their effectiveness (Goldenhar et al. 2001, Griffiths 1999). By evaluating these process and contextual issues, occupational health interventions can be improved since early adjustments can be made. Moreover, this type of evaluation allows for more meaningful interpretations as to why and how the intervention produced intended or non-intended effects.

DOES THE INTERVENTION FIT?

Well-designed and well-planned interventions sometimes fail to produce intended results. One of the reasons why, as posited by Randall and Nielsen (2012), is that intervention outcomes are largely affected by the degree of fit between the intervention and individual employees (person–intervention fit), and the degree of fit between the intervention and the environmental context in which the intervention is introduced (environment–intervention fit). They state that the degree of fit between "the active ingredients of the intervention and the required remedy for a specific presenting problem in a specific context" shapes the intervention process and then the intervention outcomes. The absence of expected outcomes, or only limited impact, result from inconsistent fit across employees and contexts.

Randall and Nielsen (2012) refer to environment–intervention fit to describe the appropriateness of the intervention in its setting. Good fit exists in a contest that supports the initial implementation of the intervention and does not interfere with or limit ongoing intervention activities. Poor fit occurs when there are changes in the top team or organizational setting (reorganization merger, downsizing) or changes in pre-intervention job design (limited opportunities for employee input, increased job demands). Person–intervention fit deals with the potential benefits the intervention has for employees in addressing a concern. Randall and Nielsen offer two processes that are helpful in determining levels of fit: first, a complete pre-intervention assessment of presenting stress-health-related problems in context. Second, intervention planning activities that involve key stakeholders in design, implementation and evaluation (participative action research).

INTERVENTION DEVELOPMENT, IMPLEMENTATION AND EVALUATION

Noblet and LaMontagne (2009) offer a seven-stage framework for undertaking an intervention project:

1. Gaining management support – all levels of management, from top to bottom, must be supportive if the intervention is to have any chance of being successful.
2. Forming a coordinating/steering team – individuals with knowledge, skill and reputation need to be identified to spearhead and champion the investigation into organizational health.
3. Conducting a needs assessment and issue determination – identifying issues affecting employee health and well-being, the seriousness of these effects, and factors likely to be sources of them.
4. Setting goals and priorities – identifying issues that need to be addressed and those having higher priority and then using this to establish intervention goals.
   Developing interventions and action plans – the interventions that will be undertaken specific steps needed to move them forward, the timing of them and who is

data collection.

These seven steps need to be integrated, and in some cases, though presented in a linear fashion, the steering/coordinating team may have to retrace some steps.

LaMontagne et al. (2012) identify five barriers at the process level to organizational-level intervention, building on their seven stages described above, along with suggestions on how to address them. They consider both internal barriers and external barriers. Internal barriers include gaining management support, articulating the need for comprehensive worker- and work-direct interventions, establishing participatory processes, early detection of opportunities and threats in the implementation of the intervention. External barriers refer to the broader labour market, the local or international economy, national cultures, political conditions, regulatory and other policy influences.

Although organizational-level interventions are known to be complex to implement, there is a growing body of research demonstrating that when they are properly developed and implemented, they can be relevant to reduce exposure to stressors and improve well-being. However, issues remain regarding their evaluation. The following section describes some of these issues in evaluating stress and well-being interventions.

Issues in Designing and Undertaking Organizational Interventions

A NEED TO CONSIDER BOTH PROCESS AND CONTEXT

Process refers to how the intervention was carried out – who did what, when, why and to what effect. Context refers to the larger job, workplace and organizational environment in which the intervention is carried out (Kompier 2004). There is a need to collect both qualitative and quantitative data to monitor the progress of the intervention as well as evaluate the success of the intervention (Semmer 2006). There is a need to consider both the reduction of workplace stressors and the reduction of adverse health consequences along with a focus on positive outcomes such as flourishing, courage and engagement. There is a need for an integrated approach to addressing psychosocial issues in the workplace that include both positive and negative job characteristics, individual differences, and personal lifestyle behaviours – both occupational and non-occupational factors.

ADDRESSING THE NEGATIVE, ENHANCING THE POSITIVE, OR BOTH?

Much of the early writing on employee well-being focused on the experience of stress and a number of potentially negative outcomes associated with the stress experience. These included job dissatisfaction depression, psychosomatic symptoms, intent to quit, absenteeism and counterproductive work behaviours. More recently attention has been focused on positive individual and organizational outcomes at work such as joy, virtues, generosity, psychological capital and peak performance (Cameron et al. 2003). Stress interventions, at both individual and organizational levels, lessen negative outcomes, positive interventions at both individual and organizational levels increase positive outcomes. Biron et al. (2012) review these approaches, providing examples of both types of interventions. Although these two approaches can be conceptually distinguished, both want to bring about changes in the workplace that improve individual and organizational well-being. They are complementary, each has potential advantages and disadvantages, and they can be combined.

INDIVIDUAL DIFFERENCES IN RESPONSE TO CHANGE

Organizational changes can make life difficult for some or many employees. Mergers, acquisitions and downsizings have been shown to have adverse impacts on employee well-being in addition to failing to meet their stated financial objects in at least half and perhaps more of their undertakings. In fact one can see a workplace intervention as increasing levels of stress for some employees. Individuals react differently to sources of stress and engage in different coping responses (Cartwright and Whatmore 2005). Organizational changes more broadly, and interventions to improve employee and organizational well-being, affect individual employees on a personal level. These personal impacts are likely to influence their perceptions of and reactions to change content and change processes. Tvedt and Saksvik (2012) suggest at least three individual difference factors, or personal resources, are predictive of readiness, openness and resilience to change: a disposition for life changes – being open to and experiencing many life changes; optimism – generally expecting good things to happen; and self-efficacy – believing one is able to successfully deal with the changes one meets. They describe a three-step "therapeutic instrument" for dealing with individual resistance to change. It begins with assessing an individual's resistance to change as well as levels of resilience The second step involves determining the impact of the change on the individual as he/she sees it. The third step involves "therapy" for individuals scoring high on resistance and negative impact of the change, a kind of counselling and coaching session in which the individual explores beliefs about what the change would entail and the impact it would have and what has actually happened so far.

LEARNING FROM FAILURES

Implementing an organizational-level intervention is a complex and difficult undertaking. Failures are as likely to occur as successes; but failures are rarely written or talked about. We can learn a lot about effective intervention from analysing why they fail. Biron et al. (2010) describe a failed intervention carried out in a unit (205 employees) of a much larger organization in the UK. They used both qualitative data (field observations and interview notes) and quantitative data (surveys). This department, following UK legislation requiring an assessment of risk factors associated with employee health and being, was required to take part. The department was given a structured package to

Changing the work itself and the work environment is more effective than changing people in improving employee health and organizational performance. Changing the work itself and the workplace addresses the sources of stress whereas changing people deals with the consequences of exposure to stress. There is a need to better "fit" the intervention and its implementation to the workplace and organizational context. Randall and Nielsen (2012) define it "as the degree of fit between the active ingredients of the intervention and the required remedy for a specific presenting problem in a specific context." Fit will influence both the intervention process and the intervention outcomes. We can learn a lot from interventions that fail (Biron et al. 2010).

CHARACTERISTICS OF SUCCESSFUL ORGANIZATIONAL-LEVEL INTERVENTIONS

Karanika-Murray et al. (2012) identified several elements of successful organizational-level interventions. These were:

1. They combine individual and organizational-level actions with a primary, secondary and tertiary focus across all organizational levels – they are integrative and comprehensive.
2. They use a multi-disciplinary approach – micro and macro – and employ multi-modal interventions with the commitment of major stakeholders.
3. They use a participatory approach.
4. They employ a well-developed implementation framework.
5. They assess the needs of the organization and its employees; solutions are tailored to the specific organizational context.
6. They integrate solutions into day-to-day business practices.
7. They incorporate external factors in supporting and driving change at the organizational level.

8. They employ clear concepts and definitions to support both implementation and evaluation.

WHAT WE KNOW ABOUT ORGANIZATIONAL INTERVENTIONS

1. Interventions should integrate more than one level – the individual level and the organizational level.
2. Interventions can target both skills and processes.
3. They are complex and should target both short and long term;
4. Resources are needed – both time and money.
5. No one size fits all. There is no approach that will work in every instance.
6. Leadership commitment is vital, and so is employee participation.
7. There is a need to take action and learn from these actions.
8. A "burning platform" may be helpful in increasing motivation to act.
9. Outside assistance is often valuable.
10. There is a need to collect "data" from people on site to both understand the issues and enlist commitment and support as well as ideas.
11. There is a need to build skills, cooperation and trust within the intervention steering team.
12. There is a need to collect both subjective and objective data to best understand their implementation, progress and consequences.
13. There is a need for more evaluation of workplace interventions. There is a gap between research and practice. It is difficult to undertake high quality evaluation research here but recent developments brings examples of better research designs in these evaluations.

The currently available data on the benefits of organizational interventions to reduce workplace stress and its negative consequences and improve organizational performance show mixed results but this could be because researchers were mainly concerned with effectiveness questions such as "did the intervention work?," instead of process and context factors such as "how and why did the intervention work?."

COUNTRY LEGISLATION SUPPORTING WORKPLACE INTERVENTIONS

Some countries (e.g. the UK, the Netherlands, Sweden and Norway) have passed legislation requiring the audit of workplaces for risks to employee health and well-being (Leka et al. 2011, Mackay et al. 2012, Daniels et al. 2012, Weyman 2012). This legislation supports prevention of risk and promotion of health. Interventions then follow from required audits, and evaluation of interventions is endorsed to improve their effectiveness. Having such legislation, however, is no guarantee that workplace interventions will always be carried out as suggested and be successful.

Conclusion

There are hundreds of academic journals and thousands of books that address content on improving individual and organizational performance. We know that transformational

leadership is associated with higher levels of employee ⋯
2010), that Total Quality Management (TQM) is associated ⋯
service (Garcia et al. 2012), that higher levels of civility ⋯
with higher levels of employee psychological health and lo⋯
et al. 2001), and that individuals scoring higher on self-ef⋯
burned out and perform at higher levels (Bandura 1997). It ⋯
of this wisdom in the form of individual- and organizatio⋯
calls for increasing researcher and practitioner collaboratio⋯
the good news is that this is increasingly being done.

Acknowledgements

Preparation of this chapter was supported in part by York Un⋯

References

Anderson, N. 2007. The practitioner–researcher divide revisited: Strat⋯ of two psychologies. *Journal of Occupational & Organizational Psy*⋯

Antoniou, A.-S. and Cooper, C.L. 2005. *Research Companion to Organizational Health Psychology*. Cheltenham: Edward Elgar.

Antoniou, A.-S. and Cooper, C.L. 2011. *New Directions in Organizational Psychology and Behavioral Medicine*. Farnham: Gower Publishing.

Argyris, C. 1970. *Intervention Theory and Method: A Behavioral Science View*. Reading, MA: Addison-Wesley.

Balogun, J. and Hope Hailey, V. 2004. *Exploring Strategic Change*. 2nd editon. London: Prentice Hall.

Bambra, C., Gibson, M., Snowden, A.J., Wright, K., Whitehead, M. and Petticrew, M. 2009. Working for health? Evidence from systematic reviews on the effects on health and health inequalities of organizational changes to the psychosocial work environment. *Preventative Medicine*, 43, 454–61.

Bandura, A. 1997. *Self-efficacy: The Exercise of Control*. New York, NY: Freeman and Co.

Barling, J., Kelloway, K. and Frone, M. 2005. *Handbook of Work Stress*. Thousand Oaks, CA: Sage Publications.

Biron, C., Cooper, C.L. and Bond, F.W. 2009. Mediators and moderators of organizational interventions to prevent occupational stress. In *The Oxford Handbook of Organizational Well-being*, edited by S. Cartwright and C.L. Cooper. Oxford: Oxford University Press, 441–65.

Biron, C., Cooper, C.L. and Gibbs, P. 2011. Stress interventions versus positive interventions: Apples and oranges? In *Oxford Handbook of Positive Organizational Scholarship*, edited by K.S. Cameron and G.M. Spreitzer. New York, NY: Oxford University Press, 938–50.

Biron, C., Gatrell, C. and Cooper, C.L. 2010. Autopsy of a failure: Evaluating process and contextual issues in an organizational-level work stress intervention. *International Journal of Stress Management*, 17, 135–58.

Biron, C., Karanika-Murray, M. and Cooper, C.L. 2012. *Organizational Stress and Well-being Interventions: Addressing Process and Context*. London: Routledge.

Briner, R.B. and Reynolds, S. 1999. The costs, benefits, and limitations of organizational-level stress interventions. *Journal of Organizational Behavior*, 20, 647–64.

Burke, R.J. 2010. Work hours, work intensity and work addiction: Weighing the costs. In *Risky Business: Psychological, Physical and Financial Costs of High Risk Behavior in Organizations*, edited by R.J. Burke and C.L. Cooper. Farnham: Gower Publishing, 62–106.

Burke, R.J. 2011. Tragic duty. In *Risky Business: Psychological, Physical and Financial Costs of High Risk Behavior in Organizations*, edited by R.J. Burke and C.L. Cooper. Farnham: Gower Publishing, 287–322.

Burke, R.J. and Cooper, C.L. 2008. *The Long Work Hours Culture: Causes, Consequences and Choices*. Bingley: Emerald Publishing.

Burke, R.J. and Cooper, C.L. 2010. *Risky Business*. Farnham: Gower Publishing.

Burke, R.J. and Cooper C.L. 2012. *The Fulfilling Workplace: The Organization's Role in Achieving Individual and Organizational Health*. Farnham: Gower Publishing.

Burke, R.J., Tomlinson, E.C. and Cooper, C.L. 2011. *Crime and Corruption in Organizations: Why it Occurs and What to Do About it*. Farnham: Gower Publishing.

By, R.T. 2005. Organizational change management: A critical review. *Journal of Change Management*, 5, 369–80.

Cameron, K.S. 2007. Forgiveness in organizations. In *Positive Organizational Behavior*, edited by D.L. Nelson and C.L. Cooper. Thousand Oaks, CA: Sage Publications, 129–42.

Cameron, K.S., Dutton, J.E. and Quinn, R.E. 2003. *Positive Organizational Scholarship: Foundations of a New Discipline*. San Francisco, CA: Berrett-Koehler.

Cartwright, S. and Cooper, C.L. 2005. Individually targeted interventions. In *Handbook of Work Stress*, edited by J. Barling, E.K. Kelloway and M.R. Frone. Thousand Oaks, CA: Sage Publications, 607–22.

Cartwright, S. and Cooper, C.L. 2009. *The Oxford Handbook of Occupational Well-being*. Oxford: Oxford University Press.

Cartwright, S. and Whatmore, L.C. 2005. Stress and individual differences: implications for stress management. In *Research Companion to Organizational Health Psychology*, edited by A. Antoniou and C.L. Cooper. Chichester: Edward Elgar, 163–73.

Caulfield, N., Chang, D., Dollard, M.F. and Elshaug, C. 2004. A review of occupational stress interventions in Australia. *International Journal of Stress Management*, 11, 149–66.

Cooper, C.L., Quick, J.C. and Schabracq, M.J. 2009. *International Handbook of Work and Health Psychology*. New York, NY: John Wiley.

Cortina, L.M., Magley, V.J., Williams, J.H. and Langhout, R.D. 2001. Incivility in the workplace: Incidence and impact. *Journal of Occupational Health Psychology*, 6, 64–80.

Cox, T., Karanika, M., Griffiths, A. and Houdmont, J. 2007. Evaluating organizational-level work stress interventions: Beyond traditional methods. *Work & Stress*, 21, 348–68.

Csikszentmihalyi, M. 1990. *Flow: The Psychology of Optimal Experience*. New York, NY: HarperCollins.

Csikszentmihalyi, M. 2003. *Good Business: Leadership, Flow, and the Making of Meaning*. New York, NY: Viking.

Daniels, K., Karanika-Murray, M., Mellor, N. and Van Veldhoven, M. 2012. Moving policy and practice forward: Beyond prescriptions for job characteristics. In *Improving Organizational Interventions for Stress and Well-being: Addressing Process and Contest*, edited by C. Biron, M. Karanika-Murray and C.L. Cooper. London: Routledge, 313–32.

Davidson, M.J. and Burke, R.J. 2011. *Women in Management Worldwide: Progress and Prospects*. Farnham: Gower Publishing.

Davis, L.E. and Cherns, A.B. 1975. *The Quality of Working Life*. Vols 1 and 2. New York, NY: Free Press.

Dollard, M.F. 2012. Psychosocial safety climate: A lead indicator of workplace psychological health and engagement and a precursor to intervention success. In *Organizational Stress and Well-being*

Interventions: Addressing Process and Context, edited by C. Biron, M. Karanika-Murray and C.L. Cooper. London: Routledge, 77–101.

Drago, R.W. 2007. *Striking a Balance: Work, Family, Life*. Boston, MA: Dollars and Sense.

Fredrickson, B.L. 2003. Positive emotions and upward spirals in organizations. In *Positive Organizational Scholarship*, edited by K.S. Cameron, J.E. Dutton and R.E. Quinn. San Francisco, CA: Berrett-Koehler, 163–75.

Friedman, S.D. and Greenhaus, J.H. 2000. *Work and Family – Allies or Enemies? What Happens When Business Professionals Confront Life Choices?* New York, NY: Oxford University Press.

Galt, V. 2012. When management needs a tune-up-or an overhaul: Plagued by low levels of trust among its executives, Union Gas turned around its corporate culture and improved efficiency. *Globe and Mail*, March 1, B10.

Garcia, E., Salanova, M., Grau, R. and Cifre, E. 2012. How to enhance service quality through organizational facilitators, collective work engagement, and relational service competence. *European Journal of Work and Organizational Psychology*, 22, 113–26.

Giga, S.L., Noblet, A., Faragher, B. and Cooper, C. 2003. Organizational stress management interventions: A review of UK-based research. *The Australian Psychologist*, 38, 158–64.

Goldenhar, L.M., LaMontagne, A.D., Katz Heaney, C. and Landsbergis, P. 2001. The intervention research process in occupational safety and health: An overview from NORA Intervention Effectiveness Research Team. *Journal of Occupational and Environmental Medicine*, 43, 616–22.

Griffiths, A. 1999. Organizational interventions: Facing the limits of the natural science paradigm. *Scandinavian Journal of Work and Environmental Health*, 25, 589–96.

Hackman, J.R. and Oldham, G.R. 1976. Motivation through the design of work: Test of theory. *Organizational Behavior and Human Performance*, 16, 259–79.

Hackman, J.R. and Oldham, G.R. 1980. *Work Redesign*. Reading, MA: Addison-Wesley.

Hewlett, S.A. and Luce, C.B. 2006. Extreme jobs: The dangerous allure of the 70-hour work week. *Harvard Business Review*, December, 49–59.

Hoffman, D.A. and Tetrick, L. 2003. *Health and Safety in Organizations*. San Francisco, CA: Jossey-Bass.

Hogan, R. and Hogan, J. 2001. Assessing leadership: A view of the dark side. *International Journal of Evaluation and Assessment*, 9, 540–51.

Hurrell, J.J. 2005. Organizational stress interventions. In *Handbook of Work Stress*, edited by J. Barling, E.K. Kelloway and M.R. Frone. Thousand Oaks, CA: Sage Publications, 623–45.

Isaac, F.W. and Ratzan, S.C. 2013. Corporate wellness programs: Why investing in employee health and well-being is an investment in the health of the company. In *The Fulfilling Workplace: The Organization's Role in Achieving Individual and Organizational Health*, edited by R.J. Burke and C.L. Cooper. Farnham: Gower Publishing, 301–13.

Katzenbach, J.R. 2000. *Peak Performance: Aligning the Hearts and Minds of Your Employees*. Boston, MA: Harvard Business School Press.

Karanika-Murray, M., Biron, C. and Cooper, C.L. 2012. Concluding comments: Distilling the elements of successful organizational intervention implementation. In *Improving Organizational Interventions for Stress and Well-being: Addressing Process and Context*, edited by C. Biron, M. Karanika-Murray and C.L. Cooper. London: Routledge, 353–61.

Kompier, M. 2004. Commentary: Does the "Management Standards" approach meet the standard? *Work & Stress*, 18, 137–39.

Kotter, J. 1995. Leading change: Why transformation efforts fail. *Harvard Business Review*, March–April, 59–67.

Kusy, M. and Holloway, E. 2009. *Toxic Workplace! Managing Toxic Personalities and Their Systems of Power*. New York, NY: John Wiley.

LaMontagne, A.D., Noblet, A.J. and Landsbergis, P.A. 2012. Intervention development and implementation: Understanding and addressing barriers to organizational-level interventions. In *Organizational Stress and Well-being Interventions: Addressing Process and Context*, edited by C. Biron, M. Karanika-Murray and C.L. Cooper. London: Routledge, 21–37.

LaMontagne, A.D., Keegel, T., Louie, A.M., Ostry, A. and Landbergis, P.A. 2007. A systematic review of the job-stress intervention evaluation literature, 1990–2005. *International Journal of Occupational and Environmental Health*, 13, 268–80.

Langan-Fox, J., Cooper, C.L. and Klimoski, R.J. 2007. *Research Companion to the Dysfunctional Workplace*. Cheltenham: Edward Elgar.

Lawler, E.R. 2003. *Treat People Right: How Organizations and Individuals can Propel Each Other into a Virtuous Spiral of Success*. San Francisco, CA: Jossey-Bass.

Leiter, M.P., Laschinger, H.K.S., Day, A. and Gilin-Oore, D. 2011. The impact of civility interventions on employee social behavior, distress, and attitudes. *Journal of Applied Psychology*, 96, 1258–74.

Leka, S., Jain, A., Iavicoli, S., Vartia, M. and Ertel, M. 2011. The role of policy for the management of psychosocial risks at the workplace in the European Union. *Safety Science*, 49, 558–64.

Lencioni, P. 2012. *The Advantage: Why Organizational Health Trumps Everything Else*. San Francisco, CA: Jossey-Bass.

Lester, D. 2011. Violence in the workplace. In *Occupational Health and Safety*, edited by R.J. Burke and C.L. Cooper. Farnham: Gower Publishing, 179–95.

Likert, R.J. 1961. *New Patterns of Management*. New York, NY: McGraw Hill.

Likert, R.J. 1967. *The Human Organization: Its Management and Value*. New York, NY: McGraw Hill.

Loehr, J. and Schwartz, T. 2003. *The Power of Full Engagement: Managing Energy, Not Time is the Key Top High Performance and Personal Renewal*. New York, NY: Free Press.

Lowe, G. 2011. *Creating Healthy Organizations: How Vibrant Workplaces Inspire Employees to Achieve Sustainable Success*. Toronto: University of Toronto Press.

Lubit, R.H. 2004. *Coping with Toxic Managers, Subordinates … and Other Difficult People*. Upper Saddle River, NJ: Prentice-Hall.

Mackay, C., Palferman, D., Saul, H., Webster, S. and Packham, C. 2012. Implementation of the management standards for work related stress in Great Britain. In *Organizational Stress and Well-being interventions: Addressing Process and Context*, edited by C. Biron, M. Karanika-Murray and C.L. Cooper. London: Routledge, 285–311.

Mainero, L.A. and Sullivan, S.E. 2006. *The Opt-out Revolt: Why People are Leaving Companies to Create Kaleidoscope Careers*. Mountain View, CA: Davies-Black.

Maslach, C., Leiter, M.P. and Jackson, S.E. 2011. Making a significant difference with burnout interventions: Researcher and practitioner collaboration. *Journal of Organizational Behavior*, 33, 296–300.

Nielsen, K., Randall, R., Holten, A.L. and Rial-Gonzalez, E. 2010. Conducting organizational-level occupational health interventions: What works? *Work & Stress*, 24, 234–59.

Noblet, A. and LaMontagne, A.D. 2009. The challenges of developing, implementing, and evaluating interventions. In *The Oxford Handbook of Organizational Well-being*, edited by S. Cartwright and C.L. Cooper. Oxford: Oxford University Press, 466–96.

Passmore, J. and Anagnos, J. 2009. Organizational coaching and mentoring. In *The Oxford Handbook of Organizational Well-being*, edited by S. Cartwright and C.L. Cooper. Oxford: Oxford University Press, 497–521.

Peters, T.J. and Waterman, R.H. 1982. *In Search of Excellence: Lessons from America's Best-run Companies*. New York, NY: Harper.

Pfeffer, J. 2010. Building sustainable organizations: The human factor. *Academy of Management Perspectives*, 24, 34–45.

Quick, J.C. and Tetrick, L.E. 2011. *Handbook of Organizational Health Psychology*. Washington, DC: American Psychological Association.

Quick, J.C., Quick, J.D., Nelson, D. and Hurrell, J. 1997. *Preventative Stress Management in Organizations*. Washington, DC: American Psychological Association.

Randall, R. and Nielsen, K.M. 2012. Does the intervention fit? An explanatory model of intervention success and failure in complex organizational environments. In *Organizational Stress and Well-being Interventions: Addressing Process and Context*, edited by C. Biron, M. Karanika-Murray and C.L. Cooper. London: Routledge, 120–134.

Report on Business. 2012. More wellness, better performance. *Globe and Mail*, May.

Sashkin, M. and Burke, W.W. 1987. Organization development in the 1980s. *Journal of Management*, 13, 393–417.

Schabracq, M.J., Winnubst, J.A.M. and Cooper, C.L. 2003. *The Handbook of Work and Health Psychology*. New York, NY: John Wiley.

Schein, E.H. 1978. *Career Dynamics*. Reading, MA: Addison-Wesley.

Schwartz, T. 2007. Manage your energy, not your time, *Harvard Business Review*, October, 63–73.

Seldman, M. and Seldman, J. 2008. *Executive Stamina: How to Optimize Time, Energy and Productivity to Achieve Peak Performance*. New York, NY: John Wiley.

Seligman, M. 2002. *Authentic Happiness: Using the New Positive Psychology to Realize Your Potential for Lasting Fulfillment*. New York, NY: Free Press.

Semmer, N.K. 2006. Job stress interventions and the organization of work. *Scandinavian Journal of Work and Environmental Health*, 32, 515–27.

Sisodia, R., Wolfe, D.B. and Sheth, J. 2007. *Firms of Endearment: How World-class Companies Profit from Passion and Purpose*. Upper Saddle River, NJ: Wharton School Publishing.

Tepper, B.J. 2000. Consequences of abusive supervision. *Academy of Management Journal*, 43, 178–90.

Tetrick, L.E., Quick, J.C. and Gilmore, P.L. 2012. Research in organizational interventions to improve well-being: Perspectives on organizational change and development. In *Organizational Stress and Well-being Interventions: Addressing Process and Context*, edited by C. Biron, M. Karanika-Murray and C.L. Cooper. London: Routledge, 59–75.

Tims, M., Bakker, A.B. and Xanthopoulou, D. 2010. Do transformational leaders enhance their followers' daily work engagement? *Leadership Quarterly*, 22, 121–31.

Tvedt, S.D and Saksvik, P.O. 2012. Perspectives on the intervention process as a special case of organizational change. In *Improving Organizational Interventions for Stress and Well-being: Addressing Process and Context*, edited by C. Biron, M. Karanika-Murray and C.L. Cooper. London: Routledge, 102–19.

Weyman, A. 2012. Evidence based practice – its contribution to learning in managing workplace health risks. In *Improving Organizational Interventions for Stress and Well-being: Addressing Process and Context*, edited by C. Biron, M. Karanika-Murray and C.L. Cooper. London: Routledge, 333–50.

Wittenberg-Cox, A. and Maitland, A. 2008. *Why Women Mean Business: Understanding the Emergence of Our Next Economic Revolutions*. San Francisco, CA: Jossey-Bass.

PART 1

Creating Healthy Workplaces: Models and Approaches

2 *The WHO Global Approach to Protecting and Promoting Health at Work*

EVELYN KORTUM

Abstract

This chapter outlines the policy instruments of the World Health Organization (WHO) in the area of workers' health. It explains the WHO global approach to developing, implementing and evaluating healthy workplace programmes which address the protection and promotion of workers' health in an integrated manner, and explains its rationale for such programmes. Psychosocial workplace hazards and work-related stress are set on an equal footing with the physical working environment in the WHO approach, which also covers the personal health resources and the interface with the community in which the company is active to ensure the application of business social responsibility. The chapter underlines the importance of psychosocial hazards with respect to a comprehensive policy framework, and explains why it has such a high relevance. Lastly, some good practice examples are outlined that comply with the principles of the WHO healthy workplace approach.

Introduction: The WHO Action Plan to Address the Health of Workers Worldwide

The WHO action on protecting and promoting the health of workers is mandated by the Constitution of the Organization, as well as a number of resolutions passed at various sessions of the World Health Assembly, the supreme decision-making body of the Organization with its 194 Member States to date.

The workplace is an ideal setting for early interventions and for protecting and promoting the health of workers and their families (Ivanov and Kortum 2008). Currently, and despite the existence of effective interventions to prevent occupational diseases and injuries, there are still major gaps in the health status of workers between and within countries. The current global policy instrument in place, the WHO Global Plan of Action on Workers' Health (WHO 2007), 2008–2017, provides a political framework for the development of policies, infrastructure, technologies and partnerships to achieve a basic level of health protection in all workplaces throughout the world. The prime principle underpinning the Global Plan is that workers should be able to enjoy the highest attainable standard of physical and mental health including favourable working conditions for their health and for sustaining their livelihoods. Therefore, it is important that the workplace is not detrimental to health and well-being. Participation in processes to improve workers' health should include workers and employers and their representatives. WHO Member States are, at the same time, requested to develop national plans and strategies for implementing the Global Plan of Action and to work towards protecting and promoting workers' health through essential interventions and basic services for prevention of occupational diseases and injuries. These instruments recognize the close link between occupational or workers' health on the one hand, and public health on the other. This is important as it lifts the workers' health issues out of a silo and transfers these onto the public health agenda where they can get the prominent attention they deserve.

Traditionally, Ministries of Health take a general public health approach in developing their policies and strategies, while Ministries of Labour take an approach that focuses primarily on occupational health, as they may differ in terms of their priorities and actions in relation to issues such as work, employment and workplace risks and their prevention. The Global Plan brings these approaches together, and attempts to deal with a large array of aspects of workers' health, including the importance of building a solid evidence base, and the benefits from focused international networking. To enhance the capacity for implementation of a large number of activities related to the Global Plan, the WHO is supported by its large Global Network of Collaborating Centres for Occupational Health (CCs) and other partners. The WHO also has formal relations with three non-governmental organizations (ICOH, IOHA and IEA). Work with relevant intergovernmental organizations, such as the ILO, is key to success.

The WHO Healthy Workplace Initiative (WHO 2010a) is only one way towards achieving these aims, but its advantage is that it combines all approaches that have proven to be successful in building healthy and safe workplaces. The model, which was developed and published after extensive intra- and extramural consultations, has the advantage of being applicable to all sizes of workplaces, and has been well accepted by the international scientific and business community as a useful approach to ensuring workers' health. Particularly, the global approach to dealing with well-being issues, including psychosocial risks, was a much needed variation from the traditional occupational health and safety (OHS) approach (Kortum 2012).

The WHO Healthy Workplace Initiative

The WHO Healthy Workplace model and framework include both content and process, which are made very explicit within the model. The model may be implemented in any workplace of any size, in any country. There is no "one-size-fits-all" and each enterprise must adapt good practices to their own workplace, their own culture and their own country legislatory framework. The WHO model and framework brings together the principles and common factors that appear to be universally supported in the literature and in the perceptions of experts and practitioners in the fields of health, safety and organizational health (WHO 2010a). The main criteria applied to the development, implementation and evaluation of healthy workplace programmes include a number of core values that reply to business ethics including personal and social codes of behaviour, the business case to include costs of prevention versus those resulting from accidents or occupational diseases, and the legal case where business liability will enhance adherence to legislation and policies. Further criteria of the model refer to management commitment, worker involvement, an approach to learning from others and working towards sustainability. Sustainability strategies are means for businesses to engage themselves in improving health protection and health promotion, as well as social well-being for workers and their communities. Therefore, the healthy workplace initiative promotes a comprehensive approach to workplace health promotion and protection.

The definition of a healthy workplace is strongly based on the WHO definition of health which is enshrined in its Constitution and which stresses an integrated approach to health: "A state of complete physical, mental and social well-being, and not merely the absence of disease," as well as evidence from best practice in the research and practice literature. The WHO definition of a healthy workplace (WHO 2010b) encompasses the four avenues of influence where actions can best take place and the most effective processes by which employers and workers can take action:

A healthy workplace is one in which workers and managers collaborate to use a continual improvement process to protect and promote the health, safety and well-being of all workers and the sustainability of the workplace by considering the following, based on identified needs:

- *health and safety concerns in the physical work environment;*
- *health, safety and well-being concerns in the psychosocial work environment including organization of work and workplace culture;*
- *personal health resources in the workplace provided by the employer; and*
- *ways of participating in the community to improve the health of workers, their families and other members of the community.*

The four avenues of influence found in the model in Figure 2.1 refer to content and not process. Each of the four avenues is explained further below, including some examples of interventions to make the workplace healthier and safer. Important to note is that all four avenues interact and have common denominators.

The physical work environment is the part of the workplace facility that can be detected by human or electronic senses, including the structure, air, machines, furniture, products, chemicals, materials and processes that are present or that occur in the workplace, and which can affect the physical or mental safety, health and well-being of workers. If the worker performs his or her tasks outdoors or in a vehicle, then that location is the physical work environment. Examples of interventions include:

- eliminating a toxic chemical or substituting with one less hazardous;
- installing machine guards or local exhaust ventilation;
- training workers on safe operating procedures; and
- providing personal protective equipment such as respirators or hard hats.

Personal health resources in the workplace encompass the supportive environment to include health services, information, resources, opportunities and flexibility an enterprise

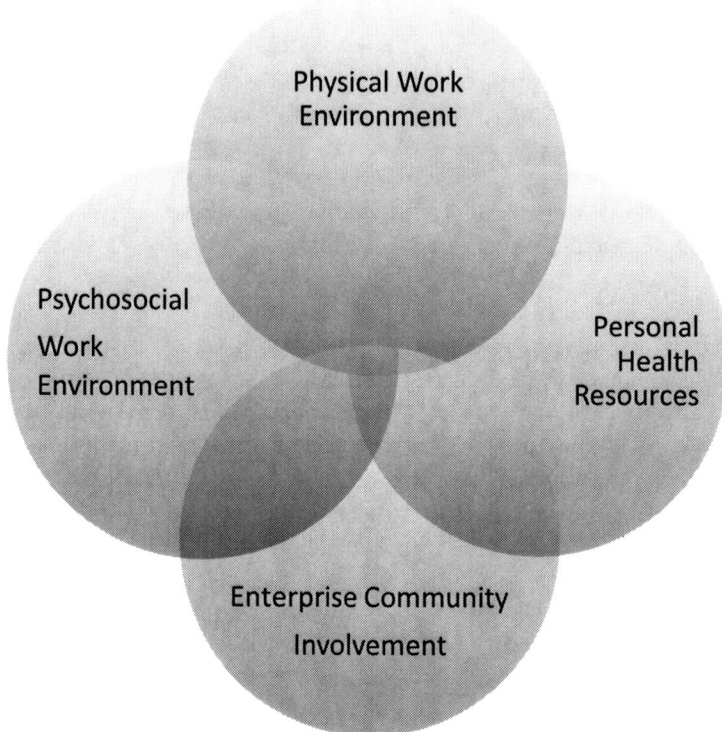

Figure 2.1   The four avenues of influence

provides to workers to support or motivate their efforts to improve or maintain healthy personal lifestyle practices. They also enable them to monitor and support their ongoing physical and mental health. Examples include:

- providing fitness facilities, classes or equipment for workers;
- providing healthy food choices in the cafeteria and vending machines;
- putting no-smoking policies in place, and providing smoking cessation assistance;
- providing information about alcohol and drugs, and employee assistance counselling;
- providing confidential medical services such as health assessments, medical examinations, medical surveillance and medical treatment if not accessible in the community (e.g. antiretroviral treatment for HIV).

The psychosocial work environment includes the organization of work and the organizational culture; the attitudes, values, beliefs and practices that are demonstrated on a daily basis in the enterprise/organization, and which affect the mental and physical well-being of employees. These are sometimes generally referred to as workplace stressors, which may cause emotional or mental stress to workers. Examples of interventions include:

- reallocating work to reduce workload;
- enforcing zero tolerance for harassment, bullying or discrimination;
- allowing flexibility in how and when work is done to respect work–family balance;

- recognizing and rewarding good performance appropriately;
- allowing meaningful worker input into decisions that affect them.

Enterprise community involvement or business responsibility comprises the activities, expertise and other resources an enterprise engages in or provides to the social and physical community or communities in which it operates; and which affect the physical and mental health, safety and well-being of workers and their families. It includes activities, expertise and resources provided to the immediate local environment, but also the broader global environment. Examples include:

- providing free or affordable primary health care to workers, and including access for family members, small- and medium-sized enterprise (SME) employees and informal workers;
- providing free supplemental literacy education to workers and their families;
- providing leadership and expertise related to workplace health and safety to SMEs in the community;
- implementing voluntary controls over pollutants released into the air or water;
- allowing workers to volunteer for non-profit organizations during work hours;
- providing financial support to worthwhile community causes without an expectation of concomitant enterprise advertising.

Clearly every enterprise may not have the need or capacity to address each of these four avenues at the same time. The way an enterprise addresses the four avenues must be

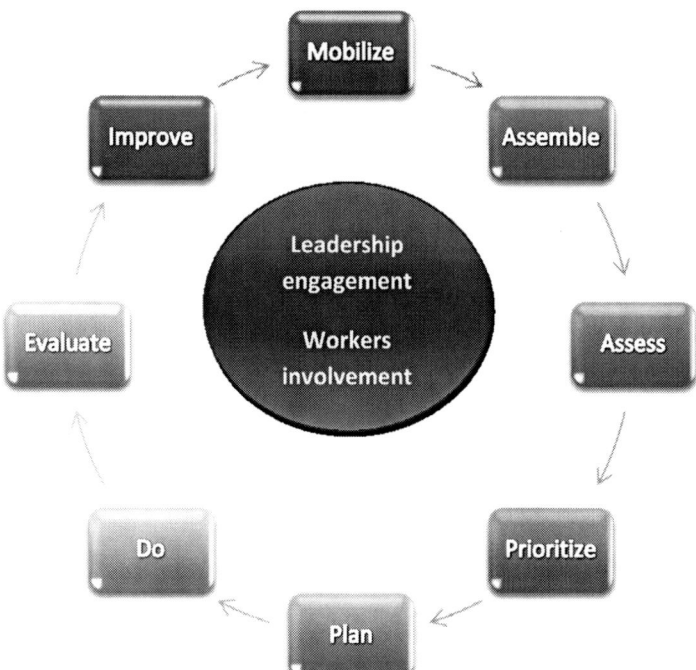

Figure 2.2   Leadership engagement

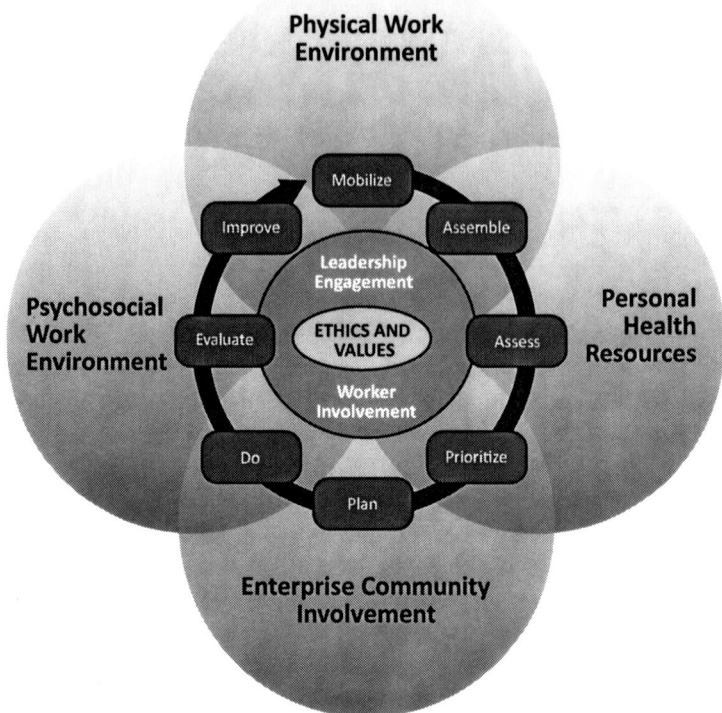

Figure 2.3   The WHO Healthy Workplace model and framework

based on the needs and priorities identified through an assessment process that involves extensive consultation with workers and their representatives through health and safety committees. Implementing a healthy workplace programme that is sustainable and effective in meeting the needs of workers and employers requires more than knowing what kinds of issues to consider. To successfully create such a healthy workplace, an enterprise must follow a process that involves continual improvement (a management systems approach). This is graphically represented by the continual improvement loop of mobilize, assemble, assess, prioritize, plan, do, evaluate and improve demonstrated in Figure 2.2 on the previous page.

Two core principles that underlie this model are featured in the centre of Figure 2.3. These principles of management commitment and workers' involvement are not merely steps in the process, but are ongoing circumstances or conditions that must be tapped into at every stage of the process.

Good Practices are Slowly Spreading

It is encouraging to observe that there are good practices at the international level that address the four avenues of influence in a comprehensive manner. An example from the automotive industry (Panter 2012) in South Africa was the implementation of an organizational health management programme to ensure the health and well-being of its employees. An HIV prevention programme is in place in 44 workplaces and has 24 community peer educators. It concentrates on information about HIV and AIDS and other sexually transmitted diseases, distributing condoms, promoting voluntary counselling and testing. This is a comprehensive programme integrating physical and psychosocial working environment management systems with health promotion, HIV/AIDS programmes and community health activities. The success factors were the involvement of employee representatives in the initial design and implementation of the programme. Support was provided by the corporation's leadership and parent company. Also the involvement in the implementation process of trade unions and other employee formations was successful. The programme now enjoys ongoing support from the middle management.

An example from India (Venkatesh 2006) is a programme that promotes health and safety, while implementing activities for the prevention of incidents and diseases. In most developing countries the physical working environment is still the number one risk area. This programme is also comprehensive with activities to improve all avenues of influence of the WHO model. It includes, for instance, a system to increase recognition, arrangements of working hours, means to balance work and family demands, activities to prevent infectious and non-communicable diseases, a programme to support the community in health, education, women empowerment, protection of the environment, as well as the continuous improvement of the physical working environment. Systems for recognition or awards and permissible flexible shifts including clear job descriptions were provided. Good measurable results were so far obtained for injury reduction.

In a Kenyan hotel chain, the management implemented a comprehensive wellness programme addressing HIV/AIDS and non-communicable diseases (Lutalo 2007). The wellness programme provides hotel employees with a comprehensive set of activities, ranging from the provision and promotion of health information to the provision of medical facilities and a sexual harassment policy. It also includes occupational health and safety provisions and covers non-communicable and communicable diseases. Benefits were registered from the surrounding community and hotel guests who benefit from access to programmes and facilities.

Of Equal Importance But Still Largely Ignored: The Psychosocial Working Environment

The psychosocial working environment is one of four avenues of influence of the WHO policy framework for healthy workplaces. Psychosocial hazards have nowadays become a global phenomenon but are still ignored by many. Signs are that many workplaces around the world experience repeated reorganizing, downsizing and expanding of organizations. This has become very common and is related to established health effects among workers and employees (Ferrie et al. 2007, Theorell et al. 2003, Westerlund et al. 2004a, 2004b). In addition, the experience of job insecurity has been associated with poorer physical and mental health outcomes (Ferrie et al. 1998, Metcalfe et al. 2003, Ostry and Spiegel 2004, Pollard 2001, Virtanen et al. 2005). According to a recent European survey, job insecurity is one of the ten highest psychosocial hazards in the EU (EU-OSHA 2007). Sustained job insecurity due to precarious labour market situations has been linked to poor health behaviours by way of declines in specific coping mechanisms. Some evidence shows

Table 2.1   Global financial and mental health impact of work-related stress

Type of cost	Country	Estimated cost	Source
Work-related health loss and associated productivity loss	Globally	4–5% of the GDP	Takala 2002
Occupational diseases and accidents	Commonwealth	10 million disability-adjusted life years (DALYs) lost	CDPP 2007
Work-related stress and related mental health problems	EU (15 Member States)	On average between 3% and 4% of the GNP = €265 billion/year	Gabriel and Liimatainen 2000
Stress at work	UK	Estimate 5–10% of the GNP/year costing employers around €571 million	Worrall and Cooper 2006
Sick leave due to stress and mental strain	Sweden	€2.7 billion	Koukoulaki 2004
Stress-related illnesses	France	Between €830 and €1,656 million	EU-OSHA 2009

that temporary employment is associated with increased death from alcohol-related causes and smoking-related cancers (Kivimäki et al. 2003). The WHO estimated that 400 million people around the world suffer from mental or neurological disorders, or from psychosocial problems such as those related to smoking, drinking and drug abuse (WHO 2001). A number of health outcomes, particularly from work-related stress, have been well documented in the literature, such as heart disease, depression and musculo-skeletal disorders. Also the interlinkages of these health outcomes and work-related stress have been studied broadly (WHO 2010b).

Absences caused by occupational injuries or ill health affect not only workers' lives but also businesses and their communities. This is of growing concern globally in addition to, in particular, absenteeism due to work-related mental health problems and the rising cost to economies, employers and insurances. Statistics show that in many industrialized countries, 35–45 per cent of absenteeism from work is due to mental health problems (WHO 2003) and that about 40 per cent of employee turnover is due to stress at work. Table 2.1 demonstrates the national and global business cases through the economic impact on countries and at the global level of absences due to work-related stress resulting in mental health problems.

Generally, statistics from industrialized countries show that the collective cost of work-related stress is high, having potentially major impacts on national economies. And the link between health and productivity has been recognized for centuries as the cornerstone for a healthy economy (Goetzel et al. 2002, Oxenburgh et al. 2004, Stewart et al. 2004). The high costs of the impact of work-related stress, which is evidenced by national statistics, have facilitated public dialogue, and the issuance of many studies that attempt to address the causes and origins of work-related stress. It has been estimated that the cost of the work-related health loss and associated productivity loss is estimated to reach around 4–5 per cent of the GDP (Takala 2002).

Usually, workers in industrialized countries enjoy a welfare system that provides a public "safety net," as a result of which the burden of unemployment is shared by the

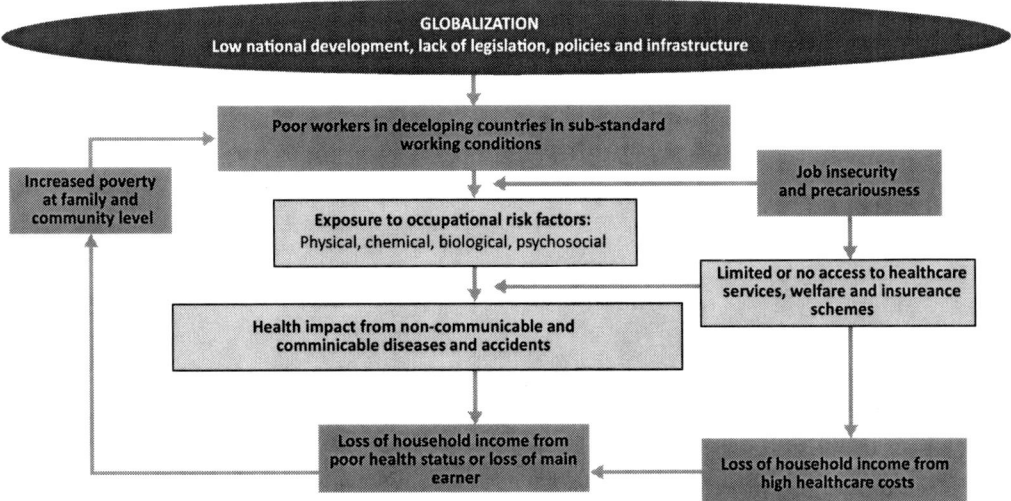

Figure 2.4   The link between poverty and sub-standard working conditions in developing countries

government (Dewa et al. 2007). In the absence of a welfare system that may protect individuals who are unable to work, for example, as a result of their mental illness, workers in developing countries are likely to continue to work despite their disability (Dewa et al. 2007) and the impact on workplace productivity is, therefore, even magnified, and goes beyond the direct costs as a result of impairment in the workplace.

In many low-income countries without welfare systems, workers and their families have no choice but to further plunge into poverty when they get sick due to the high cost of healthcare. Figure 2.4 describes this vicious cycle of poverty, which needs to be addressed together by businesses in terms of respecting OHS legislation and taking their business responsibilities towards their workers seriously and governments who need to develop and enforce legislation for the protection and for the promotion of workers' health. Politicians, policy-makers, labour unions and employers need to be convinced of the importance of occupational health and safety (Rantanen et al. 2004). This is an effort to bridge research and practical application.

Although research (EU-OSHA 2007) indicates that impact from psychosocial hazards reaches beyond the workplace, Nuwayhid (2004) argues that the internal domain of occupational health research, such as focusing on workplace hazards, work organization, exposure-disease spectrum, etc., works well in industrialized countries. He stresses that the focus should be first on the external-contextual domain, followed by a more focused, specific workplace approach, which would help build consensus amongst occupational health researchers and other disciplines (e.g. economists, social scientists, unionists, women's organizations). Such research may facilitate the creation of political mechanisms responsive to occupational health needs in developing countries and also facilitate the adoption of psychosocial risks and work-related stress into the national policy frameworks addressing workers' health. The issue seems to be that it does currently not translate well to developing countries due to the lack of mechanisms and risk assessment processes, which is why the issues need to be addressed at policy level while including the causal

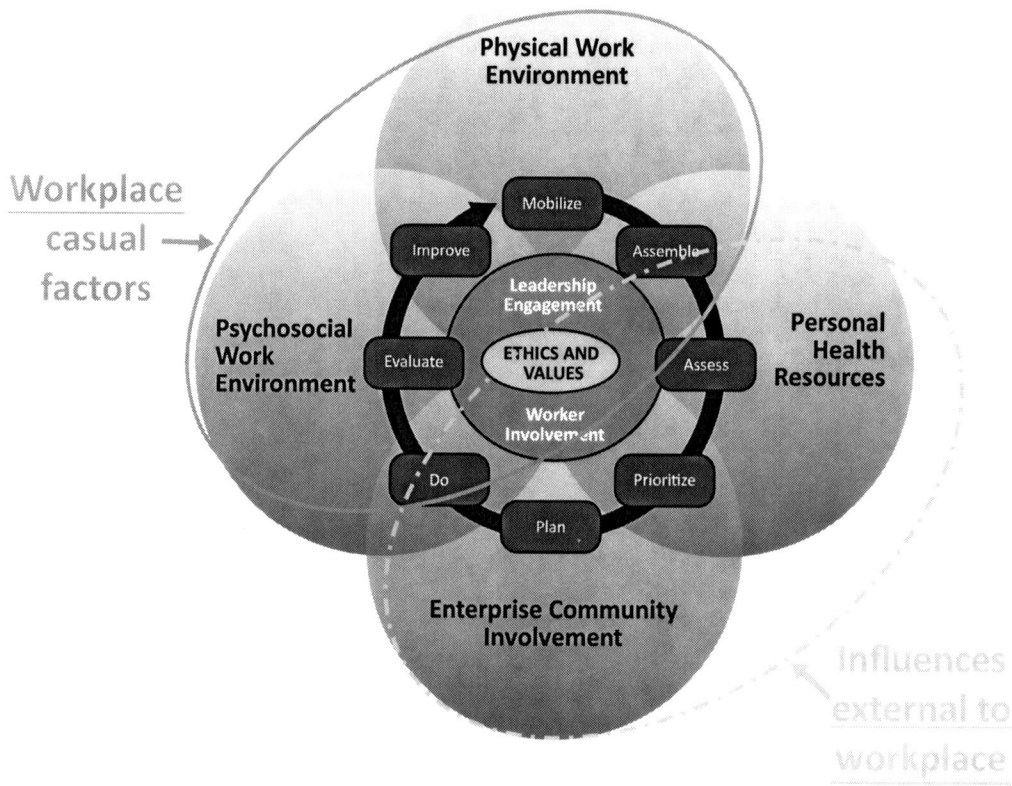

Figure 2.5   Influences on health according to the WHO 'Healthy Workplace' model (Macdonald, 2011)

factors of workplace as well as the influences external to the workplace as graphically demonstrated in Figure 2.5.

Figure 2.5 clearly outlines the importance of addressing the interplay between work and non-occupational diseases that include communicable and non-communicable diseases, which underlines that it is important that the area of the health of workers needs to develop towards a public health approach (WHO 2007). In 2011, the UN Secretary General reported regarding the prevention of non-communicable diseases that "part of the burden of non-communicable diseases is attributable to occupational risk factors including exposure to chemical, physical, biological, ergonomic and psychosocial hazards at work" (UN 2011). This also provides fertile ground for greater business responsibility and putting in place interventions at the workplace.

Even in Europe, where the policy context of health and safety is considered to be more advanced, initiatives have not had the impact anticipated both by experts and policy-makers so far, which may mainly be due to the gap between policy and practice (Levi 2005). For example, findings from the PRIMA-EF European-wide stakeholder survey (Leka and Cox 2008) also show that not only in developing countries but also in Europe there is low prioritization of work-related stress, particularly by employers. Trade unions and employer organizations believe that, a priori, there is lack of awareness about the issue of work-related stress, and governments primarily blame low prioritization of psychosocial issues in general for the lack of initiatives in this area (Natali et al. 2008).

Conclusion

Overall, psychosocial hazards and work-related stress are not prioritized by policy-makers anywhere in the world. And although no clear patterns of evaluation and impact of policy-level interventions have been reported so far, possibly due to lack of resources (time and money) and confounding variables, stakeholders highlight the need for more long-term evaluation. The development of clear key messages to reach policy-makers, accompanied by clear communication structures, is essential to obtaining impact (Leka et al. 2010). The Healthy Workplace model provides such a clear format for messages to policy-makers and is easy to adopt and to adapt. A large number of multi-nationals and development agencies have so far adopted the model as part of their business strategy towards sustainability.

Hence, addressing psychosocial hazards globally and in unison with other workplace hazards should be recognized as an important objective, particularly when statistics show that the collective cost of stress is high for national economies and at the global level, in particular for mental health problems. In many past country frameworks, this important element has often been forgotten and in current frameworks that address psychosocial hazards and mental and physical health at work, the relevant action is still lagging behind. The public health impact from psychosocial risks, added to the traditional health impact, is enormous and cannot be ignored any longer at a global level and, indeed, a call to employers, worker representatives, researchers and policy-makers to include these globally emerging issues within comprehensive and broad approaches to occupational health and national plans for occupational health, is a call for attention to occupational health per se. This is particularly the case considering the economic burden and human suffering involved. To respond to many of the new challenges posed by globalization for the health of workers, the United Nations Human Rights Council issued a report by the Special Rapporteur on the right of everyone to the enjoyment of the highest attainable standard of physical and mental health (United Nations 2012). The report rightly states that "the right to a healthy workplace environment is an integral component of the right to health."

References

Commonwealth Director of Public Prosecutions (CDPP) 2007. Annual report 2007–2008. Available at: http://www.cdpp.gov.au/Publications/Annual-Reports/ [accessed 13 August 2013].

Dewa, C.S., McDaid, D. and Ettner, S.L. 2007. An international perspective on worker mental health problems: Who bears the burden and how are costs addressed. *Canadian Journal of Psychiatry*, 52, 346–56.

European Agency for Safety and Health at Work (EU-OSHA) 2007. *Expert Forecast on Emerging Psychosocial Risks Related to Occupational Safety and Health*. Luxembourg: Office for Official Publications of the European Communities.

European Agency for Safety and Health at Work (EU-OSHA) 2009. *Expert Forecast on Emerging Chemical Risks Related to Occupational Safety and Health*. Luxembourg: Office for Official Publications of the European Communities.

Ferrie, J.E., Westerlund, H., Oxenstierna, G. and Theorell, T. 2007. The impact of moderate and major workplace expansion and downsizing on the psychosocial and physical work environment and income in Sweden. *Scandinavian Journal of Public Health*, 35(1), 62–9.

Ferrie, J.E., Shipley, M.J., Marmot, M.G., Stansfeld, S.A. and Davey Smith, G. 1998. An uncertain future: The health effect threats of employment security in white-collar men and women. *American Journal of Public Health*, 88(7), 1030–36.

Gabriel, P. and Liimatainen, M.-R. 2000. *Mental Health in the Workplace*. Geneva: International Labour Office.

Goetzel, R.Z., Ozminkowski, R.J., Sederer, L.I. and Mark, T.L. 2002. The business case for quality mental health services: Why employers should care about the mental health and well-being of their employees. *Journal of Occupational and Environmental Medicine*, 44, 320–30.

Ivanov, I. and Kortum, E. 2008. WHO strategies and action to protect and promote the health of workers. In *Quality of Working Life*. Liber Amicorum, 1–4.

Kivimäki, M., Vahtera, J., Virtanen, M., Elovainio, M., Pentti, J. and Ferrie, J.E. 2003. Temporary employment and risk of overall and cause specific mortality. *American Journal of Epidemiology*, 158, 663–8.

Kortum, E. 2012. Editorial: A need to broaden our perspective to address workers' health effectively in the 21st Century. *Industrial Health*, 50, 71–2.

Koukoulaki, T. 2004. Stress prevention in Europe: Trade union activities. In *Stress at Work in Enlarging Europe*, edited by S. Iavicoli, P. Deitinger, C. Grandi, M. Lupoli, A. Pera and M. Petyx. Rome: ISPESL, 17–27.

Leka, S. and Cox, T. 2008. The future of psychosocial risk management and the promotion of well-being at work in the European Region: A PRIMA time for action. In *The European Framework for Psychosocial Risk Management*, edited by S. Leka and T. Cox. Nottingham: I-WHO, 174–84.

Leka, S., Jain, A., Iavicoli, S., Vartia, M. and Ertel, M. 2010. The role of policy for the management of psychosocial risks at the workplace in the European Union. *Safety Science*, 49(4), 558–64.

Levi, L. 2005. Working life and mental health – a challenge to psychiatry? *World Psychiatry*, 4(1), 53–7.

Lutalo, M. 2007. The wellness program of Serena Hotels, Kenya – a case study. World Bank Global HIV/AIDS Program. Washington, DC. Available at: http://siteresources.worldbank.org/INTHIVAIDS/Resources/375798-1132695455908/GRSerenaFinalOct0307.pdf [accessed September 2013].

Macdonald, W. 2011. Need for development aid to promote occupational health. Australian Council for International Development (ACFID) Universities Linkage Network Conference on *An Australian Approach to Development? People, Practice and Policy*, Deakin University, Melbourne, 12–13 December 2011. Available at: http://www.deakin.edu.au/arts-ed/shss/events/acfid/#program [accessed September 2013].

Metcalfe, C., Davey Smith, G., Sterne, J.A., Heslop, P., Macleod, J. and Hart, C. 2003. Frequent job change and associated health. *Social Science and Medicine*, 41, 210–16.

Natali, E., Deitinger, P., Rondinone, B. and Iavicoli, S. 2008. The European Framework for Psychosocial Risk Management: PRIMA-EF. In *The European Framework for Psychosocial Risk Management*, edited by S. Leka and T. Cox. Nottingham: Institute for Work, Health and Organisations, 79–114.

Nuwayhid, I.A. 2004. Occupational health research in developing countries: A partner for social justice. *American Journal of Public Health*, 94, 1916–21.

Ostry, A.S. and Spiegel, J.M. 2004. Labor markets and employment security: Impacts of globalization on service and healthcare-sector workforces. *International Journal of Occupational and Environmental Health*, 10, 368–74.

Oxenburgh, M., Rapport, N. and Oxenburgh, P.M. 2004. *Increasing Productivity and Profit through Health and Safety: The Financial Returns from a Safe Working Environment*. Boca Raton, FL: CRC Press LLC.

Panter, C. 2012. OSH and corporate social responsibility in Mercedes Benz South Africa. In *Occupational Safety and Health and Corporate Social Responsibility in Africa: Repositioning Corporate Responsibility Towards National Development*, edited by A. Jain, B.B. Puplampu, K. Amponsah-Tawiah and N.J.A. Andreou. Bedfordshire: Cranfield Press, 152–69.

Pollard, T.M. 2001. Changes in mental well-being, blood pressure and total cholesterol levels during workplace reorganization: The impact of uncertainty. *Work & Stress*, 15(1), 14–28.

Rantanen, J., Lehtinen, S. and Savolainen, K. 2004. The opportunities and obstacles to collaboration between the developing and developed countries in the field of occupational health. *Toxicology*, 198(1–3), 63–74.

Stewart, W.F., Ricci, J.A. and Leotta, C. 2004. Health-related lost productive time (LPT): Recall interval and bias in LPT estimates. *Journal of Occupational and Environmental Medicine*, 46, S12–S22.

Takala, J. 2002. Life and health are fundamental rights for workers (interview). *Labour Education*, 1, 1–7.

Theorell, T., Oxenstierna, G., Westerlund, H., Ferrie, J., Hagberg, J. and Alfredsson, L. 2003. Downsizing of staff is associated with lowered medically certified sick leave in female employees. *Occupational and Environmental Medicine*, 60(9), 1–5.

United Nations 2011. *Political Declaration of the High-level Meeting of the General Assembly on the Prevention and Control of Non-Communicable Diseases*. UN General Assembly, Sixty-sixth session, Agenda item 117, 16 September 2011.

United Nations 2012. *Report of the Special Rapporteur on the Right of Everyone to the Enjoyment of the Highest Attainable Standard of Physical and Mental Health*, Anand Grover. A/HRC/20/15. 10 April 2012.

Venkatesh, H. 2006. *Promotion of Health & Safety and Prevention of Incidents and Diseases*. India.

Virtanen, M., Kivimäki, M., Joensuu, M., Virtanen, P., Elovainio, M. and Vahtera, J. 2005. Temporary employment and health: A review. *International Journal of Epidemiology*, 34, 610–22.

Westerlund, H., Theorell, T. and Alfredsson, L. 2004a. Organizational instability and cardiovascular risk factors in white-collar employees: An analysis of correlates of structural instability of workplace organization on risk factors for coronary heart disease in a sample of 3,904 while collar employees in the Stockholm region. *European Journal of Public Health*, 14(1), 37–42.

Westerlund, H., Ferrie, J., Hagberg, J., Jeding, K., Oxenstiera, G. and Theorell, T. 2004b. Workplace expansion, long-term sickness absence, and hospital admission. *Lancet*, 363(9416), 1193–7.

World Health Organization 2001. *Mental Health in Europe*. Regional Office for Europe, Copenhagen: WHO.

World Health Organization 2003. *Investing in Mental Health*. Geneva: WHO.

World Health Organization 2007. *Global Plan of Action for Workers' Health, 2008–2017*. Geneva: WHO.

World Health Organization 2010a. *WHO Healthy Workplace Framework and Model: Background and Supporting Literature Review*. Geneva: WHO.

World Health Organization 2010b. *Healthy Workplaces: A Model for Action. For Employers, Workers, Policy-makers and Practitioners*. Geneva: WHO.

Worrall, L. and Cooper, C.L. 2006. *The Quality of Working Life: Managers' Health and Well-being*. Executive Report, Chartered Management Institute.

3 Work and the Dynamics of Development: An Integrated Model

ANTONELLA DELLE FAVE and MARTA BASSI

Abstract

This chapter addresses the topic of work from a theoretical perspective, within a general framework encompassing cultural dimensions as well as the role of work in the process of individual development. To this purpose, work will be discussed as a basic component of the social and personal life, taking into account the variety of its structural features and the quality of experience individuals associate with it, in relation to their goals and expectations, health conditions, and specific skills and potentials.

Introduction: Culture and Work

Culture is a crucial element to consider when analysing work and its impact on the life of individuals. As clearly stated by Jablonka and Lamb (2005), human evolution is grounded into factors and mechanisms that go far beyond genetics. Epigenetic variations, much more frequent than genetic mutations, take place at various levels as non-random consequences of environmental pressures on the organism, enhancing its plasticity. Among these variations, emerging at progressively higher levels of complexity in the individual–environment interaction, the prominent ones are differences in gene expression and in cell structure; body-to-body routes of materials' transmission (such as breastfeeding); social learning; and symbolic information and artefacts. The role of individual intentionality and direction becomes growingly crucial across these levels (Jablonka and Lamb 2007).

Culture originally developed within the constraints of ecological niches. Specific climatic conditions, available sources of food and of raw materials gave rise to highly diversified societies. Cultural differentiation occurring through the various types of epigenetic variations played the same role that the creation of new species and subspecies plays in biology: it improved flexibility and it increased the amount of information and survival strategies which humans could adopt to cope with the environmental demands (Newson et al. 2007). Ultimately, cultural inheritance influenced the features and history of human communities much more than biology (Richerson and Boyd 2005, Delle Fave et al. 2011a).

Culture can thus be understood as information originally produced and stored in the human brain, and transmitted through various mechanisms (Henrich et al. 2008). In particular, besides its individual intrasomatic localization, culture is embedded in material and symbolic artefacts, from books to buildings, from frescoes to utensils. The storage of cultural information in extrasomatic carriers has important implications for the survival of individuals and groups. In addition, cultural products outlive their biological creators, thus representing a repository of information that can be transmitted across generations as effectively – or even more effectively – than genes.

Artefact production, the social structure of each human community, the labour division between genders and across social classes represent outcomes of the cultural information developed and transmitted within a specific ecological niche. The material and intellectual products of the daily work of each member of a community – be they carpenters, scientists, architects, poets, farmers, office employees or economists – are the substantial core of this cultural creation and transmission process. Some of these products are perishable goods, others are long-lasting objects, others are even difficult to destroy once produced (plastics or nuclear waste, for example). Some of them, like pharmaceutical or medical products, become vital for the biological survival of the members of a society, or even of the whole species; others, especially in post-industrial societies, are temporarily fashionable but exposed to quick extinction; others represent symbols of values and meanings developed within a specific cultural context: this category comprises for example law provisions, artworks, educational systems and their contents.

Rules, norms and values formalized in laws and in institutionalized normative systems, such as the national constitutions, represent the visible outcomes of the shared understanding of reality and the meaning attribution process ceaselessly performed by cultural groups in time (Baumeister 2005, Leung and Morris 2001). In particular, constitutions comprise the basic social values, which can be defined as assumptions on what is desirable for the individual and for the group in a specific culture (Rokeach 1974, Schwartz and Bilsky 1987). They represent sets of cultural instructions or attempted solutions to universal human problems, such as biological survival, development and maintenance of relationships, transmission of information across generations (Kluckhohn and Stroedbeck 1961). The solutions to these problems can however vary across human communities. Variation can concern the specific strategies identified to solve a problem, or the priority attributed to a given problem in relation to the other ones to be solved. Different solutions and priorities are the foundations of cultural diversity.

Moving from the analysis of constitutional texts from different Western and non-Western countries, Massimini and Calegari (1979) systematized universal human problems and related solutions into 11 units of information, operationalizing them as components of a cultural network. These units can be grouped into overarching categories, representing more general issues each community has to deal with: biocultural reproduction (comprising the unit's work, income and property); the structure and organization of the society, and its contents' transmission mechanisms (including education, information exchange, participation, status, decision-making, legal system); and the core values on which the culture is built (comprising individual and social values). The whole set of units has been defined 'network' because its elements interact with one another, and are connected by mutual influences (Massimini and Delle Fave 2000).

Work is listed as the first unit of the network, and its role appears essential for the biological and cultural survival of any community. Work as a core component of society is also recurrent in the great majority of national constitutions in the world. Its importance for human life is therefore hardly underestimated. Nowadays, work is considered such a relevant component of the individual and social life that the World Health Organization has included it in its definition of mental health. According to this definition, mental health is "a state of well-being in which the individual realizes his or her own abilities, can cope with the normal stresses of life, can work productively and fruitfully, and is able to make a contribution to his or her community" (WHO 2004).

Work from the Subjective Perspective: Optimal Experience, Goals and Meanings

The complexity of human behaviour can hardly be captured through simplified paradigms, and the current state of the art in psychological studies does not allow for a unified perspective yet. Individuals and societies, like any other living system, show dynamic features, tending towards progressively higher levels of complexity through the exchange of information with the environment that leads to a ceaseless system's transformation (Pribram 1996). In particular, the acquisition of new information is transformed into stable neural connections at the biological level; into personal skills, beliefs, meanings and purposes at the psychological level; and into codified rules, norms and values at the social level. However, with time the integration of new information at the biological, psychological and social levels leads to progressive changes in the whole system configuration (Delle Fave 2007).

As bio-cultural entities, humans inherit a genotype, and build their *culturetype* by acquiring cultural information throughout life (Boyd and Richerson 1985), progressively attaining higher levels of complexity. Throughout their lives individuals interact with the information available in their daily environment by means of the process of *daily psychological selection* (Csikszentmihalyi and Massimini 1985). They differentially replicate subsets of this information, thus actively contributing to the horizontal and vertical transmission of culture. Psychological selection results from the person's preferential investment of attention and resources on a limited amount of the environmental opportunities for action and engagement. In particular, when daily activities are perceived as sources of well-being and positive experiences, they are more likely to be preferentially replicated and cultivated in the long run. Their selective cultivation provides the individual with increasingly complex competences and skills, fostering personal growth and development. More generally, psychological selection affects both the developmental trajectory of each individual and the survival and transmission of cultural information, norms and values (Massimini and Delle Fave 2000).

Several cross-cultural studies (summarized in Delle Fave et al. 2011a) showed that two core elements play a key role in guiding psychological selection. The first one is the association of specific activities with optimal experience, or flow (Csikszentmihalyi 1975/2000), characterized by engagement, skill investment, involvement and enjoyment. The second component is the long-term meaning individuals attribute to the activities available to them in their daily environment (Hicks and King 2009, Schlegel et al. 2011, Steger et al. 2008). Meaning-making represents the way in which people actively organize their own experience in time around their values and beliefs through goal setting, definition of priorities, and action strategies. As previously outlined, with time individuals can attribute different meanings to the same situation, according to progressively more complex principles in organizing experience (Kegan 1994, Kunnen and Bosma 2000). Through the attribution of meaning to specific life activities and domains, individuals pursue goals they deem as relevant, as well as consistent with social values and others' needs.

The potential association of daily activities with optimal experience and meaning is influenced by the level of complexity and challenge characterizing them. In particular, as concerns the work domain, the enrolment in repetitive, simplified and merely executive tasks strongly limits the availability of occasions for optimal experiences. In a recent study conducted on a large sample of adult participants belonging to different cultures (Delle Fave et al. 2011a), traditional work activities such as handicrafts, cooking, farming and gardening were quoted as occasions for flow much more frequently than office and factory tasks. This is not surprising, since technology and automation have led to the standardization of behaviour and of its products, to restrictions in individual initiative, to a decrease in the variation of activity structure and outcomes, with energy and artefact consumption prevailing on the creative transformation and effective use of environmental resources (Oskamp 2000). Nevertheless, work does represents a substantial opportunity for optimal experience, and considering that it accounts for a substantial portion of individuals' daily lives in most countries, understanding its potential in promoting well-being is a crucial issue for researchers and practitioners.

Research recurrently highlighted that optimal experience during work activities is characterized by high cognitive investment, as theoretically expected, but also by below average scores of affect and intrinsic motivation, however counterbalanced by the perception of high stakes in the task at hand, and long-term goals. This finding was obtained with the most diverse groups of professionals, including white-collar workers, managers, assembly-line workers (Csikszentmihalyi and Lefevre 1989, Haworth and Hill 1992), physicians (Delle Fave and Massimini 2003) and teachers (Bassi and Delle Fave in press). All participants constantly reported an experience of engagement, concentration and clear goals while working, but they did not describe the task as highly enjoyable or intrinsically rewarding. This phenomenon – not surprising, given the compulsory nature of work – was labelled work paradox (Csikszentmihalyi and Lefevre 1989).

The profile of optimal experience at work contrasts with the one associated with leisure activities such as sports and hobbies, which are characterized by intrinsic motivation, cognitive engagement, but by below-average goals and stakes in the activity. This finding leads to some considerations in light of the other core dimension of psychological selection, namely meaning. The immediate pleasure and reward derived from an activity can be related to intrinsic motivation (Ryan and Deci 2000), while the perception of long-term goals refers to broader pursuits that include professional achievements, potentially – though not necessarily – related to extrinsic motives such as career development and salary.

The human tendency to set and pursue goals, and its role in supporting the achievement of developmental tasks were largely investigated (Gollwitzer 1999, Sheldon et al. 2004). The perception of long-term goals and meanings during optimal experience at work was detected in different samples (Delle Fave and Massimini 2005, Demerouti 2006, Rheinberg et al. 2007, Salanova et al. 2006), suggesting the functional value of the work paradox in promoting competence development and professional commitment. These findings were further confirmed by a more recent study, conducted among adult participants from seven Western countries with the aim to investigate different components of perceived well-being, including meanings and goals (Delle Fave et al. 2013). The percentage of participants reporting work as a goal and a meaningful thing was second only to the percentage of those reporting family. Job related goals and meaningful things referred primarily to work as a value per se, and to its role in promoting individual well-being in terms of satisfaction, competence development and self-actualization. Other studies highlighted that the work domain can offer meaningful occasions for proving one's worth, sustaining self-esteem, and ultimately contributing to a full life (Page and Vella-Brodrick 2009, van den Heuvel et al. 2009). Recently Linley et al. (2010) provided an overview of the ways people can find meaning and purpose in work settings, by giving coherence and direction to their activities and by aligning organizations' strategies to support personal fulfillment.

Work and Social Integration

The potential of work as an opportunity for skill implementation and goal pursuit is especially evident when dealing with populations who live in disadvantaged conditions, such as immigrants or people with disabilities. To this purpose, it is important to notice that disadvantage is not only related to individual impairments and limitations; it also depends on the cultural attitude towards such limitations. As often emphasized, the so-called disadvantaged groups are only disadvantaged in an environment in which their condition brings about disadvantageous consequences (Saravanan et al. 2001).

As concerns migration, a study conducted among participants who had moved to Italy from India, Eastern Europe, South America and Africa (Delle Fave and Bassi 2009) showed that job opportunities represent one of the crucial components of adaptation (Zlobina et al. 2006), and that it can promote the retrieval of optimal experiences in the new country. Of course, participants with higher education levels and engaged in professionally qualified jobs were advantaged in this endeavour. However, also women employed as nurses, housemaids and caregivers reported their job as the prominent opportunity for optimal experience and goal pursuit in daily life, even though many of them had attained college education in their homeland. Nevertheless, working in families allowed them to develop competencies and to mobilize resources at different levels: the professional one, the relational one, and the cultural one, supporting and facilitating their integration process.

As concerns disability, studies were conducted among people with blindness or motor impairments (Delle Fave and Massimini 2004). Work was prominent among their perceived opportunities for optimal experiences, with variations according to the kind and time of onset of disability. People born blind mostly referred to handicrafts and manual activities such as knitting, working as physiotherapists or bookbinders, bottoming chairs with straw. On the contrary, people with acquired blindness mostly quoted activities based on intellectual skills, such as teaching. These competencies, developed during school years, turned to be a resource for their integration into productive life, counterbalancing the difficulties faced in performing practical tasks. Participants with motor disabilities (prominently medullary lesions due to trauma) highlighted the multifaceted role of work: an opportunity to improve personal skills and performance, but also a way to participate into social life. Similar findings were obtained among Nepalese youth with motor disabilities. Work accounted for almost half of the activities they associated with optimal experience; it included both traditional activities, such as woodcarving and tailoring, and modern jobs, such as computer programming and printing. Again, the association of optimal experience with work facilitated both personal skill development and the person's active contribution to society.

Finally, work proved to be a key resource for women and men with achondroplasia (Cortinovis et al. 2011): it was prominently associated with optimal experience, and it represented a major present challenge and future goal for these participants. These empirical evidences support the "normalization" function of employment (Adelson 2005) in circumstances implying diversity at any level.

Work, Cultural Meanings and Daily Life

The WHO definition of mental health reported in the first section of this chapter, apparently very broad and inclusive, is however strongly focused on performance, and it endorses a clearly Western pragmatist perspective: well-being means not only doing, working, making, but also being fruitfully productive (Delle Fave and Fava 2011). This approach entails a dangerous discriminatory attitude towards some categories of citizens. Can we consider a person unable to work because of quadriplegia as lacking mental health? Can we assume that after retirement people get suddenly affected by some kind of mental disorder? Should a hermit who devotes his life to meditation, rituals and prayer be classified as a mentally disturbed person? And – on the contrary – what about people with Down syndrome or mild cognitive impairment, diagnosed as mentally retarded, who work productively and fruitfully?

The identification of work as a substantial component of mental health, within an allegedly universal definition, derives in fact from a culture-bound perspective. In particular, Western societies emphasize performance more than experience, quantification of time and activities more than their quality. This is true of all life domains, including work. The conceptualization of productivity itself is influenced by cultural features: in industrial and post-industrial societies it is quantified on the basis of well-defined material and economic standards. In other contexts, on the contrary, its relevance and role can widely vary according to the typology of activity and to the family and community needs and resources (Haworth 1997). For example, in rural contexts where agriculture and handicrafts are the prominent sources of subsistence, work activities are an integral part of the daily life, they are often shared with other family members, they represent opportunities for socialization and exchange of information within the community, and their role and meaning in individuals' life is manifold, in that these activities can satisfy a variety of basic and higher-order needs, from biological survival to personal skill development and self-expression (Delle Fave and Massimini 1988, 1991). For the disadvantaged categories of citizens exemplified in the previous pages work is not prominently relevant in terms of productivity, but – besides satisfying survival needs – it is perceived as an opportunity for social inclusion and interpersonal interactions. For this reason these categories of citizens do not focus on job performance to achieve optimal experiences, but on the social context of work.

Overall, the conceptualization of work as a specific and separated domain of daily life is typical of industrial and post-industrial societies (Csikszentmihalyi 1990, 1997), in which a clear distinction is made between the time devoted to work and the time devoted to all the other daily activities, in particular leisure and family/social interactions (Rojek 2000).

Moreover, the relationship between work and leisure is not homogeneous within a given society, but it depends on people's occupation. To this purpose, Parker (1997) proposed three patterns of work–leisure relationship, which can be related to the main theories developed in this field. The *extension* pattern derives from the spillover approach, that posits mutual influences in terms of skill development and levels of satisfaction between different areas of life (Leiter and Durup 1996, Staines 1980). This pattern is typical of people involved in creative and autonomy-supporting jobs, such as artists, scientists and specialized professionals. The *opposition* pattern, also originating from the spillover approach, applies to people enrolled in risky and damaging jobs, who compensate through leisure the frustrations and constraints of work. The *separation* pattern, stemming from Dubin's theory of segmentation (1956), considers work and leisure as two independent life domains, with no mutual influences, and it applies to the great majority of today's workers, employed in neither particularly creative nor dangerous jobs. Considering that in post-industrial societies the extension pattern is confined to a minority of jobs, not available to a large portion of the population, the negative consequences of the dichotomized perspective embedded in the opposition and separation patterns are clearly reflected in the psychological literature on work, that emphasizes the need for fostering workers' motivation and for countering the levels of stress and burnout associated with productive activities (Le Blanc et al. 2008, Schaufeli et al. 2009).

Unfortunately, many jobs do not provide workers with the opportunities for concentration, control, engagement and personal initiative that allow for the emergence of optimal experience. Similarly, not all jobs share the same richness in meaning and the same potential for personal growth and development. Work conditions can be widely improved, however this is not always feasible to the extent that job tasks themselves become engaging and enjoyable. In these circumstances, finding involvement, goals and meaning in other life domains such as hobbies, cultural and social interests, family and interpersonal relations represents a more practicable way to foster well-being and to provide people with both opportunities for individual development and social participation (Rojek 2000). To address this issue, researchers have recently focused on the benefits derived from pursuing balance in life as a whole and across domains.

Harmony and Balance in Daily Life: Unexplored Dimensions of Well-being

A historical and interdisciplinary overview of harmony as a concept related to quality of life (Delle Fave 2014) has recently highlighted that, beyond philosophical and conceptual differences, in all cultural traditions harmony has been directly or indirectly related to well-being. The concepts of *ataraxia* (Epicure, ancient Greece), *anasakti* (Hinduism), *viraaga* (Buddhism) imply evenness of judgement, peace and quietness of mind, and the ability to preserve balance and serenity in both enjoyable and challenging times.

Harmony represents an indicator of a good quality of life, at both the individual and social levels. At the psychological level, it has been variously defined as a dynamic state of existence, related to satisfaction or contentment, agency, spiritual enrichment and a positive outlook (Lu 2001); a condition of self-acceptance and positive relationships

(Muñoz Sastre 1998); a component of wisdom, together with self-love, good judgement, appreciation and purpose in life (Jason et al. 2001); a dimension of spirituality, together with meaning and purpose, social connectedness, peacefulness, gratefulness, forgiveness and self-discipline (Koenig 2008). A recent international survey has highlighted that, when invited to define happiness in their own words, people very often refer to inner harmony and balance (Delle Fave et al. 2011b). At the social level, harmony is a prominent value of collectivistic societies, and it can be considered a goal per se – according for example to the Confucian perspective – or a means to pursue consensus, conformity and conflict avoidance (Leung et al. 2002). More recently, it has been identified as a dimension of social quality, together with social-economic security, social inclusion, social cohesion and empowerment to develop individual potential (Ho and Chan 2009).

Particularly interesting for the purpose of this chapter is the concept of cross-domain balance, emerged in the WHO's definition of quality of life as a multi-componential construct (WHOQOL Group 2004). Sirgy and Wu (2009) recently explored balance across life domains as a core component of well-being. Their theoretical approach moves from the assumption that, in order to satisfy the full spectrum of human needs, people have to engage in multiple life domains. A balanced life derives from fulfilment in various important domains, combined with little or no negative affect in other ones. Sirgy and Wu articulate their theory in three basic postulates. The first one states that people who derive life satisfaction from multiple life domains (balance) are likely to experience higher levels of subjective well-being than those whose life satisfaction stems from a single domain. According to the second postulate, balance contributes to subjective well-being because through involvement in a single life domain people can satisfy only a limited part of the full spectrum of developmental needs. Finally, the third postulate states that, since subjective well-being can only be attained when both survival and growth needs are met, balance across life domains is necessary to well-being. A very important aspect of this model consists in the salience of the domains considered. The attainment of well-being must concern relevant life domains, that are deemed as meaningful by the individual and that allow for the satisfaction of both basic and growth needs.

As a consequence, people experiencing moderate levels of satisfaction from multiple – and salient – life domains are likely to report higher levels of subjective well-being than people experiencing high levels of satisfaction stemming from a single – though salient – domain. In other words, it is better to be moderately happy and satisfied in multiple life domains, than to experience extremely positive feelings in one single domain, to the detriment or neglect of the others.

The concept of cross-domain balance has important implications for the work context. Several studies have highlighted that impairments or limitations in one life domain do not significantly modify the general quality of life level, which is relatively stable (Cummins and Nistico 2002). As discussed above, especially in contemporary urban contexts many people are forced to devote a relevant amount of their daily life to jobs that are low in complexity, challenge variation and opportunities for self-expression and skill development. Nevertheless, investing resources and attention in multiple life areas can allow them to compensate for the dissatisfaction with work, thus achieving higher levels of well-being.

Moreover, considering the spillover effect, previous studies have shown that the importance individuals attach to single life domains is crucial in determining the weight of a domain in contributing to overall well-being (Diener et al. 2000). In order to preserve well-being in spite of a job that does not provide opportunities for optimal experiences, meaning-making and skill development, downplaying its importance can be an adaptive strategy (Wu 2009).

Conclusion: An Integrated Model of Development

As outlined in the previous pages, work is one, albeit crucial, component of human life. Its structure and relevance are shaped by the ecosystem as well as by the cultural context. Individual workers, in their turn, build their job experience according to the intrinsic complexity and challenges of the activity – on the one hand – and to their personal skills, goals and meanings – on the other hand. Figure 3.1 summarizes the various dimensions of the individual–environment interplay highlighting their dynamic relationship. Work can be considered a specific case of this dynamic interplay.

The research findings illustrated in this chapter suggest that individual development can take place in multiple life domains – education, work, leisure, relationships – provided that people are exposed to complex and challenging, as well as personally and socially meaningful, opportunities for action. Individuals should be encouraged to follow their own path toward complexity through the effective use of their competencies and skills, and through the cultivation of activities associated with lifelong and socially relevant meanings, as diversified as possible across the different life domains.

This process has also long-term cultural implications. Through their psychological selection, individuals can actively influence the complexity of their cultural system. The

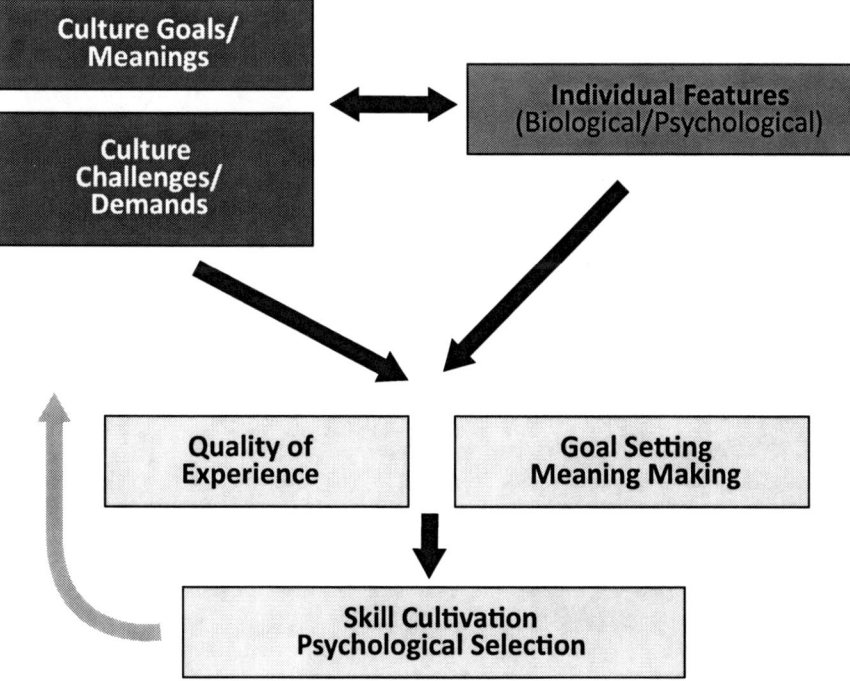

Figure 3.1   An integrated model of development

contents and values of a given culture are the result of the combined action of individuals who find meaning and opportunities for self-expression in the most varied activities and tasks. These people come from different backgrounds, have different access to job opportunities in their life, experience hardship or affluence, live in metropolitan areas or in small villages, cope with extraordinary circumstances or with the little troubles of ordinary life. In order to give the adequate relevance to such a variety of perspectives, a paradigm centred on harmony, as the dynamic integration of various aspects and components of a whole, can promote the development of interventions and social policies respectful of differentiation and diversity.

References

Adelson, B.M. 2005. *Dwarfism: Medical and Psychosocial Aspects of Profound Short Stature*. Baltimore, MD: Johns Hopkins University Press.

Baumeister, R.F. 2005. *The Cultural Animal*. New York, NY: Oxford University Press.

Bassi, M. and Delle Fave, A. in press. Optimal experience among teachers: New insights into the work paradox. *The Journal of Psychology*.

Boyd, R. and Richerson, P.J. 1985. *Culture and the Evolutionary Process*. Chicago, IL: Chicago University Press.

Cortinovis, I., Luraschi, E., Intini, S., Sessa, M. and Delle Fave, A. 2011. The daily experience of people with achondroplasia. *Applied Psychology: Health And Well-Being*, 3, 207–27.

Csikszentmihalyi, M. 1975/2000. *Beyond Boredom and Anxiety*. San Francisco, CA: Jossey-Bass.

Csikszentmihalyi, M. 1990. *Flow: The Psychology of Optimal Experience*. New York, NY: Harper & Row.

Csikszentmihalyi, M. 1997. Activity, experience and personal growth. In *Physical Activity in Human Experience: Interdisciplinary Perspectives*, edited by J. Curtis and S. Russell. Champaign: Human Kinetics, 59–88.

Csikszentmihalyi, M. and Lefevre, J. 1989. Optimal experience in work and leisure. *Journal of Personality and Social Psychology*, 56, 815–22.

Csikszentmihalyi, M. and Massimini, F. 1985. On the psychological selection of bio-cultural information. *New Ideas in Psychology*, 3, 115–38.

Cummins, R.A. and Nistico, H. 2002. Maintaining life satisfaction: The role of positive cognitive bias. *Journal of Happiness Studies*, 3, 37–69.

Delle Fave, A. 2007. Theoretical foundations of the ESM. In *Experience Sampling Method: Measuring the Quality of Everyday Life*, edited by J. Hektner, J. Schmidt and M. Csikszentmihalyi. New York, NY: Sage, 15–28.

Delle Fave, A. 2014. Harmony. In *Encyclopedia of Quality of Life and Well-Being Research*, edited by A. Michalos. Dordrecht: Springer.

Delle Fave, A. and Bassi, M. 2009. Sharing optimal experiences and promoting good community life in a multicultural society. *Journal of Positive Psychology*, 4, 280–89.

Delle Fave, A. and Fava, G.A. 2011. Positive psychotherapy and social change. In *Positive Psychology as Social Change*, edited by R. Biswas-Diener. Dordrecht: Springer, 267–92.

Delle Fave, A. and Massimini, F. 1988. Modernization and the changing contexts of flow in work and leisure. In *Optimal Experience: Psychological Studies of Flow in Consciousness*, edited by M. Csikszentmihalyi and I. Csikszentmihalyi. New York, NY: Cambridge University Press, 193–213.

Delle Fave, A. and Massimini, F. 1991. Modernization and the quality of daily experience in a Southern Italy village. In *Contemporary Issues in Cross-Cultural Psychology*, edited by N. Bleichrodt and P.J.D. Drenth. Amsterdam: Swets & Zeitlinger B.V., 110–19.

Delle Fave, A. and Massimini, F. 2003. Optimal experience in work and leisure among teachers and physicians: Individual and bio-cultural implications. *Leisure Studies*, 22, 323–42.

Delle Fave, A. and Massimini, F. 2004. Bringing subjectivity into focus: Optimal experiences, life themes, and person-centered rehabilitation. In *Positive Psychology in Practice*, edited by P.A. Linley and S. Joseph. New York, NY: Wiley, 581–97.

Delle Fave, A. and Massimini, F. 2005. The investigation of optimal experience and apathy: Developmental and psychosocial implications. *European Psychologist*, 10, 264–74.

Delle Fave, A., Massimini, F. and Bassi, M. 2011a. *Psychological Selection and Optimal Experience across Cultures: Social Empowerment through Personal Growth*. Dordrecht: Springer.

Delle Fave, A., Brdar, I., Freire, T., Vella-Brodrick, D. and Wissing, M.P. 2011b. The Eudaimonic and hedonic components of happiness: Qualitative and quantitative findings. *Social Indicators Research*, 100, 185–207.

Delle Fave, A., Wissing, M.P., Brdar, I., Vella-Brodrick, D. and Freire, T. 2013. Cross-cultural perceptions of meaning and goals in adulthood: Their roots and relation with happiness. In *The Best Within Us: Positive Psychology Perspectives on Eudaimonia*, edited by A. Waterman. Washington, DC: American Psychological Association, 227–48.

Demerouti, E. 2006. Job characteristics, flow, and performance: The moderating role of conscientiousness. *Journal of Occupational Health Psychology*, 11, 266–80.

Diener, E., Napa Scollon, C.K, Oishi, S., Dzokoto, V. and Suh, E.M. 2000. Positivity and the construction of life satisfaction judgments: Global happiness is not the sum of its parts. *Journal of Happiness Studies*, 1, 159–76.

Dubin, R. 1956. Industrial workers' worlds: A study of the central life interest of industrial workers. *Social Problems*, 3, 358–90.

Gollwitzer, P.M. 1999. Implementation intentions: Strong effects of simple plans. *American Psychologist*, 54, 493–503.

Haworth, J.T. (ed.). 1997. *Work, Leisure, and Well-being*. London: Routledge.

Haworth, J. and Hill, S. 1992. Work, leisure and psychological well-being in a sample of young adults. *Journal of Community and Applied Social Psychology*, 2, 147–60.

Henrich, J., Boyd, R. and Richerson, P.J. 2008. Five misunderstandings about cultural evolution. *Human Nature*, 19, 119–37.

Hicks, J.A. and King, L.A. 2009. Meaning in life as a subjective judgment and a lived experience. *Social and Personality Psychology Compass*, 3/4, 638–53.

Ho, S.S.M. and Chan, R.S.Y. 2009. Social harmony in Hong Kong: Level, determinants and policy implications. *Social Indicators Research*, 91, 37–58.

Jablonka, E. and Lamb, M.J. 2005. *Evolution in Four Dimensions: Genetic, Epigenetic, Behavioural and Symbolic Variations in the History of Life*. Cambridge, MA: MIT Press.

Jablonka, E. and Lamb, M.J. 2007. Précis of "Evolution in Four Dimensions." *Behavioral and Brain Sciences*, 30, 353–92.

Jason, L.A., Reichler, A., King, C., Madsen, D., Camacho, J. and Marchese, W. 2001. The measurement of wisdom: A preliminary effort. *Journal of Community Psychology*, 29, 585–98.

Kegan, R. 1994. *In Over Our Heads*. New York, NY: Cambridge University Press.

Kluckhohn, F. and Stroedbeck, F. 1961. *Variants in Value Orientation*. New York, NY: Row Peterson.

Koenig, H.G. 2008. Concerns about measuring "spirituality" in research. *The Journal of Nervous and Mental Disease*, 196, 349–55.

Kunnen, E.S. and Bosma, H.A. 2000. Development of meaning making: A dynamic systems approach. *New Ideas in Psychology*, 18, 57–82.

Le Blanc, P., de Jonge, J. and Schaufeli, W. 2008. Job stress and occupational health. In *An Introduction to Work and Organizational Psychology: A European Perspective*, edited by N. Chmiel. Malden, MA: Blackwell, 119–47.

Leiter, M. and Durup, M. 1996. Work, home, and in-between: A longitudinal study of spillover. *Journal of Applied Behavioral Science*, 32, 29–47.

Leung, K. and Morris, M.W. 2001. Justice through the lens of culture and ethnicity. In *Handbook of Law and Social Sciences: Justice*, edited by J. Sanders and V.L. Hamilton. New York, NY: Plenum, 343–78.

Leung, K., Tremain Koch, P. and Lu, L. 2002. A dualistic model of harmony and its implications for conflict management in Asia. *Asia Pacific Journal of Management*, 19, 201–20.

Linley, P.A., Harrington, S. and Garcea, N. (eds) 2010. *Oxford Handbook of Positive Psychology and Work*. New York, NY: Oxford University Press.

Lu, L. 2001. Understanding happiness: A look into the Chinese folk psychology. *Journal of Happiness Studies*, 2, 407–32.

Massimini, F. and Calegari, P. 1979. *Il Contesto Normativo Sociale [The Social Normative Context]*. Milan: Angeli.

Massimini, F. and Delle Fave, A. 2000. Individual development in a bio-cultural perspective. *American Psychologist*, 55, 24–33.

Muñoz Sastre, M.T. 1998. Lay conceptions of well-being and rules used in well-being judgements among young, middle-aged, and elderly adults. *Social Indicators Research*, 47, 203–31.

Newson, L., Richerson, P.J. and Boyd, R. 2007. Cultural evolution and the shaping of cultural diversity. In *Handbook of Cultural Psychology*, edited by S. Kitayama and D. Cohen. New York, NY: Guilford Press, 454–76.

Oskamp, S. 2000. A sustainable future for humanity? *American Psychologist*, 55, 496–508.

Page, K.M. and Vella-Brodrick, D.A. 2009. The "what," "why" and "how" of employee well-being. *Social Indicators Research*, 90, 441–58.

Parker, S. 1997. Work and leisure futures: Trends and scenarios. In *Work, Leisure, and Well-being*, edited by J. Haworth. London: Routledge, 180–91.

Pribram, K.H. 1996. Interfacing complexity at the boundary between the natural and social sciences. In *Evolution, Order and Complexity*, edited by E.L. Khalil and K.E. Boulding. New York, NY: Routledge, 40–60.

Rheinberg, F., Manig, Y., Kliegel, R., Engeser, S. and Vollmeyer, R. 2007. Flow bei der Arbeit, doch Glück in der Freizeit. Zielausrichtung, Flow und Glücksgefühle [Flow during work but happiness during leisure time: Goals, flow experience and happiness]. *Zeitschrift für Arbeits- und Organisationspsychologie*, 3, 105–15.

Richerson P.J. and Boyd R. 2005. *Not by Genes Alone: How Culture Transformed Human Evolution*. Chicago, IL: University of Chicago Press.

Rojek, C. 2000. *Leisure and Culture*. Basingstoke: Macmillan.

Rokeach, M. 1974. *The Nature of Human Values*. New York, NY: Free Press.

Ryan, R. and Deci, E. 2000. Self-determination theory and the facilitation of intrinsic motivation, social development, and well-being. *American Psychologist*, 55, 68–78.

Salanova, M., Bakker, A. and Llorens, S. 2006. Flow at work: Evidence for an upward spiral of personal and organizational resources. *Journal of Happiness Studies*, 7, 1–22.

Saravanan, B., Manigandam, C., Macaden, A., Tharion, G. and Bhattacharji, S. 2001. Re-examining the psychology of spinal cord injury: A meaning centered approach fom a cultural perspective. *Spinal Cord*, 39, 323–6.

Schaufeli, W.B., Leiter, M. and Maslach, C. 2009. Burnout: 35 years of research and practice. *Career Development International*, 14, 204–20.

Schlegel, R.J., Hicks, J.A., King, L.A. and Arndt, J. 2011. Feeling like you know who you are: Perceived true self-knowledge and meaning in life. *Personality and Social Psychology Bulletin*, 37(6), 745–56.

Schwartz, S.H. and Bilsky, W. 1987. Toward a theory of the universal structure and content of values: Extensions and cross-cultural replications. *Journal of Personality and Social Psychology*, 58, 878–91.

Sheldon, K.M., Elliot, A.J., Ryan, R.M., Chirkov, V., Kim, Y., Wu, C., Demir, M. and Sun, Z. 2004. Self-concordance and subjective well-being in four cultures. *Journal of Cross-cultural Psychology*, 35, 209–23.

Sirgy, M.J. and Wu, J. 2009. The pleasant life, the engaged life, and the meaningful life: What about the balanced life? *Journal of Happiness Studies*, 10, 183–96.

Staines, G. 1980. Spillover versus compensation: A review of the literature on the relationship between work and nonwork. *Human Relations*, 33, 111–29.

Steger, M.F., Kashdan, T.B., Sullivan, B.A. and Lorents, D. 2008. Understanding the search for meaning in life: Personality, cognitive style, and the dynamic between seeking and experiencing meaning. *Journal of Personality*, 76, 199–228.

van den Heuvel, M., Demerouti, E., Schreurs, B.H.J., Bakker, A.B. and Schaufeli, W.B. 2009. Does meaning-making help during organizational change? Development and validation of a new scale. *Career Development International*, 14, 508–33.

WHOQOL Group 2004. Can we identify the poorest quality of life? Assessing the importance of quality of life using the WHOQOL-100. *Quality of Life Research*, 13, 23–34.

World Health Organization. 2004. *Promoting Mental Health: Concepts, Emerging Evidence, Practice*. Summary report. Geneva: WHO.

Wu, C. 2009. Enhancing quality of life by shifting importance perception among life domains. *Journal of Happiness Studies*, 10, 37–47

Zlobina, A., Basabe, N., Paez, D. and Furnham, A. 2006. Sociocultural adjustment of immigrants: Universal and group predictors. *International Journal of Intercultural Relations*, 30, 195–211.

4 ACT: A Third Wave Behavioural-Cognitive Approach to Creating Healthy Workplaces

JULIE MÉNARD and BRENT BERESFORD

Abstract

This chapter examines the efficacy of Acceptance and Commitment Therapy (ACT) as an intervention tool in organizations in order to promote employees' health and wellness. Based on a number of different studies, the authors present how the literature has shown a significant impact of this approach on the psychological health of employees and on different organizational outcomes such as workloads, presenteeism and absenteeism. The authors expose a brief example of such an intervention and conclude with an agenda for future research and implications for practice.

Introduction

Robert Frost once said: "the difference between a job and a career is the difference between forty and sixty hours a week." We spend countless hours at work in order to financially support our families and ourselves. For many, most of the day's sunlit hours are spent sitting behind a desk. Thus, it is obvious why there is an importance in being and feeling well in the workplace. This imperative need and its important link to performance in daily work tasks have helped to create a place for the psychologist in organizational settings. It is now well recognized that psychological distress in employees leads to more sick leave and presenteeism and performance decrease due to health problems (Lim et al. 2000). Most work settings past the pre-contemplation stage of change regarding their view on mental health amongst employees (see Prochaska and DiClemente 1984 for more details on their model of change). However, most organizations are still at the contemplation phase. Hence, they are conscious of the consequences of poor mental health but are still focusing on the costs. Many 'loops' will be necessary to properly deal with employees' health and wellness.

In order to act in a proactive way to both prevent (i.e. early intervention) and decrease employee distress in organizations there is a need to instate interventions promoting mental health, directly in work settings. Most worksite interventions – called stress management intervention (SMI) and referring to any organizational activity or that focuses on reducing the presence of work-related stressors or on supporting individuals to minimize the negative outcomes of exposure to these stressors (Ivancevich et al. 1990) – lie within secondary and tertiary level of interventions (Richardson and Rothstein 2008), when consequences of distress are already visible. Organizational-level interventions have been proved efficient to decrease employee distress, stress and improve performance (Bambra et al. 2007, Egan et al. 2007). However, those interventions aim first at changing the work environment to reduce inherent psychosocial risks (e.g. decrease demand, improve control) but do not provide direct tools for workers. As highlighted by Bond and Hayes (2002), three main issues still remain when using solely psychosocial risk management techniques in work settings:

1. some sources of workplace stress are not completely avoidable;
2. the ability of a worker to modify his/her workplace stress can be impeded by a lack of ability to find an appropriate fashion to manage their stress; and
3. people may have behaviour patterns that are linked to an increase in stress reactions outside of the workplace, and these reactions can be related to their experience of stress at the workplace.

Based on these observations, we believe that there is a place for individual-focused interventions within a comprehensive stress management programme that provide both direct tools to employees and manage psychosocial risks through work reorganization, which could be used both as an early intervention in order to prevent stress as well as a secondary and tertiary stress prevention intervention. Job stress can be prevented and controlled effectively using a systems approach that integrates primary, secondary and tertiary intervention (LaMontagne et al. 2007).

Many theories and phenomena studied in psychology have been adapted and applied to the workplace. One such model is part of a third wave of cognitive-behavioural therapies called *Acceptance and Commitment Therapy* (or ACT, pronounced as spelled and not as Aay-See-Tee). It is a relatively recent approach that has its roots in mindfulness-based therapies, and it is proving to be a very valuable tool in many areas of mental health such as depression or depressive symptoms (e.g. Bohlmeijer et al. 2011, Forman et al. 2007, Zettle and Hayes 1986, Zettle and Rains 1989), anxiety (e.g. Dalrymple and Herbert 2007, Roemer et al. 2008), chronic pain (e.g. Dahl et al. 2004, McCracken et al. 2007) and, more interestingly, work stress (e.g. Bond and Bunce 2000) (see Ruiz 2010 for a review). Meta-analyses showed ACT significantly reduces distress symptoms with medium to large effect sizes (Hayes et al. 2006, Öst 2008). ACT was developed to help individuals live with difficult emotional and cognitive experiences (Hayes et al. 2006).

In this chapter, the basic principles of ACT will be introduced in order to clarify its unorthodox take on human functioning. Afterwards, several research studies using ACT and its underlying processes will be presented in order to exemplify its place for creating healthy workplaces. Finally, in order to help practitioners use ACT within their practice, a typical ACT intervention will be described.

The Origins of Acceptance Commitment Therapy

ACT and other third-wave therapies came about partly because of the current popular scientific views regarding psychiatric diseases. Some therapists were discontented by the large presence of emphasis on *form* of mental health issues versus the lack of focus on the *function* of these problems (Hayes et al. 1999, 2012). Others raised issues with the apparent disconnection between recommended clinical treatments and the *context* in which these mental health problems played out in everyday life (Hayes et al. 1999, 2012). According to Hayes and collaborators (1999), human suffering comes, at least partly, from the following tendencies:

1. avoidance behaviours, which aim to change, control or avoid painful thoughts, emotions or bodily sensations. Those behaviours actually contribute to maintain or worsen suffering;
2. fusion with mental content (i.e. thought and emotions), leading the person to perceive mental content as the reality or as an extent of the Self;
3. an orientation towards the past or the future instead of a mindful internal and external experience of the present moment; and, finally
4. actions or behaviours that are not in line with one's values.

ACT authors looked to take a different approach, and integrated these concerns into their new perspective on human suffering that takes into consideration both the function of mental content and behaviours and the context in which they unfold.

The model underlying ACT is based on a particular philosophy of science in combination with a view of cognition from a language perspective called *functional contextualism*. This view defines psychological events such as thoughts, beliefs, feelings and emotions as the interactions of whole organisms in and with a context considered both historically and situationally (Hayes et al. 2012). In this view, behaviour, be it overt or covert, is combined with its context into one organizing unit that becomes the focus of all interventions. Therefore, no behaviour is evaluated as good or bad on its own without considering its function and its effect. Taking this view of psychological events helps ACT's clinical and academic community to address certain issues that have been raised due to the current emphasis on the form and not the function of these same events. Their answer lies in seeing the two elements functioning together as one, thus never overlooking either.

Additionally to functional contextualism, the ideas about cognition used by ACT practitioners come from the *relational frame theory* (RFT). According to RFT, at the basis of language abilities and other high level cognitive processes is an ability to learn and apply relational frames (Hayes et al. 2012). Relational framing is a learned operant behaviour in which the relationship between one thing and another comes to be known (Hayes et al. 1999). This process is ubiquitous, allowing us to constantly create relationships between anything and everything without ever having to follow any specific rules or guidelines. Once an event is compared to another, a relational frame is created, and we assume to have learnt something about these events based on this newly derived relationship. According to Hayes et al. (1999, 2012), the main issue lies in the fact that these relational frames that we learn are arbitrarily applicable and thus it is difficult to determine how and when events are related (e.g. if we say "good," you'll probably think "bad"; if we say "what doesn't kill you ...," you'll probably think "makes you stronger"). As we learn and develop, there are more and more relations of this type, expanding our relational frame web (i.e. the collection of all relational frames that we have ever learnt). Consequently, this can lead to the derivation of harmful and unhelpful relations that, once learnt, cannot be "un-learnt." The bright side is that many useful problem-solving processes are carried out via the information held in these webs. Also, without relational framing there would be no language! However, relation frames hold a strong control over our behaviour (Hayes et al. 1999, 2012). Moreover, since relational frames are rule-governed they are therefore not necessarily applicable in all types of situations, and can lead to psychological rigidity, which is associated with suffering and maladaptive functioning (Hayes et al. 1999, 2012). Hence, according to ACT theorists, one of the root causes of psychological rigidity (and human suffering) is related to the way in which we learn to use and apply language.

ACT's Six Processes and Psychological Flexibility

Third-wave practitioners have integrated this view of language into a functional-contextual model, ACT, that aims at developing *psychological flexibility* (process at the centre of Figure 4.1), a learnt and developable ability that helps individuals exert a voluntary distanciation from one's difficult thoughts and emotions in order to live them fully while making sure to strive for goals that are fully related to one's values (Hayes et al. 2006). In other words, it is the ability to be consciously in touch with the present moment, accepting it for what it truly is, and then adapting one's behaviour as needed in the light of one's personally chosen and constructed values (Hayes et al. 1999, 2012). Individuals who are psychologically inflexible tend to suppress their undesired thoughts and negative emotions, therefore using experiential avoidance. Such a strategy has been shown to be counterproductive since it increases distress and reduces the ability to take action when facing difficult events (Gross 1998).

According to the ACT model, six core processes lead to psychological flexibility. They aim at developing three response styles: open, centered and engaged. The core processes are illustrated in the Hexaflex (Figure 4.1).

ACCEPTANCE AND DEFUSION

The first two processes, acceptance and defusion, constitute the core process dyad leading to an open response style. Acceptance is accepting instead of controlling one's difficult or hurtful thoughts, emotions and bodily sensations, while reducing avoidance behaviours. Cognitive defusion is untangling from one's thoughts, emotions and bodily sensations without trying to change them.

When a person refuses to encounter his/her negative psychological experiences head-on, using techniques that seek to change these events (e.g. have positive thoughts; see the positive side of the situation) or push them away (e.g. avoid thinking), is called experiential avoidance, which could lead to detrimental consequences when avoidance behaviours take the person away from his/her valued course of action (Hayes et al. 2004). It is not necessarily the avoidance strategy itself that is troublesome. A little bit of

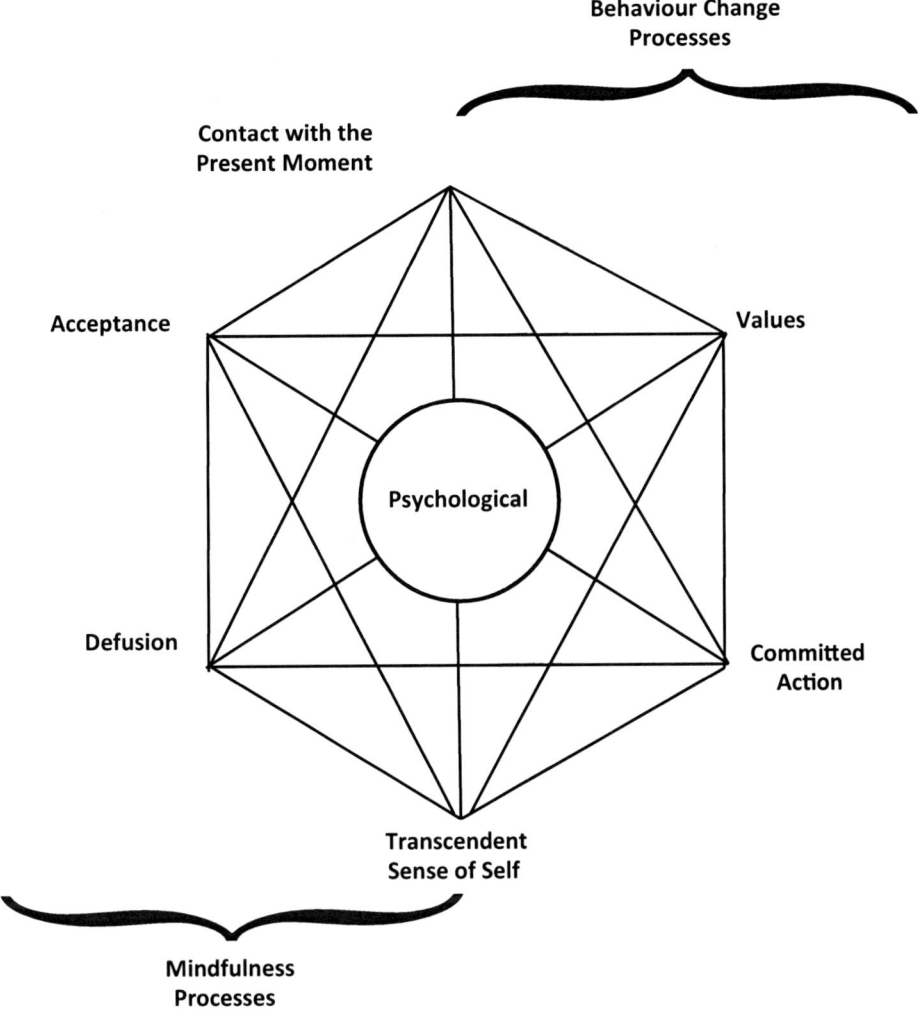

Figure 4.1   The Hexagon Model of Psychological Flexibility (Bond et al. 2006)

experiential avoidance, in the short term, can be self-protective in some specific situations such as not showing disgust when tasting an unsavoury dish cooked by one's partner, or not showing signs of stress during a job interview. The problem is its application as one's main coping strategy in the long term. Experiential avoidance is the core mechanism in the development and maintenance of psychological distress (Kashdan et al. 2006). It is the fact that it comes to be applied in unnecessary contexts that can lead to problems of adaptability (Bonanno et al. 2004). "Experiential avoidance becomes a disordered process when it is applied rigidly and inflexibly such that enormous time, effort, and energy is devoted to managing, controlling, or struggling with unwanted private events" (Kashdan et al. 2006: 1302). Paradoxically, attempting to control unpleasant thoughts, feelings and sensations (i.e. through thought suppression) actually increases both their frequency and related distress (Gross 1998).

Acceptance processes seek to encourage "the adoption of an intentionally open, receptive, flexible, and non-judgmental posture with respect to moment-to-moment experience" (Hayes et al. 2012: 77). The application of this type of process encourages appetitive-driven instead of aversive-driven behaviours. Being in contact with the present moment in a way that is focused, flexible and voluntary is an important part of the base of support for being open and engaged (Hayes et al. 2012). This type of awareness can be learnt through mindfulness exercises and thus help to improve one's ability to be centred (Baer 2003).

Cognitive fusion occurs when our thoughts control our behaviour, regardless of the context – it is a difficulty in separating the processes and products of thinking from objective reality (Hayes et al. 2012). For example, if one thinks "I am not interesting," the process of cognitive fusion lead her/him to think this thought defines her/him as an individual. The thought "not interesting" becomes an extension of the Self (i.e. "I AM not interesting"). When mental content (i.e. thought, belief, emotion) is perceived as a true reflection of the reality or an extension of the Self, there is a risk of acting according to this thought or belief, therefore allowing mental content to control one's action. In that particular case, it is likely that the person's action would be not to talk (because she/he is not interesting), than actually acting in a "not interesting" way.

Thus, defusion methods seek to increase one's ability to tell the difference between the content of one's thoughts and the processes that are incessantly occurring in our minds, allowing for a more open view of reality (Hayes et al. 2012).

MINDFULNESS AND TRANSCENDENT-SELF

The third process – mindfulness – is deliberately noticing one's internal and external experience as it unfolds in the present moment, without any judgement. Mindfulness techniques are based on Buddhist meditation and consist of bringing one's complete attention to the present experience on a moment-to-moment basis (Kabat-Zinn 1994) and have been shown to efficiently help a broad range of individuals to cope with stress and other clinical and nonclinical problems (Grossman et al. 2004). The fourth process, perspective change (also called Transcendent Self or Self-as-context), can be simply defined by developing a new/different perspective on one's issues. Developing perspective change aims at differentiating oneself and the context in which the Self evolves. Mindfulness and transcendent Self are the core process dyad leading to a centred (towards the Self instead of towards mental content) response style.

As we grow and mature, our tendency to use relational framing causes us to constantly name, categorize and evaluate events. In doing so, we become fused to these processes, and the stories and descriptions that we create about ourselves come to define who we are in our minds (Hayes et al. 1999, 2012). Thus, the Self that we come to know is a conceptualized Self (or Self-as-content). In contrast, the core process of having a view of the Self-as-context involves being able to view the Self from different perspectives regarding time, place and person. This view of the Self involves an element of social compassion because our perspective is only gained in relation to the perspective of another. One comes to understand the concept of here only in relation to the concept of there; the concept of now is understood in relation to then; and the concept of I is understood in relation to the concept of you. This "socially interconnected sense of self" allows for behavioural responses that come from a more centred state of mind (Hayes et al. 2012).

VALUES AND COMMITTED ACTION

The last two processes, values and committed action, are the core process dyad leading to an engaged response style. Values are clarified and actions in concordance with one's values – committed actions – are taken (Hayes et al. 2006, 1999, 2012).

The goal of the core process of valuing is to get away from behaviour that is controlled by a problem-solving mode of mind (e.g. pleasing others, reacting) and encourage choosing behaviours that are anchored in freely chosen personal values. In order to do this, one must take committed actions that directly aim at pursuing a valued life, instead of adopting impulsive and avoidant behaviours. These committed actions serve the purpose of maintaining moment-to-moment devotion to a higher value-based purpose, thus encouraging a more engaged behaviour pattern (Hayes et al. 1999, 2012).

According to ACT principles, in order to maintain psychological flexibility, a balance must be found between the three response styles, and thus within the six core processes (Hayes et al. 2006, 2012). What ACT interventions do concretely is to use and develop all of these processes with individuals (e.g. employees, patients) in order to foster the creation of psychological flexibility in the service of diminishing suffering.

With its roots in all of the above-described processes, the ACT approach is meant to promote a more open, centred and engaged approach to living (Hayes et al. 2012). Hence, it is for this reason that ACT can be used for more than only the treatment of mental health disorders and should be used to promote healthier living. With the fast-paced living that most of Western society is engaged in, stressors abound. It is only reasonable to believe that if approaching life with more mindfulness and acceptance, one could improve one's situation and come out feeling better. Several researchers have held to this idea, and seek to show how the core processes described in the psychological flexibility model are actual predictors of change in real-life situations such as the workplace.

ACT for Creating Healthy Workplaces: State of Science

THE ROLE OF PSYCHOLOGICAL FLEXIBILITY IN HEALTHY WORKPLACES

In a long-term comparative study, Bond and Bunce (2003) worked with employees of a customer service centre to examine the effect of one's level of acceptance on job satisfaction, work performance and mental health levels. Even after controlling for negative affectivity and locus of control, they found that higher levels of psychological flexibility in employees (as measured by the Acceptance and Action Questionnaire) still predicted better mental health and work performance. Furthermore, psychological flexibility actually mediated the predictive effect between perceived job control (i.e. perception of one's ability to exert influence over one's work environment) and work performance and job control and mental health. Low levels of job control have been shown to correlate with undesirable employee and organizational outcomes such as job dissatisfaction and increased use of sick leave (Bond and Bunce 2000). Also, increasing control over one's environment has been shown to make it more rewarding and less threatening to individuals (Ganster 1989). Thus, these findings are of particular interest because they show that psychological flexibility, one of the core process developed through ACT interventions, act as a mechanism to reduce stress and increase performance in work settings.

Bond and Flaxman (2006) found similar results in a three-wave panel design over three months carried out amongst call centre employees (n=448). They showed that the ability to learn a new work-related computer program, work performance and mental health predicted both the synergy between previously measured levels of job control and psychological flexibility. Indeed, the beneficial effects of job control on learning, performance and mental health are enhanced when people have higher levels of flexibility. This study supports the idea that encouraging the development of psychological flexibility in employees could facilitate change implementation.

In a study of a work reorganization intervention amongst call centre employees, Bond et al. (2008) considered the effect of previously measured levels of psychological flexibility on employee performance 14 months after the intervention. As in the previously described studies, the authors found that psychological flexibility levels moderated the effects on an employee's level of psychological distress and work absenteeism rates in this type of situation.

Finally, in another comparative study, Donaldson-Fielder and Bond (2004) sought to support the measurement of psychological flexibility processes in the workplace, over and above the use of emotional intelligence scales. Whereas principles of ACT encourage the acceptance of internal events, emotional intelligence promotes the regulation of one's thoughts and feelings. This study showed that levels of acceptance predicted general mental health and physical well-being in the workplace, whereas emotional intelligence did not significantly predict either.

All of these studies support the suggestion that the core processes of psychological flexibility ought to be developed in employees in order to create healthier work environments, promote improved productivity levels, and encourage superior levels of well-being in general. At the time this chapter was submitted for publication, there was 58 published randomized-controlled trials comparing ACT to diverse control conditions such as treatment as usual and waitlist. The following studies actually put ACT to the test by using group ACT interventions in the workplace and studying their effects.

Hence, ACT was adapted to a nine-hour workshop on the worksite, over a 14-week period (Bond 2004, Bond and Bunce 2000, Bond and Hayes 2002, Flaxman and Bond 2006). The goal was to use ACT as a brief, group-based, worksite stress management intervention (Bond and Hayes 2002). The typical 2+1 ACT protocol is described in details in two manuals (Bond 2004, Bond and Hayes 2002).

Bond and Bunce's (2000) study compared an ACT intervention, considered an emotion-focused intervention, to an Innovation Promotion Program (IPP), considered a problem-focused stress management intervention, and to a waitlist control group, each comprised of 30 employees in a media organization. The goal of the ACT group was to improve the acceptance of negative psychological states that result from unavoidable work-related situations, whereas the goal of the IPP group was to help individuals identify and change causes of occupational strain. Both ACT and IPP interventions improved employees' general mental health and innovativeness, compared to the waiting list. However, they showed that distressed employees who took part in ACT workshops were significantly less distressed than those from both the IPP and the waiting list (i.e. control group) after a 23-week period. Change in the ACT group was explained by the acceptance of undesirable thoughts even if there was no change in the frequency of those thoughts and no modification of work stressors.

In their 2008 study amongst call centre employees, Bond and collaborators showed that compared to a control group, a control-enhancing work reorganization intervention that included both task reorganization and ACT workshops led to improvements in mental health. They also showed that those who took part in a worksite stress intervention had lower sick leave rates than those in a control group after a 14-month period.

Another study had the goal of comparing a stress inoculation training (SIT), a common intervention in organizations that is derived from a "second-wave" cognitive behavioural therapy (CBT), to an ACT intervention (Flaxman and Bond 2010a). Whilst acknowledging that mediators of change in SIT are poorly understood, Flaxman and Bond (2010a) hypothesized that psychological flexibility would mediate the effect of the ACT intervention on mental health. Working with 107 employees from local government organizations they found that even though both interventions (SIT vs. SIT+ACT) proved to be equally effective in reducing psychological distress, the mechanism for the SIT was not found, whilst psychological flexibility was found to mediate the effects of the ACT treatment group.

In a study involving two governmental organizations in the UK (n=311), Flaxman and Bond (2010b) compared employees who took part in a stress management training session based on ACT to a control waitlist group. Results revealed that amongst those who were distressed at baseline, those who attended the three workshops reported less distress over a six-month period than those in the control group. Indeed, amongst those who were already distressed, 69 per cent showed clinical improvement. Hence, based on this result, ACT seems especially relevant amongst employees who are already showing symptoms of distress.

Other studies similarly used ACT for brief stress management interventions in health care settings. In working with 30 substance abuse counsellors, an all-day ACT seminar was found to significantly reduce levels of prejudice and stigma that these counsellors directed towards substance abusers (Hayes et al. 2004). The intervention was also shown to reduce symptoms of burnout at three-month follow up.

Using a comparison with a waitlist control group, it was shown that four one-hour ACT intervention sessions given over the course of one month amongst 106 social workers significantly reduced levels of stress and symptoms of burnout and increased mental health for those with mid to high levels of stress (Brinkborg et al. 2011).

Dahl et al.'s study (2004) took an interesting twist by applying their ACT intervention with public health service employees suffering from chronic pain and/or stress. These workers were considered to be at risk of using high amounts of sick leave or requiring early retirement because of their medical conditions. The goal of this study was to add an ACT intervention to the regular medical treatment and compare results to the control group, which would receive medical treatment as usual. The authors purported that the ACT interventions would create a more open and accepting attitude within the persons towards their negative physical experiences. They also believed that the emphasis on life goals would help to provide an incentive to working towards new ways of alleviating symptoms. The ACT interventions consisted of four one-hour individual sessions given over the span of one month, each focusing on a single core process of psychological flexibility, namely values, defusion, commitment and acceptance. The results indicated that health services employees in the treatment group used significantly less sick leave and medical services. Again their reported levels of physical symptoms and beliefs about work causing these symptoms remained similar to those in the control group. These results supported the author's hypothesis that the ACT intervention would shift the focus away from negative symptoms towards more values-based behaviours. However, because of the used method, no moderation effects could be defined. Nonetheless, these results add considerable support for the use of ACT interventions with workers suffering from chronic pain.

Thus, ACT interventions are not only proving useful in the workplace, but the psychological flexibility model is also lending itself well to empirical study in that its mechanisms seem to be working as the theories suggest.

ACT Intervention Protocol: A Brief Overview

Typically, an ACT worksite intervention (Bond 2004, Bond and Hayes 2002, Flaxman and Bond 2006) is comprised of two half-day workshops in small groups (about 10 employees) over a two-week period, during normal working hours. A booster session follows these two workshops, two or three months afterwards (2+1 method), leading to a total of nine hours of worksite training. Using ACT in small groups during working hours is logistically advantageous and allows people to share and consolidate learning and to find social support. In the following section, we will briefly describe the intervention we are currently using with workers in a career transition process, following a job loss due to downsizing or job termination. The format was inspired by Bond (2004) Bond and Hayes (2002) and Flaxman and Bond (2006) and adapted to our participants' specific context and needs.

Based on the six processes described earlier, the two main aims of the workshops are to bring employees:

1. to be in contact their internal and external experience, as it unfolds in the present moment, with acceptance and without judgement; and
2. to engage in values-driven actions.

As in the work of Bond and collaborators, the first session starts with a discussion on stress, its sources and consequences, with a special highlight on job loss in order to meet our participants' specific needs and address their current concerns. Various metaphors and mindfulness exercises based on Hayes et al. (1999, 2012) and Harris (2008, 2009) are used to develop defusion, acceptance and contact with the present moment and perspective change (Self-as-context). Many of these exercises are available on the Association for Contextual Behavioral Science (ACBS) website (www.contextualpsychology.org). Mindfulness exercises consist of eye-closed "contact with the present moment" exercises lasting between five and 10 minutes. Since most workers are not familiar with these types of exercises, we recommend not exceeding 10 minutes, at least at the first session. The session ends with homework – practising mindfulness for 10 minutes a day until the next session (one-week practice). Since mindfulness is seen as a skill, it needs practice in order to fully develop. We recommend providing participants with a CD or podcast featuring one or two mindfulness exercises in order to practice mindfulness exercises between sessions. Again, some are available on the ACBS website. In order to introduce the process of values and committed actions, we also give a value-clarification exercise to be completed by the participants.

To reinforce the importance of practice, the second session starts and ends with a mindfulness exercise. We ask participants to share their experience of mindfulness practice over the last week and encourage tips-sharing between participants. The process of values is introduced, along with the notion of committed action. Metaphors are used to highlight the detrimental consequences of fusion and experiential avoidance. Participants are invited to voluntarily share the content of their homework on value-clarification. Afterwards, participants complete a "committed actions" plan based on the values they identified. Based on the implementation intentions theory (Gollwitzer and Sheeran 2006) and outcome-based simulations, they specify their intentions of implementing value-based action in their plan, along with possible barriers to actions and solutions to resolve them. Again, the session ends with homework – 10-minutes of mindfulness practice a day until the next session (in two or three months). We also suggest participants programme a weekly reminder on their smartphone (or diary) to check:

1.   if their actions are aligned with their committed actions plan; and
2.   as a reminder of the importance of mindfulness practice.

Conclusion

Finally, the booster session starts and ends with a mindfulness exercise. Participants share their experience of mindfulness practice and commitment to the action plan over the last weeks (i.e. challenges, issues, good tips, observed outcomes). All six processes are reviewed using metaphors and exercises. At the end of the booster session, participants are invited to integrate mindfulness practice and commitment to action plan to their everyday lives.

Based on current evidence, ACT looks promising in order to create healthier workplaces. The approach has the advantage to provide employees with tools that can be used both at work and in their personal lives. The format (i.e. short workshops during normal work hours) is seen as logistically convenient for organizations. Also, ACT is a cognitive-behavioural intervention (CBI). As shown in Richardson and Rothstein's (2008) meta-analysis of 36 interventions, CBIs (which included an ACT study and a mindfulness study) had the strongest effect size on reducing distress (d=1.164). However, more research is needed to find out about the long-term efficacy of ACT more specifically and it would be useful to know how long the effects last. We still don't know how each one of the six processes plays a role in developing psychological flexibility and if some process are more efficient than others. Moreover, little is known about the effect of ACT among diverse types of workers since most studies were amongst health care, social workers and telemarketers.

There currently is a debate on whether interventions need to be broad and comprehensive to successfully prevent and manage stress. As stated by Leka and Cox (2008) intervention strategies "should address the root causes of work-related stress (primary prevention); provide training to managers and employees on stress management in order to reduce its impact (secondary prevention); and, for those that have suffered ill health as a result of work-related stress, provide them with resources to manage and reduce their respective effects (tertiary prevention)." Hence, ACT could be used as a complementary tool within a larger psychosocial risk management intervention at the organizational level. However, Richardson and Rothstein's (2008) meta-analysis did not support the multimodal intervention approach. Surprisingly, they even found that effect sizes were reduced when multiple techniques (i.e. relaxation and organizational intervention) were used. They explained these results by the difficulty to implement many components at the same time, which could work to the detriment of the more complex individual components of stress. Unfortunately, organizational interventions remain scarce so their meta-analysis was based on only five studies, which did not report effect on absenteeism and performance (Richardson and Rothstein 2008). Therefore, there is a need to remain cautious.

Hence, based on the current state of the debate, to create healthy workplaces, we recommend using ACT workshops as part of an organizational stress management intervention that takes into consideration the organizational readiness to change (Leka and Cox 2008) and actual capacity to implement such a programme.

References

Baer, R.A. 2003. Mindfulness training as a clinical intervention: A conceptual and empirical review. *Clinical Psychology: Science and Practice*, 10, 125–43.

Bambra, C., Egan, M., Thomas, S., Petticrew, M. and Whitehead, M. 2007. The psychosocial and health effects of workplace reorganization-2. A systematic review of task restructuring interventions. *Journal of Epidemiology and Community Health*, 61, 1028–37.

Bohlmeijer, E.T., Fledderus, M., Rokx, T.A.J.J. and Pieterse, M.E. 2011. Efficacy of an early intervention based on acceptance and commitment therapy for adults with depressive symptomatology: Evaluation in a randomized controlled trial. *Behaviour Research and Therapy*, 49(1), 62–7.

Bonanno, G.A., Papa, A., LaLande, K., Westphal, M. and Coifman, K. 2004. The importance of being flexible: The ability to both enhance and suppress emotional expression predicts long-term adjustment. *Psychological Science*, 15, 482–7.

Bond, F.W. 2004. ACT for stress. In *A Practical Guide to Acceptance and Commitment Therapy*, edited by S.C. Hayes and K.D. Strosahl. New York, NY: Springer, 275–93.

Bond, F.W. and Bunce, D. 2000. Mediators of change in emotion-focused and problem-focused worksite stress management interventions. *Journal of Occupational Health Psychology*, 5, 156–63.

Bond, F.W. and Bunce, D. 2003. The role of acceptance and job control in mental health, job satisfaction, and work performance. *Journal of Applied Psychology*, 88(6), 1057–67.

Bond, F.W. and Flaxman, P.E. 2006. The ability of psychological flexibility and job control to predict learning, job performance, and mental health. *Journal of Organizational Behavior Management*, 26, 113–30.

Bond, F.W. and Hayes, S.C. 2002. ACT at work. In *Handbook of Brief Cognitive Behaviour Therapy*, edited by F.W. Bond and W. Dryden. Chichester: John Wiley and Sons, 119–40.

Bond, F.W., Flaxman, P.E. and Bunce, D. 2008. The influence of psychological flexibility on work redesign: Mediated moderation of a work reorganization intervention. *Journal of Applied Psychology*, 93(3), 645–54.

Brinkborg, H., Michanek, J., Hesser, H. and Berglund G. 2011. Acceptance and commitment therapy for the treatment of stress among social workers: A randomized control trial. *Behaviour Research and Therapy*, 49, 389–98.

Dahl, J., Wilson, K.G. and Nilsson, A. 2004. ACT and the treatment of persons at risk for long-term disability resulting from stress and pain symptoms: A preliminary randomized trial. *Behavior Therapy*, 35, 785–801.

Dalrymple, K.L. and Herbert, J.D. 2007. Acceptance and commitment therapy for generalized social anxiety disorder: A pilot study. *Behavior Modification*, 31, 543–68.

Donaldson-Fielder, E.J. and Bond, F.W. 2004. The relative importance of psychological acceptance and emotional intelligence to workplace well-being. *British Journal of Guidance and Counselling*, 32(2), 187–203.

Egan, M., Bambra, C., Thomas, S., Petticrew, M., Whitehead, M. and Thomson, H. 2007. The psychosocial and health effects of workplace reorganization: A systematic review of organizational-level interventions that aim to increase employee control. *Journal of Epidemiological Community Health*, 61, 945–54.

Flaxman, P.E. and Bond, F.W. 2006. Acceptance and commitment therapy in the workplace. In *Mindfulness-Based Interventions: A Clinician's Guide*, edited by R.A. Baer. London: Elsevier, 377–402.

Flaxman, P.E. and Bond, F.W. 2010a. A randomized worksite comparison of acceptance and commitment therapy and stress inoculation training. *Behaviour Research and Therapy*, 48(8), 816–20.

Flaxman, P.E. and Bond, F.W. 2010b. Worksite stress management training: Moderated effects and clinical significance. *Journal of Occupational Health Psychology*, 15(4), 347–58.

Forman, E.M., Herbert, J.D., Moitra, E., Yeomans, P.D. and Geller, P.A. 2007. A randomized controlled effectiveness trial of acceptance and commitment therapy and cognitive therapy for anxiety and depression. *Behavior Modification*, 31(6), 772–99.

Ganster, D.C. 1989. *Measurement of Worker Control. Final Report to the National Institute of Occupational Safety and Health* (Contract No. 88-79187). Cincinnati, OH: National Institute of Occupational Safety and Health.

Gollwitzer, P.M. and Sheeran, P. 2006. Implementation intentions and goal achievement: A meta-analysis of effects and processes. *Advances in Experimental Social Psychology*, 38, 69–119.

Gross, J.J. 1998. The emerging field of emotion regulation: An integrative review. *Review of General Psychology*, 2, 271–99.

Grossman, P., Niemannb, L., Schmidt, S. and Walach, H. 2004. Mindfulness-based stress reduction and health benefits: A meta-analysis. *Journal of Psychosomatic Research*, 57(1), 35–43.

Harris, R. 2008. *The Happiness Trap: How to Stop Struggling and Start Living*. London: Constable and Robinson.

Harris, R. 2009. *ACT Made Simple: An Easy-To-Read Primer on Acceptance and Commitment Therapy*. Oakland, CA: New Harbinger.

Hayes, S.C., Strosahl, K. and Wilson, K.G. 1999. *Acceptance and Commitment Therapy: An Experiential Approach to Behavior Change*. New York, NY: Guilford Press.

Hayes, S.C., Strosahl, K. and Wilson, K.G. 2012. *Acceptance and Commitment Therapy: The Process and Practice of Mindful Change*. 2nd edition. New York, NY: Guilford Press.

Hayes, S.C., Luoma, J.B., Bond, F.W., Masuda, A. and Lillis, J. 2006. Acceptance and commitment therapy: Model, processes and outcomes. *Behaviour Research and Therapy*, 44(1), 1–25.

Hayes, S.C., Bissett, R., Roget, N., Padilla, M., Kollenberg, B.S., Fisher, G., Masuda, A., Pistorello, J., Rye, A.K., Berry, K. and Niccolls, R. 2004. The impact of acceptance and commitment training on stigmatizing attitudes and professional burnout of substance abuse counselors. *Behaviour Therapy*, 35, 821–36.

Ivancevich, J.M., Matteson, M.T., Freedman, S.M. and Phillips, J.S. 1990. Worksite stress management interventions. *American Psychologist*, 45(2), 252–61.

Kabat-Zinn, J. 1994. *Wherever You Go, There You Are: Mindfulness Meditation in Everyday Life*. New York, NY: Hyperion.

Kashdan, T.B., Barrios, V., Forsyth, J.P. and Steger, M.F. 2006. Experiential avoidance as a generalized psychological vulnerability. *Behaviour Research and Therapy*, 44, 1301–20.

LaMontagne, A.D., Keegel, T. and Vallance, D.A. 2007. Protecting and promoting mental health in the workplace: Developing a systems approach to job stress. *Health Promotion Journal of Australia*, 18, 221–8.

Leka, S. and Cox, T. 2008. *Best Practices in Work-related Stress Management Interventions*. PRIMA-EF, Institute of Work, Health and Organizations.

Lim, D., Sanderson, K. and Andrews, G. 2000. Lost productivity among full-time workers with mental disorders. *The Journal of Mental Health Policy and Economics*, 3(3), 139–46.

McCracken, L.M., MacKichan, F. and Eccleston, C. 2007. Contextual cognitive-behavioral therapy for severely disabled chronic pain sufferers: Effectiveness and clinically significant change. *European Journal of Pain*, 11, 314–22.

Öst, L. 2008. Efficacy of the third wave of behavioral therapies: A systematic review and meta-analysis. *Behaviour Research and Therapy*, 46(3), 296–321.

Prochaska, J.O. and DiClemente, C.C. 1984. *The Transtheoretical Approach: Crossing Traditional Boundaries of Therapy*. Homewood, IL: Dow Jones-Irwin.

Richardson, K.M. and Rothstein, H.R. 2008. Effects of organizational stress management interventions programs: A meta-analysis. *Journal of Occupational Health Psychology*, 13(1), 69–93.

Roemer, L., Orsillo, S.M. and Salters-Pedneault, K. 2008. Efficacy of an acceptance-based behavior therapy for generalized anxiety disorder: Evaluation in a randomized controlled trial. *Journal of Consulting and Clinical Psychology*, 76, 1083–9.

Ruiz, F.J. 2010. A review of Acceptance and Commitment Therapy (ACT) empirical evidence: Correlational, experimental psychopathology, component and outcome studies. *International Journal of Psychology and Psychological Therapy*, 10(1), 125–62.

Zettle, R. and Hayes, S.C. 1986. Dysfunctional control by client verbal behavior: The context of reason-giving. *The Analysis of Verbal Behavior*, 4, 30.

Zettle, R. and Rains, J. 1989. Group cognitive and contextual therapies in treatment of depression. *Journal of Clinical Psychology*, 45(3), 436–45.

PART 2

Work–Life Balance and Physical Health Interventions

5

Improving Employee Health and Work–Life Balance: Developing and Validating a Coaching-based ABLE (Achieving Balance in Life and Employment) Programme

ARLA DAY, LORI FRANCIS, SONYA STEVENS, JOSEPH J. HURRELL, JR. and PATRICK MCGRATH

Abstract

There is a growing need for organizations to better manage stress and conflict-related issues in the workplace. However, although much is known about the antecedents and consequences of these issues, very few studies have tried to validate broad-based intervention programmes designed to mitigate the impact of stress and conflict. The main goal of the first study was to examine the efficacy of the six-month job stress and work–life balance workshop to improve employee outcomes. The main goals of the second study were to develop and validate a stress and conflict phone-based coaching programme that would be more accessible and tailored.

Introduction

To date, we have a good understanding of the work-related antecedents of work–life conflict and job stress as well as their consequences (see Sonnentag and Frese 2003 for an overview). Both work factors (e.g. poor collegial relationships; Leiter et al. 2012) and individual factors (e.g. negative coping styles; Day and Livingstone 2001) are associated with reduced employee well-being and increased conflict and stress. An inability to manage work and non-work demands as well as conflicts can exacerbate stress and increase individual strain, leading to negative repercussions for organizations (e.g. turnover; Kossek and Ozeki 1999) as well as for employees, their families and society (Parasuraman and

Greenhaus 1997). The resulting stress can increase negative physiological strain outcomes (e.g. cardiovascular problems, high blood pressure, ulcers; Kristensen 1996, Wager et al. 2003) and negative behavioural outcomes (e.g. absenteeism; Kahn and Byosiere 1992; drug use; Frone et al. 1994). The financial costs to organizations are also substantial (see for example, Cooper and Dewe 2008, Duxbury and Higgins 2003).

These relationships of stress and conflict with negative employee outcomes indicate a need for organizations to identify and offer programmes, interventions, and resources to reduce job stress, support positive work–life balance, and in turn, improve worker health and well-being. However, despite the large body of research on the predictors of stress, conflict and health, there is a surprising lack of quantitative research and few validated, broad-based intervention programmes designed to mitigate these issues (DeRango and Franzini 2003, Kelloway et al. 2008, Kenny and Cooper 2003). That is, despite the knowledge we have accumulated about work conflict and stress, despite the individual and organizational costs associated with these issues, and despite the acknowledged need for such programmes and interventions, little quantitative, longitudinal intervention research has been conducted to identify and evaluate initiatives to reduce conflict and improve balance.

Hurrell (2005) argued that more research is needed to identify and evaluate the individual and organizational interventions that can be used to reduce employee stress and conflict. Therefore, in Study 1, we created and validated a pilot programme of workshops to help employees identify stressors, set goals, manage stress, develop coping strategies, and maintain work–life balance. Learning from the results and feedback obtained in Study 1, we developed and validated a phone-based coaching programme based on the same materials, the results of which are described in Study 2.

Job Stress and Work–Life Balance

Job stress is a widely reported phenomenon: it negatively affects employees, their families and, ultimately, their communities (Kelloway and Francis 2006), and it has been listed as one of the top 10 leading causes of work-related death (Sauter et al. 1990). Similarly, work–family conflict is associated with negative health costs for individuals, in terms of increased depression and alcohol use, and decreased physical health (Adams et al. 1996, Duxbury and Higgins 2003, Frone 2003). This conflict also impairs family functioning (Adams et al. 1996, Duxbury and Higgins 2003). Many illnesses, including diabetes and cardiovascular concerns, are preventable with early interventions aimed at nutrition, exercise and stress management (Hu et al. 2001). However, many organizations and employees do not know how to reduce these stressors and improve work-life balance.

Job stress may arise directly from the demands of the job, interactions with others at work, and conflicts among work and non-work priorities. Researchers have done a good job of identifying the antecedents of stress and work–life conflict. For example, time factors (e.g. work overload, scheduling of work and family demands) are associated with negative health, stress, and work–family conflict (Day and Livingstone 2001, Rogers et al. 1991, Williams 2003), job withdrawal and decreased life satisfaction (Bellavia and Frone 2005, Frone et al. 1997, Hammer et al. 2004, Totterdell 2005). Interpersonal relationships both at work and home also affect individuals' well-being (Hain and Francis 2006), such that a lack of social support predicts decreased well-being and increased strain (see Viswesvaran et al. 1999 for a meta-analysis). Low job control and unclear role expectations are associated with decreased job and life satisfaction and increased stress (Adams and Jex 1999, Carayon 1995, Day and Jreige 2002, Peiró et al. 2001, Spector 1986). Conversely, supportive supervisors (Thomas and Ganster 1995), high work commitment (Day and Chamberlain 2006) and schedule flexibility (Thomas and Ganster 1995) are associated with less work–family conflict.

Job stress and conflict models may help provide a framework to organize these antecedents, and help identify ways to improve employee well-being. These models typically identify three major stress components (Hurrell et al. 1998, Lazarus and Folkman 1984):

1. stressors or demands are external events or elements that have the potential to create stress;
2. perceived stress occurs only if the individual appraises these events as being negative (i.e. "stressful");
3. strain refers to the potential immediate and long-term health outcomes (Barling 1990) that occur when an individual perceives stressors as negative and is unable to cope with them (Lazarus 1995, Quick et al. 1992).

Individuals differ in what they view as stressful and how they react to stress depending on their personal resources (Lazarus and Folkman 1984). Therefore, these models suggest that we may be able to improve employee health through interventions that reduce the demands, provide supports to help employees reassess the negativity of the demands, and reduce strain outcomes.

Work Stress and Conflict Interventions

The fact that occupational stress has negative effects on many aspects of one's life has created a need for research on interventions that can "treat, manage, and prevent" this stress (Caulfield et al. 2004: 151). We can categorize stress and conflict interventions as primary, secondary, or tertiary in nature (see for example, Hurrell 2005, Hurrell and Murphy 1996). Primary interventions aim to minimize or eliminate exposure to stressors (e.g. job redesign: Hurrell 2005, Quick et al. 1997). Secondary interventions focus on changing individuals' reactions to stressors and improving their resilience so that they can better cope with experienced stress (e.g. stress management programmes). Tertiary interventions focus on treating individuals who experience negative outcomes caused by exposure to stressors (e.g. employee assistance programmes: Quick et al. 1997), in terms of a "heal the wounded" perspective (Kelloway et al. 2008: 421).

Primary intervention (reducing stressors/demands) is a valuable component of stress reduction. Although primary prevention is often viewed as the ideal form of stress management, it is not feasible as the sole solution because it is impossible to eliminate all demands and because the same demand can be viewed differently (positively or negatively) by people. Moreover, regardless of the potential to realize improvements of employee health, these types of interventions often require organizational change and, therefore, may be viewed by some organizational stakeholders as being too costly or disruptive to implement (Cartwright and Cooper 2005, Hepburn et al. 1997). In general, improvements from workplace interventions are often short-lived, because individuals often revert to old habits.

Conversely, the traditional medical model typically focuses on tertiary interventions (i.e. treating health problems); however, a sole focus on these types of interventions is not optimal because they ignore prevention. Moreover, even in the face of stressors, many illnesses, including cardiovascular concerns and diabetes, are preventable with early interventions aimed at stress management, nutrition, and exercise (Hu et al. 2001, Stampfer et al. 2000). Therefore, secondary prevention (e.g. addressing individual perceptions of stressors and increasing individual coping resources) is important because it bridges the gap between primary and tertiary interventions. It acknowledges that not all stressors can be eliminated, but it also empowers individuals to minimize their negative reactions to potential stressors, and decreases reliance on tertiary treatment. Moreover, Cartwright and Cooper (2005: 618) argued that an "effective strategy for stress prevention needs to incorporate both work-related and worker-related initiatives." That is, incorporating different levels of work stress and conflict interventions may be most effective.

There are several components of an effective stress and conflict intervention, including self-insight, goal setting, and time management. Because of the individual nature of stress reactions, approaches such as insight and reflection (i.e. non-ruminative self-attentiveness: Trapnell and Campbell 1999) may be effective methods in any intervention programme to improve well-being and goal attainment (Grant 2003). Moreover, self-reflection can reduce reliance on maladaptive coping methods (Mactavish and Iwasaki 2005).

Similarly, reaching one's goals can improve mental health (Sheldon and Kasser 2001). Therefore, helping employees to identify effective goals (in terms of being specific, realistic, measurable in behavioural terms, attainable, and time-bound) should be effective in reducing stress and conflict (Locke and Latham 2002). Healthy behaviours (e.g. physical activity) and relaxation (e.g. listening to music) assist in facilitating health and are effective coping mechanisms (Moos et al. 2003).

Because feeling overloaded is a major concern for many employees (Williams 2003), incorporating training on time management and setting priorities may improve employee well-being. Extending the popular three-component model of organizational commitment (Meyer et al. 1993), we can utilize affective, normative, and continuance commitment to describe and prioritize time demands. Employees can categorize responsibilities as tasks that they "want to" complete (i.e. affective commitment to the task); those they "have to" complete (i.e. continuance commitment); and those they feel they "ought to" complete (i.e. normative commitment). Using this model, identifying and reducing the "ought to" responsibilities should reduce conflict and strain by reducing time commitments and focusing on high-priority goals.

Poor interpersonal relations are a significant source of job stress (Sauter et al. 1990, Warr 1987) and lead to job dissatisfaction and lower productivity (Hain and Francis 2006, Hodson 1997). Conversely, positive co-worker relationships lead to job satisfaction (Nielsen et al. 2000) and less job stress, strain, and burnout (Beehr et al. 2000, Hain and Francis 2006). Stress and conflict programmes should incorporate effective communication tactics, identify pitfalls to avoid, and introduce conflict-management techniques to improve relationships and reduce existing conflict (Oetzel and Ting-Toomey 2006). Similarly, helping employees develop and maintain effective support networks should help reduce strain and conflict.

A lack of control is linked to decreased job and life satisfaction and increased stress (Adams and Jex 1999, Carayon 1995, Day and Jreige 2002, Spector 1986). To a great degree, job control is a function of the job itself and may not be something that employees have the power to change. However, programmes that help participants examine ways to increase control in their jobs and life (e.g. time management/prioritization to control one's schedule, work with a supervisor to achieve more flexibility at work) or to improve the control that they do have over their job and home lives (e.g. delegate tasks at work and at home; declining offers of unnecessary and unwelcome tasks) may be more effective.

In addition to reducing stressors, using secondary prevention in terms of incorporating coping, reappraisal, and strain reduction techniques is important. Coping is defined as behavioural and psychological attempts to reduce the symptoms of a stressful situation (Lazarus and Folkman 1984). Emotion-focused coping involves changing how one feels about a stressful situation. For instance, enacting social support can be beneficial and calming (Taylor et al. 2000). Problem-focused coping involves changing the stressful situation through problem solving (Lazarus and Folkman 1984). Cognitive reappraisal involves changing the way one thinks about a stressful situation or reinterpreting the situation in a positive manner. Disengagement involves using food, alcohol or drugs or other activities to "disengage" from one's problems (Carver et al. 1989). Both emotion-focused and problem-focused coping are associated with lower distress (Violanti 1992). Typically, problem-focused coping is effective when stressors are within one's control, and emotion-focused coping and cognitive reappraisal are effective when the stressor is outside one's control. Conversely, disengagement is associated with negative outcomes (Day and Livingstone 2001, Tyler and Cushway 1995). Strain reduction techniques (e.g. yoga, meditation, progressive muscle relaxation) are aimed at minimizing the negative consequences of stress. Similar to reappraisal, these techniques tend to be most effective when employees cannot control the source of the stress. They may be able to reduce strain by invoking what control they have over their response to the stress.

The studies available on organizational stress interventions have demonstrated the effectiveness of specific work-stress life-coaching interventions in reducing stress (Bond and Bunce 2001, Grant 2003, Hurrell 2005, Mikkelsen et al. 2000, Terra 1995). Compared to health promotion research that has produced somewhat ambiguous results, there is some encouraging evidence for the effectiveness of stress management programmes (Kelloway et al. 2008), although some of the positive effects may be only short-term (see Giga et al. 2003). For example, Mikkelsen et al. (2000) found that participants in a short-term intervention (in which they focused on identifying job stressors and implementing strategies to reduce them) experienced reduced stress and demands relative to controls immediately after the intervention. Wolever et al. (2012) evaluated the effectiveness of two mind-body workplace stress reduction programmes. They found that participants who were randomly assigned to one of several programmes (i.e., a yoga based workplace stress reduction programme, an online mindfulness-based programme, or an in-person mindfulness-based programme) all showed improved levels of perceived stress, sleep quality, and the heart rhythm coherence ratio of heart rate variability relative to the control groups, with online and in-person delivery methods being equally effective. Theorell et al. (2001) found that, compared to a control group of managers, managers who received stress management training experienced lowered serum cortisol levels. However, some studies suggest that stress management training alone may not be effective. For instance, Thomason and Pond (1995) found that compared to a control group, participants in their stress management training programme did not experience improved health. However, introducing a self-management module into the training (i.e. goal setting, self-reinforcement) resulted in significant reductions in blood pressure.

Study 1: Workshop Intervention Programme

Based on the well-established stress and conflict literatures, we developed a comprehensive six-session workshop programme intended to provide information and practical solutions to employees experiencing work and job stress and conflict (ABLE Workshops: Achieving Balance in Life and Employment) to be conducted over a six-month period. We based the workshop materials on the literature demonstrating the importance of goal setting, time management, setting priorities, developing coping strategies, identifying individual and organizational resources and support, reducing demands in one's life, and balancing work and life responsibilities. We conducted a pilot study to examine the validity of the newly developed ABLE workshop. We assessed employees' physical and psychological health at two times (i.e., before starting and upon ending the six-month programme). Employees also participated in concurrent exercise and nutrition programmes offered by the organization.

Hypothesis 1: Participants in the ABLE workshop will experience:

a)   decreased work–life conflict;
b)   decreased strain and stress;
c)   decreased negative mood;
d)   decreased burnout (emotional exhaustion); and
e)   increased positive mood) from Time 1 (before workshops) to Time 2 (after workshops).

Hypothesis 2: At a group level, there will be a significant decrease in:

a)   workers' compensation claims; and
b)   bsenteeism over the period of the intervention compared to the equivalent time prior to the intervention.

Method

Sixty-three employees (61 women and 2 men) from a long-term health care facility participated in a six-month programme on job stress and work–life balance. They met once a month for six months in one of several group settings. In between the workshops, they received weekly emails providing additional information on the past session, reminders and overviews about the upcoming sessions, and useful tips to put the workshop content into action. All participants also had the opportunity to contact a member of the research team to answer any questions or deal with new issues that may have arisen in between session. A listserv also was created in which all participants could post tips for improving balance and reducing stress and for sharing success stories.

The interactive workshops were based on research on stress, goal-setting theory, coping, health, time management, and work–life balance, and they incorporated basic learning principles to help employees reduce and manage stressors and attain personal health goals.

The average age of participants was 46.29 years (SD=9.99). Participants had been in their current jobs an average of 12.37 (SD=8.62) years, and they worked an average of 41.61 hours per week (SD=6.41). The majority (50.8 per cent) worked an 8-hour day (ranging from 6- to 12-hour days), and 75 per cent of participants work a stable Monday–Friday shift. Approximately 50 per cent of participants were married or living common law with their partners; 77.4 per cent had at least one child, and 21 per cent indicated that they had eldercare responsibilities.

Because of missing data at the second data collection, the final sample when examining Time 2 variables is N=23. There were no differences between this subsample and the full sample.

Measures

DEMOGRAPHICS

Participants were asked to respond to several questions about their work (e.g. work hours, job tenure), personal demographics (e.g. age, gender, education), and family status information (e.g. marital status, children, eldercare). The organization also collected data on the number of days participants missed from work for (a) the year prior to the study; and (b) the six months of the study and the six months following the study, as well as health and safety claims made to a government compensation board (i.e. Workers' Compensation, WCB) for the same two time periods.

WORK–LIFE CONFLICT

Work–life conflict was assessed using six items developed by Day (1996, see also Day and Chamberlain 2006). Using a five-point Likert scale (1=*strongly disagree*, 5=*strongly agree*), participants were asked to indicate the extent to which agreed with the items (e.g. "It is hard to balance my role as an employee with my life outside of work"). Cronbach's alpha for Time 1 was $\alpha=0.89$, with item-total correlations ranging from r=0.58 to 0.85; Cronbach's alpha for Time 2 was $\alpha=0.94$, with item-total correlations ranging from r=0.47 to 0.92.

STRAIN

Strain was assessed with the 20-item Strain Symptoms Checklist (Bartone et al. 1989), which measures both psychological and physiological health symptoms and complaints. Using a six-point frequency scale ranging from 0 (never) to 5 (always), participants indicated the extent to which they agreed with the items (e.g. "feeling life is pointless"; "headaches"). Cronbach's alpha for Time 1 was $\alpha=0.90$, with item-total correlations ranging from r=0.58 to 0.85; Cronbach's alpha for Time 2 was $\alpha=0.84$, with item-total correlations ranging from r=0.47 to 0.76.

EMOTIONAL EXHAUSTION

Emotional exhaustion was assessed using the five-item Emotional Exhaustion subscale of the Maslach Burnout Inventory – General Survey (Maslach et al. 1996). Respondents were asked to rate each item using a seven-point frequency scale (0 = never; 6 = every day). Emotional exhaustion assesses the extent to which participants feel tired and drained from work (e.g. "I feel emotionally drained from work"). In the present sample, the internal reliability of this scale was α=0.83, with item-total correlations ranging from r=0.38 to 0.67.

MOOD

The 20-item Positive and Negative Affectivity Scale (PANAS, Watson et al. 1988) was used to measure positive and negative mood. Using a five-point Likert-type scale (1 = very slightly or not at all; 5 = extremely), respondents were asked to rate the extent to which they experienced a list of positive emotions (e.g. enthusiastic, determined) or negative emotions (e.g. distressed, hostile) within the past few weeks.[1] In the present sample, the internal reliability of this positive mood scale was α=0.89, with item-total correlations ranging from r=0.55 to 0.67. The internal reliability of the negative mood scale was α=0.91, with item-total correlations ranging from r=0.55 to 0.67.

Study 1: Workshop Intervention Results

In order to assess Hypotheses 1a–e, we conducted a paired t-test for each of the outcomes. When comparing their pre- and post-workshop scores, employees experienced significant increases in positive mood, $t(23)=-2.80$, $p<0.01$). They also experienced significant decreases in strain, $t(22)=2.50$, $p<0.05$, stress, $t(23)=2.51$, $p<0.05$, negative mood, $t(23)=3.12$, $p<0.01$, negative physical symptoms, $t(23)=2.83$, $p<0.01$, and emotional exhaustion, $t(22)=2.38$, $p<0.05$. There was no change in work–life conflict.

To test Hypothesis 2 pertaining to absenteeism rates and worker compensation claims, we examined organizational data from the 12-month period prior to the study and compared it to the 12-month period that covered the 6 months of the intervention and the 6 months following the intervention. The average number of days missed from work dropped from 8 to 6 from the 12-month period prior to the intervention to the subsequent 12-month period (6 months of the intervention and 6 months after the intervention; i.e. a 25 per cent improvement). There also was an 18 per cent reduction in worker compensation claims (i.e. average uses dropped from 0.22 to 0.18) during this same period.

Study 1: Discussion

The goal of the pilot study was to examine the efficacy of the six-month intervention programme on employees' self-rated well-being and stress indices. Participants reported significant increases in positive mood as well as significant reductions in stress, strain, negative mood, and exhaustion. Contrary to expectations, however, work–life conflict was not significantly reduced for this group. Importantly, the group also experienced a significant decrease in worker compensation claims and improvements in attendance. We collected qualitative data after the conclusion of the intervention. The majority of

---

1   Watson et al. (1988) found that the test-retest reliability for the PANAS when used to evaluate mood over the "past few weeks" (0.58) is similar to the test-retest reliability for the PANAS when used to evaluate mood in the "moment" (0.54).

participants reported that the sessions were very useful and led them to healthier lifestyle practices.

LIMITATIONS AND FUTURE RESEARCH

Although results suggests that this programme was valuable, there are several cautions that should be taken with it. The small sample size (which was exacerbated by missing data at the Time 2 data collection) within one organization limits the generalizability of the study. A true longitudinal design should include more than two data collection times (Kelloway and Francis 2012), and it would help establish the extent to which the effects of the programme could be maintained. Finally, although there were positive changes in the participants, we cannot conclude that these effects are due solely to the workshops. For example, the participants were also involved in other health-related workshops, which may have had positive effects on their well-being. Similarly, other systematic changes within the organization may have had positive effects on all employees. Having a control group with which to compare these participants would be very valuable. Therefore, future research should study the effectiveness of this programme using a larger sample over a longer time period, and including a control group.

There are several practical limitations as well: participants provided several suggestions to improve the content, process, and format of the workshops. Although the organization allowed participants to attend the workshops during work time, and although several different sessions were offered to accommodate the shifts of the participants, many of the participants indicated it was difficult to get away during busy times (e.g. when colleagues were away from work). Therefore, they suggested that a more flexible nature tailored to their individual schedules would be more effective. They also suggested decreasing the time gap between the sessions to enhance compliance and to stay more connected with the programme. Because participants were dealing with unique issues, several noted that tailoring the topics to individuals would be beneficial, make it more relevant, and be a better use of the individual participant's time. Although they noted that they felt comfortable in the workshops, several also indicated that some issues were not discussed because of concerns about privacy.

Cartwright and Cooper (2005) argued that there is a scarcity of comprehensive and regular evaluation studies examining the efficacy of stress intervention programmes. Study 1 was an important first step in developing and validating a programme to help employees reduce stress and improve work–life balance. However, it is possible to improve the programme based on the practical limitations inherent in the first study. Moreover, it is necessary to provide a more rigorous examination of the efficacy of this programme. Cartwright and Cooper (2005) also argued that few studies incorporate control groups, which are integral to examining the efficacy of intervention programmes. Moreover, stress interventions may be unsuccessful because they are too short in duration and do not allow sufficient practice time (Niven and Johnson 1989). Therefore, based on the positive results of the pilot study, the positive comments of participants, and the suggestions for future research, we conducted a second study in which we expanded the ABLE programme by incorporating the workshop content into a new, more accessible and more tailored phone-based coaching delivery format.

Study 2: ABLE Phone-Based Coaching

We modeled the ABLE Coaching Programme after a validated coaching programme format that would address participants' suggestions, while ensuring the efficacy of the content and the overall programme. ABLE is based on the delivery strategy of the Strongest Families Program, a phone-based coaching health treatment programme designed to help families who are experiencing psychological problems (McGrath et al. 2001). The model ensures its services are accessible to all families (regardless of location) and is convenient, accessible, and cost-effective (Lingley-Pottie et al. 2005).

This format addresses the practical limitations identified in the pilot study, by allowing the programme to:

1. reach employees regardless of their proximity to the research centre (e.g. include employees in both rural and urban areas);
2. reduce time demands and scheduling conflicts for participants;
3. tailor the programme to individual needs; and
4. increase feelings of confidentiality.

The content of ABLE is similar to the workshop, but it has more flexibility to tailor the programme for individuals because it involves individual coaching sessions. These sessions addressed knowledge transfer, primary interventions (reducing demands), and secondary interventions (coping, reappraisal, reducing strain), and healthy behaviours. Participants received programme materials (i.e. manual and web link) and talked to a trained coach via telephone on a weekly basis.

SUMMARY AND HYPOTHESES

Hypothesis 1: There will be significant interaction between the control and intervention groups before and after the ABLE intervention, in that, compared to the control group, the intervention group will report increased:

a) positive mood;
b) life satisfaction, as well as decreased;
c) negative mood;
d) perceived stress; and
e) hassles from Time 1 to Time 2.

Method

PROGRAMME DESCRIPTION

The Achieving Balance in Life and Employment (ABLE) is a 12-week phone-based coaching programme, targeting employee well-being and health using tailored goal setting and by helping participants to increase their personal resources. ABLE is based on a validated distance-coaching programme, and integrates the latest research on work–life balance and job stress into its programme materials.

We conducted a longitudinal intervention study with employees from 15 organizations throughout Nova Scotia, Canada. They participated in one of two sessions of the ABLE Programme, using a wait-list control design. Each participant was assigned a coach, who phoned the participant weekly at a mutually compatible time, and worked with the participant to guide them through the programme, answer questions, and modify the programme to meet their specific needs. Coaches had 24-hour access to a supervisor and registered psychologist, and they had weekly case supervision meetings. Calls were recorded to help problem solve, improve coach effectiveness, and monitor quality. The supervisor conducted a random check on the calls to ensure programme fidelity and standardization.

Therefore, participants learned about these coping and strain reduction methods, and identified problem-focused coping strategies to use for demands that are within their control and emotion-focused coping and reappraisal strategies to use for demands that are beyond their control. They also learned strategies to minimize coping through disengagement. More specifically, participants:

1. use self-reflection to identify the types of stressor encountered (i.e. controllable or uncontrollable situation) as well as one's typical personal coping styles (long term) and strategies (specific situations);
2. evaluate the effectiveness of their own coping styles;
3. identify methods for increasing functional long-term coping styles, as well as effective short-term strategies;
4. improve interpersonal functioning and social relationships; and
5. identify and implement effective strain reduction techniques as identified above (including an assessment of the individual needs and perceived effectiveness of these techniques).

INTERVENTION AND SURVEY PROCESS

The 15 organizations were recruited through our organizational contacts who had expressed an interest in such a programme. The contacts were asked to communicate with all of their employees through their regular channels (e.g. email, voice mail, or organizational mailings) to inform them of the ABLE programme and to provide the website for more information and to register.

Employees who were interested in the study completed a registration page containing several surveys designed to screen out individuals who have any clinical levels of depression and anxiety. The programme targeted employees who experience moderate levels of conflict and burnout, and therefore, any individuals who were classified as very high on depression and anxiety were excluded from the study. A clinical member of the research group contacted these individuals for further consultation and provided information to ensure that they would be able to receive the proper treatment for their symptoms.

PARTICIPANTS AND PROCEDURE

Participants (N=169: 123 women; 45 men; 1 not reported) were recruited for the ABLE programme intervention study through contacts from these 15 organizations (e.g.

health care, government, university, service) across an eastern Canadian province. Organizational contacts distributed a standardized recruitment email to their employees. All participation was voluntary and anonymous (i.e. employers did not know who was participating). Participants were divided into intervention and control groups. All participants completed a survey electronically prior to the onset of the ABLE programme and after the first session.

There were 103 participants (86 women; 16 men; 1 did not indicate) who completed the survey at both Time 1 and Time 2. Approximately half of the participants were assigned to the ABLE treatment group (N=56: 46 female; 10 male) and the rest of the participants were assigned to a wait-list control group (N=46: 40 female; 6 male). A true randomized design was not used because the entire sample of participants was not recruited in time to begin the programme. Therefore, all participants were assigned to the treatment group until this group was full and then participants were assigned to a wait-list control group.

The average age of participants was 43.56 years (SD=9.61). They had a mean tenure of 9.2 years (range: 13 weeks to 35 years) and worked an average of 41.6 hours per week (range: 17.5 to 90). Participants had a wide variety of jobs, including: Human Resources Manager, Accountant, Administrative Assistant, Interior Designer, Library Assistant, Director of Operations, Paramedic, Registered Nurse, and Lab Technologist. Participants were well educated; all but one participant completed high school and 79.0 per cent had completed at least one college or university degree. Almost three quarters (74.0 per cent) reported being married or common law and reported having at least one child. Table 5.1 shows the socio-demographics of the first study.

MEASURES

Participants completed the following measures at three times throughout the study:

Demographic, background and work information: Participants were asked to indicate their age, gender, type of job, number of hours worked, shift work.

Mood: The 10 positively-worded items from the Positive and Negative Affectivity Scale (PANAS, Watson et al. 1988) were used to measure positive mood. Using a five-point Likert-type scale (1 = very slightly or not at all; 5 = extremely), respondents were asked to rate the extent to which they experienced a list of emotions (e.g. enthusiastic, excited, determined, inspired) within the past few weeks. The internal reliability of this scale was $\alpha=0.91$ (with item-total correlations ranging from r=0.57 to 0.76) at Time 1, and $\alpha=0.92$ (with item-total correlations ranging from r=0.64 to 0.80) at Time 2.

Life satisfaction: Life satisfaction was measured using the 10-item Life Satisfaction Scale (Tepperman and Curtis 1995). Respondents used a five-point scale (1=strongly disagree; 5=strongly agree) to indicate the extent to which they were satisfied with various aspects of their life in general (e.g. "please indicate how satisfied you are with each of the following ... my relationship with my child(ren) ... my job ... my life as a whole"). In the present study, the internal reliability was $\alpha=0.79$ (with item-total correlations ranging from r=0.59 to 0.75).

Table 5.1 Study 1: Means, standard deviations and intercorrelations among the study variables

	M	SD	1	2	3	4	5	6	7	8	9	10	11	12	13	14
						Time 1							Time 2			
1. Age	46.29	9.99	—													
2. Work hours	41.61	6.41	-0.07	—												
Time 1																
3. WLC	2.72	0.91	-0.26[a]	0.23	(0.89)											
4. Positive mood	2.95	0.73	0.15	0.07	-0.18	(0.89)										
5. Negative mood	1.95	0.80	-0.02	-0.02	0.20	-0.25	(0.91)									
6. Emotional exhaustion	3.55	1.77	0.02	-0.27[a]	0.29[a]	-0.38[b]	0.43[b]	(0.54)								
7. Strain	3.01	0.80	0.33[a]	-0.23	-0.36[b]	0.31[a]	-0.11	-0.08	(0.90)							
8. Stress	3.07	0.96	-0.11	0.17	0.48[a]	-0.28[a]	0.51[c]	0.45[c]	-0.09	(0.70)						
Time 2																
9. WLC	2.45	0.95	-0.20	-0.15	0.59[b]	-0.12	-0.44[a]	0.29	-0.10	-0.08	(0.94)					
10. Positive mood	3.24	0.72	0.42[a]	0.16	-0.10	0.57[b]	-0.16	-0.20	0.02	-0.07	-0.14	(0.91)				
11. Negative mood	1.63	0.57	-0.42[a]	-0.20	0.12	-0.00	0.56[b]	0.09	0.33	0.30	-0.06	-0.33	(0.87)			
12. Emotional exhaustion	3.28	1.24	-0.35	-0.31	0.32	-0.38	0.27	0.71[c]	0.23	0.39	0.42	-0.53[a]	0.35	(0.83)		
13. Strain	3.29	0.69	0.24	-0.26	-0.06	0.25	-0.14	-0.23	0.73[c]	0.01	-0.03	0.24	0.09	0.01	(0.84)	
14. Stress	2.74	0.72	-0.34	0.13	0.55[b]	-0.23	0.06	0.30	0.02	0.37	0.57[b]	-0.47[a]	0.34	0.63[b]	-0.16	(0.43)

Notes: [a] $p<0.05$, [b] $p<0.01$, [c] $p<0.001$.

Table 5.2   Study 1: means, standard deviations and intercorrelations among study variables

	M	SD	1	2	3	4	5	6	7	8	9	10	11	12	13
1. Age	43.34	9.64													
2. Gender	—	—	0.08												
3. Group	—	—	-0.07	-0.10											
Time 1															
4. Positive mood	2.92	0.77	-0.03	-0.01	-0.19[a]										
5. Negative mood	2.29	0.82	-0.06	0.24[b]	0.10	-0.28[b]									
6. Life satisfaction	3.28	0.79	-0.06	-0.16	-0.03	0.52[c]	0.43[c]								
7. Perceived stress	3.53	0.86	-0.08	0.18[a]	-0.05	-0.24[c]	0.40[c]	-0.36[c]							
8. Daily hassles	4.62	1.28	-0.17	0.19[a]	-0.03	-0.30[c]	0.35[c]	-0.38[c]	0.46[c]						
Time 2															
9. Positive mood	2.89	0.78	0.02	0.01	-0.41[c]	0.48[c]	-0.24[a]	0.29[b]	-0.17	-0.21[a]					
10. Negative mood	2.10	0.76	-0.09	0.11	0.29[c]	0.25[b]	0.56[c]	-0.32[c]	0.30[b]	0.45[a]	-0.30[c]				
11. Life satisfaction	3.48	0.74	-0.04	-0.15	-0.29[b]	0.36[c]	-0.40[c]	0.59[c]	-0.31[c]	-0.34[c]	0.51[c]	-0.51[c]			
12. Perceived stress	3.39	0.88	0.00	0.10	0.31[c]	-0.31[c]	0.29[b]	-0.22[a]	0.47[c]	0.27[b]	-0.43[c]	0.48[c]	-0.46[c]		
13. Hassles	4.57	1.22	-0.14	0.16	0.29[c]	-0.15	0.33[c]	-0.29[b]	0.28[b]	0.71[c]	-0.35[c]	0.50[c]	-0.45[c]	0.50[v]	

Notes: [a] $p<0.05$; [b] $p<0.01$; [c] $p<0.001$.

Table 5.3   Study 1: Repeated measures means, standard deviations, and t-tests for outcomes

	Time 1			Time 2			
	M	SD	SE	M	SD	SE	T
WLC	2.66	0.789	0.16	2.45	0.95	0.20	1.26
Positive mood	2.90	0.56	0.11	3.24	0.72	0.15	-2.80[a]
Negative mood	2.07	0.82	0.17	1.63	0.57	0.12	3.12[b]
Emotional exhaustion	3.81	1.51	0.31	3.28	1.24	0.26	2.38[a]
Strain	2.96	0.80	0.16	3.29	0.69	0.14	-2.83[b]
Stress	3.21	0.90	0.18	2.74	0.72	0.15	2.51[a]

Notes: [a] $p<0.05$; [b] $p<0.01$; [c] $p<0.001$.

Perceived stress: Perceived stress was assessed using a three-item scale. Respondents used a five-point scale (1 = strongly disagree; 5 = strongly agree) to indicate the extent to which they were stressed at work, at home, and in general. In the present study, the internal reliability was $\alpha=0.86$ (with item-total correlations ranging from $r=0.49$ to 0.72).

Hassles: Hassles ware assessed using items modified from the Daily Hassles Scale. Respondents used a five-point scale (1 = strongly disagree; 5 = strongly agree) to indicate the extent to which they experience hassles in various areas. In the present study, the internal reliability was $\alpha=0.79$ (with item-total correlations ranging from $r=0.59$ to 0.75).

Results

CORRELATIONS

Table 5.3 shows the repeated measures means, standard deviations and t-tests for outcomes of this first study, whereas the means, standard deviations, reliabilities and intercorrelations among the study variables are presented opposite in Table 5.2.

ABLE AND OUTCOMES

To examine the impact of the ABLE programme on employee well-being outcomes (i.e. Hypotheses 1 a–e, a 2 (group: treatment vs. control) x 2 (time 1 vs. time 2) repeated measures MANOVA was conducted (see Table 5.4 for means and standard errors). There was a significant multivariate effect for the Group x Time interaction for negative mood ($F(1, 109)=6.79$, $p=0.01$, $^2=0.06$), life satisfaction ($F(1, 110)=12.02$, $p=0.001$, $\eta^2=0.10$), perceived stress ($F(1, 104)=6.49$, $p=0.012$, $\eta^2=0.06$), and hassles ($F(1, 105)=8.34$, $p=0.01$, $\eta^2=0.07$). That is, compared to the control group, the ABLE treatment group experienced significant increases in life satisfaction, and reductions in negative mood, perceived stress, and hassles from Time 1 to Time 2 (see Figures 5.1, 5.2, 5.3 and 5.4). There were no differences in positive mood and job satisfaction for the two groups between Time 1 and Time 2.

Table 5.4   **Study 2: Summary of the well-being means and standard errors for the ABLE treatment participants and the control group at Time 1 and Time 2**

| | ABLE treatment group | | | | Wait-list control group | | | | |
| | Time 1 | | Time 2 | | Time 1 | | Time 2 | | |
	M	SE	M	SE	M	SE	M	SE	F
Positive mood	3.16	0.09	3.25	0.09	2.72	0.10	2.59	0.10	2.32
Negative mood	2.30	0.11	1.85	0.09	2.40	0.12	2.32	0.10	6.79[b]
Life satisfaction	3.36	0.09	3.75	0.09	3.25	0.10	3.22	0.10	12.02[c]
Job satisfaction	3.43	0.14	3.62	0.13	3.08	0.15	3.02	0.14	1.68
Perceived stress	3.53	0.11	3.07	0.12	3.54	0.12	3.51	0.13	6.49[a]
Hassles	4.59	0.16	4.13	0.16	4.58	0.18	4.63	0.18	8.34[b]

Notes: [a] $p<0.05$; [b] $p<0.01$; [c] $p<0.001$.

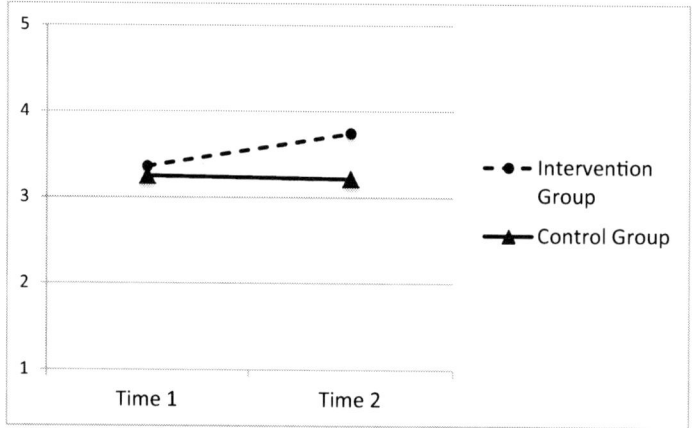

Figure 5.1   **Impact of ABLE intervention on life satisfaction**

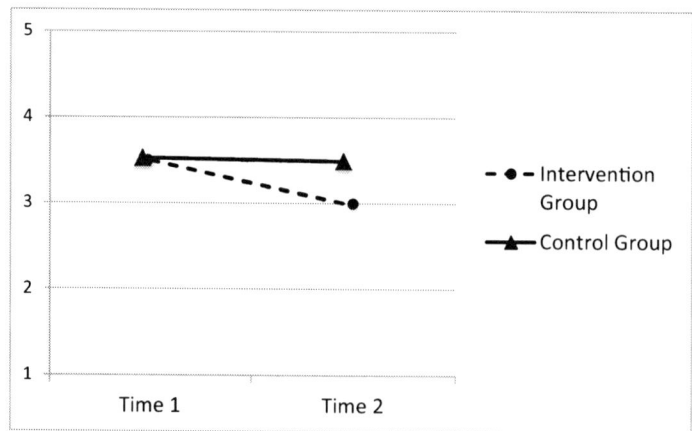

Figure 5.2   **Impact of ABLE on perceived stress**

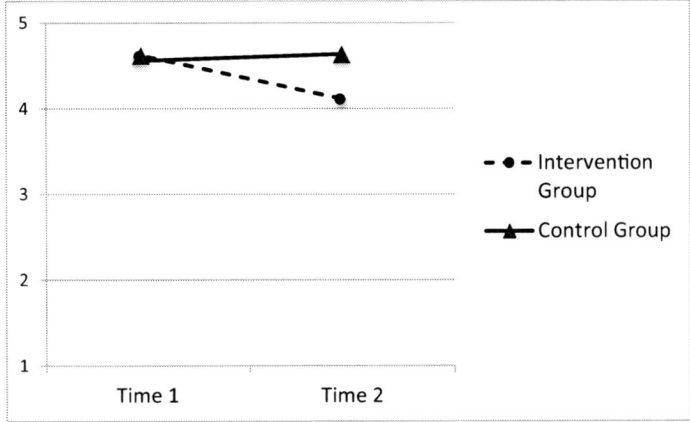

Figure 5.3   Impact of ABLE on negative mood

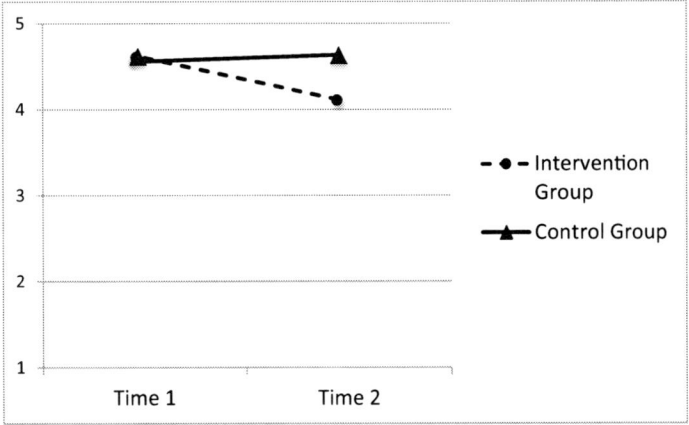

Figure 5.4   Impact of ABLE on hassles

In addition to the quantitative scales, feedback on the programme was gathered from participants after completion of the programme: 95 per cent of participants reported that the ABLE programme was beneficial and they indicated that they would encourage other employees to take it. Eighty-nine per cent of participants reported making positive changes in their life in general, with 78 per cent of participants reporting that they achieved their programme goals. Importantly, half of participants reported having someone in their life comment on the positive changes they experienced.

Study 2: Discussion – ABLE Coaching

In Study 2, we developed the ABLE Coaching Programme based on the ABLE workshops, and we examined its ability to improve the well-being of employees when compared to a control group. There were significant interactions for several outcomes, indicating that the well-being of participants in the ABLE intervention was significantly improved when compared to the control group. As hypothesized, the ABLE programme demonstrated a significant impact on several indices of well-being, including increased life satisfaction, and decreased negative mood, perceived stress, and hassles. These results support and extend the results from Study 1, and provide valuable information on the validity of stress and conflict programmes.

Limitations and Future Research

Study 2 addressed several limitations from the first study, but it still was subject to some of the same concerns. A true longitudinal study should involve more than two data collection points (Kelloway and Francis 2012). Therefore, it is important for future research to examine the long-term effects of the programme. Although we incorporated a control group into the design, future research should use a control-wait list design to demonstrate changes in the control group during a second intervention stage.

The study was originally designed to have a randomized assignment. Although we ensured that organizations still had approximately half of their participating employees in the intervention and half in the control, randomization was not possible due to organizational and time constraints. This lack of randomization may have impacted results, such as the first employees to sign up were more likely to be in the initial intervention group, and they may have felt a more pressing need for the programme, or were more committed to it. The inclusion of the control group may minimize these concerns, as we were able to compare the two groups. Nonetheless, future research should incorporate a randomized design to assess the validity of this programme.

General Discussion

Despite the negative impact of job stress and conflict on individuals and organizations, there are few validated programmes to help employees manage these demands. In fact, Kelloway et al. (2008: 419) argued that "the question of what organizations can do to avert or mitigate the negative consequences of stress – arguably the single most important question in the field – remains largely unaddressed." Therefore, in this chapter, we have reported on two programmes to address this dearth of research and to add to the literature on workplace interventions. We developed and validated a workshop and a phone-based programme to address job stress and work–life balance issues. In the development of these programmes, we heeded Quick et al.'s (1997) suggestion that primary, secondary, and tertiary efforts to reduce job stress and conflict should be integrated to improve employee health, as we tried to incorporate a variety of methodologies to address stress and conflict.

The results from both the workshop study and the ABLE coaching programme are promising, in that they demonstrated positive effects of the programmes on several important employee outcomes, even when compared to a control group (Study 2). These findings suggest that providing workshops or a more advanced, tailored coaching programme may help improve employee health and well-being. Future research should examine the efficacy of this and similar programmes with a larger sample, and examine the impact of organizational practices on individual health and performance in more detail.

LIMITATIONS AND FUTURE RESEARCH

The results are positive that such both ABLE programmes were able to contribute to employee well-being and work behaviours. However, we don't know the exact components of either ABLE programme that contributed to its effectiveness. For example, it may have been the content, the manual, and the formal support (in both studies), or the coaches and tailored topics (in Study 2), or a combination of all of these approaches/factors. Moreover, it may be inappropriate to assume that everyone's success was due to the identical factors. That is, just as everyone has different demands, resources and learning styles, different facets of the programme may have contributed to its success for different individuals. The current research identified that this type of intervention is effective, and future research may be able to disentangle its components to identify the most effective factors or the specific person by component interactions.

Although we had organizational measures of absenteeism and worker compensation claims in Study 1, we collected only self-report data in Study 2. Future research using ABLE should incorporate organizational data (supervisory reports, HR data, colleague ratings), physiological indicators of health (e.g. blood pressure), as well as self-report indices of health and well-being.

Finally, although both programmes demonstrated effectiveness, the participant feedback for the coaching programme was very positive, and both the qualitative comments and the feedback ratings indicated that the presence of the coach was a very valuable component of the programme. Future research should compare the relative effectiveness of these types of interventions, taking other cost–benefit information into consideration.

PRACTICAL IMPLICATIONS

The positive findings from both studies indicate that these types of programmes may be beneficial for both employees and organizations. Organizations may consider implementing such programmes to help improve employee well-being and employee work-related behaviours. Moreover, because Kompier and Cooper (1999) noted that the more interventions should involve a dual responsibility of both the individual and the organization to implement the programme, and based on suggestions to integrate job stress interventions within the organizational context (e.g. Hurrell 2005, Quick et al. 1997), organizations may consider incorporating programmes such as the ABLE into an overall organizational programme, maximizing the ownership of both employees and the organization.

Conclusion

The main goals of this study were to examine the ability of a job stress and work–life balance workshop to improve employee outcomes and to use this information to develop a stress and conflict programme that would be more accessible and tailored. Cooper and Dewe (2008: 523) argued that there was "growing evidence to support the case that workplace well-being interventions make good business sense." Our two studies provide further evidence of the efficacy of their statement in terms of employee health and organizational outcomes.

References

Adams, G.A. and Jex, S.M. 1999. Relationships between time management, control, work-family conflict, and strain. *Journal of Occupational Health Psychology*, 4, 72–7.

Adams, G.A., King, L.A. and King, D.W. 1996. Relationships of job and family involvement, family social support, and work-family conflict with job and life satisfaction. *Journal of Applied Psychology*, 81, 411–20.

Barling, J. 1990. *Employment, Stress, and Family Functioning*. Oxford: John Wiley and Sons.

Bartone, P.T., Ursano, R.J., Wright, K.M. and Ingraham, L.H. 1989. The impact of a military air disaster on the health of assistance workers: A prospective study. *Journal of Nervous and Mental Disease*, 177(6), 317–28.

Beehr, T.A., Jex, S.M., Stacy, B.A. and Murray, M.A. 2000. Work stressors and coworker support as predictors of individual strain and job performance. *Journal of Organizational Behavior*, 21, 391–405.

Bellavia, G.M. and Frone, M.R. 2005. Work-family conflict. In *Handbook of Workplace Stress*, edited by J. Barling, E.K. Kelloway and M. Frone. Thousand Oaks, CA: Sage Publications, 35–62.

Bond, F. and Bunce, D. 2001. Job control mediates change in a work reorganization intervention for stress reduction. *Journal of Occupational Health Psychology*, 6, 290–302.

Carayon, P. 1995. Chronic effect of job control, supervisory social support, and work pressure on office-worker stress. In *Organizational Risk Factors for Job Stress*, edited by S.L. Sauter and L.R. Murphy. Washington, DC: American Psychological Association, 357–70.

Cartwright, S. and Cooper, C.L. 2005. Individually targeted interventions. In *Handbook of Work Stress*, edited by J. Barling, E.K. Kelloway and M.R. Frone. Thousand Oaks, CA: Sage Publications, 607–22.

Carver, C.S., Scheier, M.F. and Weintraub, J.K. 1989. Assessing coping strategies: A theoretically based approach. *Journal of Personality and Social Psychology*, 56, 267–83.

Caulfield, N., Chang, D., Dollard, M.F. and Elshaug, C. 2004. A review of occupational stress interventions in Australia. *International Journal of Stress Management*, 11(2), 149–66.

Cooper, C. and Dewe, P. 2008. Well-being – absenteeism, presenteeism, costs and challenges. *Occupational Medicine*, 58(8), 522–4.

Day, A.L. 1996. *Conceptualizing Work and Family Conflict using an Inter-role Perspective: The Importance of Time, Strain, and Behaviour*. Doctoral dissertation. Retrieved from Dissertations Abstracts International.

Day, A.L. and Chamberlain, T. 2006. Committing to your work, spouse, and children: Implications for work-family conflict. *Journal of Vocational Behavior*, 68, 116–30.

Day, A.L. and Jreige, S. 2002. Using Type A behavior pattern to explain the relationship between job stressors and psychosocial outcomes. *Journal of Occupational Health Psychology*, 7, 109–20.

Day, A.L. and Livingstone, H. 2001. Chronic and acute stressors among military personnel: Do coping styles buffer their negative impact on health? *Journal of Occupational Health Psychology*, 6, 348–60.

DeRango, K. and Franzini, L. 2003. Economic evaluation of workplace health interventions: Theory and literature review. In *Handbook of Occupational Health Psychology*, edited by J.C. Quick and L.E. Tetrick. Washington, DC: American Psychological Association, 417–30.

Duxbury, L. and Higgins, C. 2003. *Work-life Conflict in Canada in the New Millennium: A Status Report (Final Report)*. Public Health Agency of Canada. Available at: http://www.phac-aspc.gc.ca/publicat/work-travail/index.html [accessed 16 February 2007].

Frone, M.R. 2003. Work-family balance. In *Handbook of Occupational Health Psychology*, edited by J.C. Quick and L.E. Tetrick. Washington, DC: American Psychological Association, 143–62.

Frone, M.R., Cooper, M.L. and Russell, M. 1994. Stressful life events, gender and substance use: An application of Tobit regression. *Psychology of Addictive Behaviors*, 8, 59–69.

Frone, M.R., Russell, M. and Cooper, M.L. 1997. Relation of work-family conflict to health outcomes: A four-year longitudinal study of employed parents. *Journal of Occupational and Organizational Psychology*, 70, 325–35.

Giga, S.I., Cooper, C.L. and Faragher, B. 2003. The development of a framework for a comprehensive approach to stress management interventions at work. *International Journal of Stress Management*, 10(4), 280–296.

Grant, A.M. 2003. The impact of life coaching on goal attainment, metacognition, and mental health. *Social Behavior and Personality*, 31(3), 253–64.

Hain, C. and Francis, L. 2006. *Coworker Relationships: Using a New Measure to Predict Health Related Outcomes*. Paper presented at Work, Stress, and Health Conference: Marking a Difference in the Workplace, Miami, FL.

Hammer, T.H., Saksvik. P.O., Nytro, K., Torvatn, H. and Bayazit, M. 2004. Expanding the psychosocial work environment: Workplace norms and work-family conflict as correlates of stress and health. *Journal of Occupational Health Psychology*, 9, 83–97.

Hepburn, C.G., Loughlin, C.A. and Barling, J. 1997. Coping with chronic work stress. In *Coping with Chronic Work Stress*, edited by B.H. Gottlieb. New York, NY: Plenum Press, 343–66.

Hodson, R. 1997. Group relations at work: Solidarity, conflict, and relations with management. *Work and Occupations*, 24, 426–52.

Hu, F.B., Manson, J.E., Stampfer, M.J. and Colditz, G. 2001. Diet, lifestyle and the risk of type 2 diabetes mellitus in in women through diet and lifestyle. *The New England Journal of Medicine*, 345, 790.

Hurrell, J.J., Jr., 2005. Organizational stress intervention. In *Handbook of Workplace Stress*, edited by J. Barling, E.K. Kelloway and M. Frone. Thousand Oaks, CA: Sage Publications, 623–45.

Hurrell, J.J. and Murphy, L.R. 1996. Occupational stress intervention. *American Journal of Industrial Medicine*, 29(4), 338–41.

Hurrell, J.J., Nelson, D.L. and Simmons, B.L. 1998. Measuring job stressors and strains: Where we have been, where we are, and where we need to go. *Journal of Occupational Health Psychology*, 3, 368–89.

Kahn, R.L. and Byosiere, P.B. 1992. Stress in organizations. In *Handbook of Industrial and Organizational Psychology*, Vol. 3, edited by M.D. Dunnette and L.M. Hough. Palo Alto, CA: Consulting Psychologists Press, 571–650.

Kelloway, E.K. and Francis, L. 2006. *Stress and Strain in Nova Scotia Organizations: Results of a Recent Province-wide Study*. Paper presented at the Nova Scotia Psychologically Healthy Workplace Conference.

Kelloway, E.K. and Francis, L. 2012. Longitudinal research and data analysis. In *Research Methods in Occupational Health Psychology*, edited by R. R. Sinclair, M. Wang, and L. E. Tetrick. New York, NY: Taylor & Francis., 374–94.

Kelloway, E.K., Hurrell, J.J. and Day, A. 2008. Workplace interventions for occupational stress. In *The Individual in the Changing Working Life*, edited by K. Näswell, M. Sverke and J. Hellgren. Cambridge: Cambridge University Press, 419–40.

Kenny, D.T. and Cooper, C. 2003. Introduction: Occupational stress and its management. *International Journal of Stress Management*, 10, 275–9.

Kompier, M. and Cooper, C.L. (eds). 1999. *Preventing Stress, Improving Productivity: European Case Studies in the Workplace*. London: Routledge.

Kossek, E. and Ozeki, C. 1999. Bridging the work-family policy and productivity gap: A literature review. *Community, Work and Family*, 2, 7–32.

Kristensen, T.S. 1996. Job stress and cardiovascular disease: A theoretical critical review. *Journal of Occupational Health Psychology*, 3, 246–60.

Lazarus, R.S. 1995. Theoretical perspectives in organizational research. In *Organizational Stress: A Handbook*, edited by R. Crandall and P.L. Perrewe. Philadelphia: Taylor & Francis, 3–14.

Lazarus, R.S. and Folkman, S. 1984. *Stress, Appraisal, and Coping*. New York: Springer.

Leiter, M.P., Day, A., Oore, D.G. and Spence Laschinger, H.K. 2012. Getting better and staying better: Assessing civility, incivility, distress, and job attitudes one year after a civility intervention. *Journal of Occupational Health Psychology*, 17(4), 425.

Lingley-Pottie, P., Watters, C., McGrath, P.J. and Janz, T. 2005. Providing family help at home. Proceedings of 38th Annual Hawaii International Conference on System Sciences (CD/ROM). 4–9 January, Computer Society Press.

Locke, E.A. and Latham, G.P. 2002. Building a practically useful theory of goal setting and task motivation: A 35 year Odessey. *American Psychologist*, 57, 705–17.

Mactavish, J. and Iwasaki, Y. 2005. Exploring perspectives of individuals with disabilities on stress-coping. *Journal of Rehabilitation*, 71(1), 20–31.

Maslach, C., Jackson, S.E. and Leiter, M.P. 1996. *Maslach Burnout Inventory Manual* (3rd edn). Palo Alto: Consulting Psychologists Press.

McGrath, P.J., Lingley-Pottie, P., Emberly, D.J., Thurston, C. and McLean, C. 2009. Integrated knowledge translation in mental health: Family help as an example. *Journal of the Canadian Academy of Child and Adolescent Psychiatry*, 18(1), 30–37.

Meyer, J.P., Allen, N.J. and Smith, C.A. 1993. Commitment to organizations and occupations: Extension and test of a three-component conceptualization. *Journal of Applied Psychology*, 78, 538–51.

Mikkelsen, A., Saksvik, P.O. and Landsbergis, P. 2000. The impact of a participatory organization intervention on job stress in a community health care institution. *Work and Stress*, 14, 156–70.

Moos, R.H., Holahan, C.L. and Beutler, L.E. (eds) 2003. Special issue on coping. *Journal of Clinical Psychology*, 59(12), 1257–403.

Nielsen, I.K., Jex, S.M. and Adams, G.A. 2000. Development and validation of scores on a two-dimensional workplace friendship scale. *Educational and Psychological Measurement*, 60, 628–43.

Niven, N. and Johnson, D. 1989. Taking the lid off stress management. *Industrial and Commercial Training*, 21(5), 8–11.

Oetzel, J.G. and Ting-Toomey, S. 2006. *The SAGE Handbook of Conflict Communication: Integrating Theory, Research, and Practice*. Thousand Oaks: Sage Publications.

Parasuraman, S. and Greenhaus, J.H. (eds) 1997. *Integrating Work and Family: Challenges and Choices for Changing World*. Westport: Quorum.

Peiró, J.M., Gonzalez-Romá, V., Tordera, N. and Mañas, M.A. 2001. Does role stress predict burnout over time among health care professionals? *Psychology and Health*, 16, 511–25.

Quick, J.C., Murphy, L.R., Hurrell, J.J. and Orman, D. 1992. The value of work, the risk of distress, and the power of prevention. In *Stress and Well-being at Work*, edited by J.C. Quick, L.R. Murphy and J.J. Hurrell. Washington, DC: American Psychological Association, 3–13.

Quick, J.C., Quick, J.D., Nelson, D.L. and Hurrell Jr, J.J. 1997. *Preventive Stress Management in Organizations*. American Psychological Association.

Rogers, S.J., Parcel, T.L. and Menaghan, E.G. 1991. The effects of maternal working conditions and mastery on child behaviour problems: Studying the intergenerational transmission of social control. *Journal of Health and Social Behaviour*, 32, 145–64.

Sauter, S.L., Murphy, L.R. and Hurrell, J.J. 1990. Prevention of work-related psychological disorders: A national strategy proposed by the National Institute for Occupational Health and Safety (NIOSH). *American Psychologist*, 45, 1146–58.

Sheldon, K.M. and Kasser, T. 2001. Goals, congruence, and positive well-being: New empirical support for humanistic theories. *Journal of Humanistic Psychology*, 41(1), 30–50.

Sonnentag, S. and Frese, M. 2003. Stress in organizations. In *Handbook of Psychology: Industrial and Organizational Psychology*, Vol. 12, edited by W.C. Borman, D.R. Ilgen and R.J. Klimoski. Hoboken: John Wiley and Sons, Inc., 453–91.

Spector, P. 1986. Perceived control by employees: A meta-analysis of studies concerning autonomy and participation at work. *Human Relations*, 39, 1005–16.

Stampfer, M.J., Hu, F.B., Manson, J.E., Rimm, E.B. and Willett,W.C. 2000. Primary prevention of coronary heart disease in women through diet and lifestyle. *The New England Journal of Medicine*, 343, 16.

Taylor, S.E., Klein, L.C., Lewis, B.P., Gruenewald, T.L., Gurung, R.A.R. and Updegraff, J.A. 2000. Behavioral responses to stress in females: Tend-and-befriend, not fight-or-flight. *Psychology Review*, 107(3), 411–29.

Tepperman, L. and Curtis, J. 1995. A life satisfaction scale for use with national adult samples from the USA, Canada, and Mexico. *Social Indicators Research*, 35, 255–70.

Terra, N. 1995. The prevention of job stress by redesigning hobs and implementing self-regulating teams. In *Job Stress Interventions*, edited by L.R. Murphy, J.J Hurrell, Jr., S.L. Sauter and G.P. Keita. Washington, DC: American Psychological Association, 265–81.

Theorell, T., Emdad, R., Arnetz, B. and Weingarten, A.M. 2001. Employee effects of an educational program for managers at an insurance company. *Psychosomatic Medicine*, 63(5), 724–33.

Thomas, L. and Ganster, D. 1995. Impact of family-supportive work variables on work-family conflict and strain: A control perspective. *Journal of Applied Psychology*, 80, 6–15.

Thomason, J.A. and Pond, S.B. 1995. Effects of instruction on stress management skills and self-management skills among blue-collar employees. In *Job Stress Interventions*, edited by L.R. Murphy, J.J. Hurrell, S.L. Sauter and G.P. Keita. Washington, DC: American Psychological Association, 7–20.

Totterdell, P. 2005. Work schedules. In *Handbook of Workplace Stress*, edited by J. Barling, E.K. Kelloway and M. Frone. Thousand Oaks: Sage, 35–62.

Trapnell, P.D. and Campbell, J.D. 1999. Private self-consciousness and the five-factor model of personality: Distinguishing rumination from reflection. *Journal of Personality and Social Psychology*, 76(2), 284–304.

Tyler, P. and Cushway, D. 1995. Stress in nurses: The effects of coping and social support. *Stress Medicine*, 11, 243–51.

Violanti, J.M. 1992. Coping strategies among police recruits in a high stress training environment. *Journal of Social Psychology*, 132, 717–29.

Viswesvaran, C., Sanchez, J.I. and Fisher, J. 1999. The role of social support in the process of work stress: A meta-analysis. *Journal of Vocational Behavior*, 54, 314–34.

Wager, N., Fieldman, G. and Hussey, T. 2003. The effect on ambulatory blood pressure of working under favourably and unfavourably perceived supervisors. *Occupational and Environmental Medicine*, 60, 468–74.

Warr, P.B. 1987. *Work, Unemployment and Mental Health*. Oxford: Clarendon Press.

Watson, D., Clark, L.A. and Tellegen, A. 1988. Development and validation of brief measures of positive and negative affect: The PANAS scales. *Journal of Personality and Social Psychology*, 54(6), 1063–70.

Williams, C. 2003. Sources of workplace stress. *Perspectives on Labour and Income*, 4, 5–12.

Wolever, R.Q., Bobinet, K.J., McCabe, K., Mackenzie, E.R., Fekete, E., Kusnick, C.A. and Baime, M. 2012. Effective and viable mind-body stress reduction in the workplace: A randomized controlled trial. *Journal of Occupational Health Psychology*, 17(2), 246–58.

6 A Case Study in the Design, Implementation and Evaluation of a Worksite Wellness Programme for Reducing Cardiovascular Risks

DOUGLAS W. ROBLIN, BRANDI E. ROBINSON and STACEY A. BENJAMIN

Abstract

Worksite wellness programmes provide a unique opportunity for employers and health plans to collaborate in creating opportunities to improve the health of employees and health plan enrollees. We describe the steps in design, implementation and prospective evaluation of a worksite wellness programme to reduce cardiovascular risks. The context for this study was a partnership between Kaiser Foundation Health Plan of Georgia, Inc. (KPGA) and a small professional services corporation in the metropolitan Atlanta area. Programme design involved extensive preparatory discussions between KPGA and the firm's leadership about organizational readiness for change, implementation policies and procedures, and involvement of a research and evaluation component. Programme participation consisted of an enrolment session, a six-month period of worksite wellness programme activities, and a disenrolment session. The evaluation design was a pre-/post-cohort design, with each participant acting as his/her control. The primary outcomes were changes in activation, lifestyle (physical activity and dietary intake), and biometric markers (body mass index, blood pressure, total serum cholesterol).

Introduction

Because employees spend a large proportion of their waking hours at their worksites, worksite wellness programmes provide an important opportunity for promoting attitudes and behaviours that can motivate practice of a healthy lifestyle, reduce chronic disease risks and improve employee productivity (Goetzel and Ozminkowski 2008, Sorensen et al. 2011). Reviews and meta-analyses of worksite wellness programmes to address cardiovascular risk factors indicate those programmes, on average, improve lifestyle behaviours (such as leisure physical activity and dietary intake) and biometric markers of cardiovascular risk (such as blood pressure levels, serum lipids and obesity) (Anderson et al. 2009, Benedict and Arterburn 2008, Chan Osilla et al. 2012, Conn et al. 2009, Engbers et al. 2005, Kuoppala et al. 2008, Pelletier 1996, 2001). In the short term, these programmes may also benefit employers in terms of reduced presenteeism, absenteeism and health care costs (Baicker et al. 2010, Kuoppala et al. 2008).

Recognizing the potential benefits of worksite wellness programme availability and participation, "Healthy People 2010" and "Healthy People 2020" recommend an "increase [in] the proportion of worksites that offer a comprehensive employee health promotion program to their employees" (Centers for Disease Control and Prevention 2012). The American Heart Association recently endorsed worksite wellness programmes as one strategy for improving cardiovascular health (Carnethon et al. 2009). Provisions of the Patient Protection and Affordable Care Act include grants for small firms to implement wellness programmes and flexibility for employers to implement financial incentives for employee participation in wellness programmes (Koh and Sebelius 2010).

Worksite wellness programmes – and related resources and activities – are neither widely nor uniformly available across employers and worksites in the United States. The 2004 National Worksite Health Promotion Survey reported that 19.6 per cent of worksites offered a physical activity programme, 22.7 per cent a nutrition education programme, 21.4 per cent a weight management programme, and 26.1 per cent a cardiovascular disease management programme (Linnan et al. 2008). Small worksites (<100 employees) are significantly less likely to offer wellness programmes or activities than large worksites (Linnan et al. 2008, Wilson et al. 1999). Few worksites, small or large, offer comprehensive, multifactorial worksite wellness programmes. In the 2004 National Worksite Health Promotion Survey, only 6.9 per cent of worksites had a programme incorporating five key elements (health education, supportive social and physical environment, integration, linkage to related programmes, and worksite screening) that defined a comprehensive programme according to Healthy People 2010 (Linnan et al. 2008). Where programmes are available, employee participation in a worksite wellness programme or activity is moderate. For example, Grosch et al. (1998) found that 32.8 per cent of employees reported participation in an "exercise programme," 57.8 per cent in a "health education programme" and 58.4 per cent in a "screening test."

Particularly for small and mid-size employers, relevant and experienced staff and other resources needed for design, implementation and evaluation of a worksite wellness programme may be limited or simply unavailable. A lack of resources was noted by 63.4 per cent of employers in a 2004 survey as one challenge in implementation of worksite wellness programmes (Linnan et al. 2008). Many integrated health care delivery systems (e.g. health maintenance organizations, physician-hospital systems) have both the expertise and resources for this purpose. These systems typically design, implement, and evaluate a range of wellness programmes germane to worksite health promotion: health education programmes for promoting physical activity, a healthy diet, and weight management; health risk appraisals; chronic disease management programmes; and, mental health and stress management programmes. These systems also often are the health care providers for current, and potentially future, employees and families of these employees. Thus, it seems there is substantial community health benefit that might be obtained from a partnership between integrated health delivery systems with expertise and resources needed to design, implement and evaluate worksite wellness programmes with employers who have a need for worksite wellness programmes without the relevant expertise and resources. In this chapter, we describe the partnership between Kaiser Permanente Georgia (KPGA), a health maintenance organization, and a small professional services company, to design, implement and evaluate a worksite wellness programme to reduce cardiovascular risks among the company's employees.

Methods

STUDY SETTING

Kaiser Permanente Georgia (KPGA) is primarily a group-model managed care organization (MCO) that, at the time of this study, provided comprehensive medical services to approximately 250,000 residents of the metropolitan Atlanta area. The employer group which participated in this worksite wellness programme was a small (<100 employees) professional corporation that was established in the early 1970s and provided insurance and risk management services both locally and regionally. The local workforce consisted of 90 employees (including support, professional and executive staff) and worked out of one primary facility in the Atlanta area. Most of the employees and their families were provided health insurance coverage through KPGA. The protocol for evaluation of the worksite wellness programme was reviewed, approved and monitored by the KPGA Institutional Review Board.

PROGRAMME DESIGN AND IMPLEMENTATION

Overall design and implementation of the worksite wellness programme followed guidelines of Kaiser Permanente's HealthWorks programme. HealthWorks is based on principles to facilitate the design and implementation of a worksite wellness programme for employers who are new to workforce health and to improve programme performance among employers with some worksite wellness programme experience. Principles which inform the HealthWorks programme generally follow the organizational practices which have been identified as supporting design and implementation of worksite wellness programmes that have been effective in improving employee health or productivity (Goetzel et al. 2007, Goetzel and Pronk 2010, Pronk and Kottke 2009, Terry et al. 2008, Weiner et al. 2009). HealthWorks resources include a series of workbooks and supporting materials to help employers recognize the importance of worksite wellness programmes, implement strategies to promote programme participation, select programme activities relevant to their programme's goals and workforce health needs, and evaluate programme impact.

ORGANIZATIONAL COMMITMENT

Commitment of an organization's leadership to a worksite wellness programme is an important antecedent of the likely success (or failure) of a worksite wellness programme (Weiner et al. 2009). For this pilot worksite wellness programme, organizational commitment was obtained through an extended negotiated process between the employer leadership team and the KPGA design, implementation and evaluation team. The negotiation process and identification of key players in both groups was somewhat fluid initially, but players and roles were more clearly defined over the course of a series of meetings during a four-month period.

Initial interest in, and assessment of commitment for, a worksite wellness programme was elicited through discussions between the employer's KPGA account manager and various members of the employer's leadership team. Once the account manager had a sense of the employer's interest in collaborating with KPGA on design and implementation of a worksite wellness programme, KPGA's manager of wellness and health promotion activities, and KPGA's manager of employer accounts and sales were brought into discussions. Those discussions allowed the KPGA manager of wellness and health promotion:

1. to explain the HealthWorks approach and KPGA's expertise and resources in health education and health promotion; and
2. to get a preliminary assessment of the employer's priorities for key components of programme design: duration of the programme, content and schedule of programme activities, location of activities, and incentives to motivate programme participation.

KPGA's research staff were then brought into discussions initially with the KPGA account manager and worksite wellness programme manager, and ultimately into discussions with the employer leadership team, when it was clear that the employer was committed not only to implementation of a worksite wellness programme but also interested in a formal evaluation of the effects of the programme on its employees.

Thus, contributions to design, implementation and evaluation of this worksite wellness programme consisted of a collaboration between the employer leadership team and a KPGA project team. The employer's leadership team included its CEO, several vice presidents, a designated contact for the worksite wellness programme, and human resources manager, the latter two who were to be responsible for local management of corporate activities related to this pilot worksite wellness programme. The KPGA project team consisted of the account manager (who served as the primary point of contact with the employer leadership team), the worksite manager (who organized and directed programme activities), and a research investigator (who directed data collection, analysis and evaluation).

EMPLOYEE ENGAGEMENT

Following the employer's leadership agreement to proceed with a worksite wellness programme, the KPGA team conducted two one-hour wellness fairs (sessions) prior to launch of the programme. Employees were encouraged by the employer to attend. The wellness fairs were held on-site during the typical lunch break, and a box lunch was provided as an incentive to attend. An outline of the worksite wellness programme, including the evaluation programme, was presented by the KPGA team. The KPGA team was introduced, and employees were encouraged to ask questions about, or comment on, both the worksite wellness programme and the evaluation programme.

PARTICIPATION INCENTIVES

Incentives, particularly monetary incentives, can be important for motivating employee participation in worksite wellness programme activities (Seaverson et al. 2009). Monetary incentives may be associated with:

1. programme outcomes – a benefit is linked to achievement of a specific target; or
2. processes – benefits are linked to participation in the programme or specific activities; however, ethical considerations often arise in who might benefit and who might not benefit from a particular incentive plan (Schmidt 2012, Schmidt et al. 2010).

In this pilot study, the employer leadership team advocated a monetary incentive linked to participation over a monetary incentive linked to an outcome. A premium was placed on a high participation rate as a programme goal, and there was a desire to avoid the appearance of differentially rewarding certain subsets of employees. For participating in the worksite wellness programme, employees received a $100 contribution from the employer to participating employees' flex-spending accounts. In addition, some non-monetary rewards were provided to participants, such as a DVD on exercise that could be used by employees outside formal programme activities.

ACCESS TO PROGRAMME ACTIVITIES

The employer provided time off for selected programme activities in order to encourage employee participation. In addition, all programme activities were located and performed at the employer's offices in Atlanta. The offices included several large meeting rooms that could accommodate the number of anticipated participants for educational seminars and limited exercise sessions. This location made it convenient for employees and was presumed likely to improve programme participation. It also allowed flexibility in scheduling activities at the beginning or end of the work day as well as around the midday lunch break.

PROGRAMME DURATION AND CONTENT

The duration of the wellness programme was six months. Duration of the worksite wellness programme took into consideration the financial resources and time that could be devoted by the employer and KPGA to a programme and an estimate of the length of time that might be needed to demonstrate change in lifestyle or health.

Programme content was selected from a menu of options (Table 6.1) and ultimately consisted of educational seminars (1/month; topics included "Prehypertension," "Art of healthy cooking," "Reading Nutrition Labels"), exercise sessions (2/month on-site; with DVD/VCR copies available for home use), and "e-blasts" (50; emails addressing nutrition and exercise). Educational seminars were conducted by experienced health educators and dietitians and exercise sessions were supervised by certified exercise instructors and personal trainers. Selection of content was partly informed by the biometric and lifestyle data collected on survey during the enrolment session. Although the health risk profile of the employer group was overall relatively positive, various measures indicated a mild to moderate elevated cardiovascular risk.

Table 6.1   Worksite wellness programme content topics

Topic	Description
Healthy Supermarket Tour	Take a virtual trip down the aisles of your average supermarket while you learn how to make healthy meal choices. Get tips on reading labels and food safety.
Taking Charge of Your Health	No more excuses. This class will focus on committing to and designing your own health improvement programme. Healthy eating, exercise, stress and work/home balance are the foci of this workshop.
Focus on Fitness	Learn the basics of exercise, get motivated for personal fitness and learn ideas for physical activity.
Healthy Heart	Hands-on heart, artery and cholesterol models help participants understand heart disease. Learn of the risk factors for heart disease and how to make lifestyle changes for better health.
Healthy Cooking	Consider your personal eating habits and how to move towards goals for healthier eating. A professional chef or dietitian will cook a healthy meal and explain the benefits of healthy eating. Participants can sample the food while learning about healthier food choices, serving sizes and the benefits of good nutrition.
Weight Management	Learn how to make healthy choices, including regular exercise, proper nutrition, and a positive attitude can help you lose weight and keep it off and/or maintain body weight.
Controlling Cholesterol	Learn about "good" and "bad" cholesterol, and how they impact your heart health. Get information on how healthier food choices and lifestyle changes can reduce your risk of heart attack and stroke.
Fast Food and Your Health	The fast food industry is offering a variety of reduced fat and lower sodium alternatives. Learn how to make healthier choices in fast food restaurants for you and your children.
Label Reading 101	This workshop helps you to understand food labels. Get information on serving size, vitamins, sugars, trans fat, percentage daily value, sodium and more.

HEALTH RISK APPRAISALS

The worksite wellness programme included an enrolment session (September 2009) during which biometric data were collected on each participant and a disenrolment session six months later (March 2010) during which the same biometric data were collected. A copy of the biometric results was provided to each participant at the enrolment session. Also, at the enrolment session, a KPGA nurse was available to provide interpretation of the biometric results (in a private room adjacent to the public room where the health fair was conducted).

TAILORING PROGRAMME CONTENT AND ACTIVITIES

Health risk appraisal results can be used to tailor worksite wellness programme activities to each participant's specific health risks. In this case, resources were not sufficient to offer individually-tailored worksite wellness programmes. However, summary results of the biometric data obtained on the enrolment session were available shortly after that session. Review of these results by the employer leadership team and the KPGA project team indicated that, although a relatively healthy employee population was profiled, sufficient indications of early cardiovascular risks justified orienting worksite wellness programme activities toward cardiovascular risk reduction

Evaluation Design

As an incentive to the employer to support the evaluation programme, the KPGA team offered evaluation reports at baseline and conclusion of the worksite wellness programme at no cost to the employer. The employer requested, and the KPGA team concurred, that participation in the worksite wellness programme would not require participation in the evaluation programme. While they recognized the value of the evaluation programme to leadership, they did not want the additional effort required of participation in the evaluation programme to discourage participation in the worksite wellness programme.

At the enrolment session, employees were offered the option of participating in an evaluation of the effectiveness of the programme. Written, informed consent was required by the KPGA Institutional Review Board. Consent involved release of the biometric data to the research team and completion of a brief written survey. For participating in the evaluation of the worksite wellness programme, consenting participants received from the KPGA research programme a $25 American Express gift certificate for completing the baseline biometric assessment and survey and an additional $25 for completing the six-month follow-up biometric assessment and survey.

CONCEPTUAL MODEL

The conceptual model for evaluation of a worksite wellness programme designed to reduce cardiovascular risks is presented in Figure 6.1 and is based on health behavioural

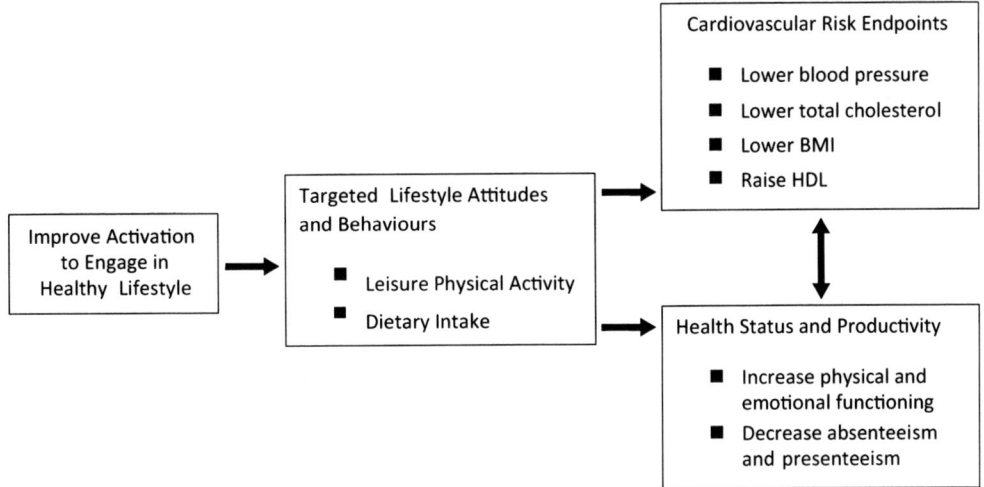

Figure 6.1   Common sense model of health behaviour and health outcomes for a worksite wellness programme to reduce cardiovascular risks

theory and well established clinical literature and consensus statements from professional societies that associate lifestyle with clinical outcomes. The "common sense model" of health behaviour suggests that individuals, when informed and motivated about health benefits and risks ("activated"), will be more likely to adhere to recommendation to practice a healthy lifestyle; and, these individuals, in turn, will experience improved health status (Leventhal et al. 2003, McAndrew et al. 2008, Meyer et al. 1985). We hypothesized that participation in the worksite wellness programme would increase average activation and that this would be associated with improved lifestyle (dietary intake and leisure physical activity) and an improved cardiovascular risk profile.

EXPERIMENTAL DESIGN

Various experimental designs were considered (Mohr 1992, Rossi et al. 1999). A randomized controlled trial would have been ideal; however, practical and logistical circumstances in this employer group made this option infeasible. The anticipated sample size was small, and, randomization would have split the sample, further weakening statistical power to detect changes in study measures. It would have been logistically difficult to prevent unwanted participation in wellness activities by employees in a control group and likely would have been perceived as unfair that some employees might benefit, and others not, as a consequence of the design of a randomized controlled trial.

Our overall evaluation design was a pre-/post-design in which participants in both baseline and six-month follow-up assessments were evaluated for significance in change on biometric and behavioural measures. Each participant acted as his/her own control, essentially (although not perfectly) controlling for other unmeasured potential confounders on the effectiveness of the worksite wellness programme.

Of the 60 programme participants who completed both the baseline biometric assessment and survey, 43 completed the biometric assessment and survey six months later. Characteristics of those who completed both the baseline and follow-up assessments were compared with those who completed only the baseline assessment to ascertain potential selection effects that might bias evaluation results.

EVALUATION MEASURES

Data included basic biometric measures related to cardiovascular risks and survey-based measures related to psychosocial factors, lifestyle, health status, and both absenteeism and presenteeism. Biometric measures were obtained by a KPGA registered nurse and a trained health educator and took approximately 15 minutes to obtain. Survey measures were based on previously validated items, most of which had been previously included in surveys administered to samples of KPGA adult enrollees. The survey was administered in written form and took approximately an additional 15 minutes to complete.

Biometric measures

Biometric data included: height obtained by stadiometer and weight obtained by a calibrated digital scale, waist circumference, systolic and diastolic blood pressure obtained by a calibrated digital blood pressure meter, and serum total cholesterol, high density lipoprotein (HDL) and glucose obtained by the finger stick method. Body mass index (BMI) was computed from height and weight and classified as obese (BMI $\geq$ 30 kg/m^2) or not.

Activation

Activation is a component of the Chronic Care Model (Wagner et al. 1996) and includes items related to information and knowledge ("I understand the nature and causes of my health problems") and self-efficacy ("I am confident that I can maintain lifestyle changes, like diet and exercise, even during times of stress") (Hibbard et al. 2004, 2005). It represents a person's engagement in health attitudes and behaviours and a general sense of self-efficacy.

Lifestyle

Three measures of dietary intake were derived from responses to the Block fat and fruit and vegetable screeners (Block et al. 2000). Percent calories from fat, F/V servings per day, and daily fibre intake (grams per day) were estimated from equations provided in Block et al. (2000) with recommended modifications (Torrin Block 2006, personal communication). Leisure physical activity was ascertained from responses to the Behavioural Risk Factor Surveillance Survey activity items and measured using two dichotomous variables: physical activity at the recommended level and physical inactivity (Centers for Disease Control and Prevention 2003 and 2005). Physical activity at the recommended level was defined as moderate physical activity (leisure activities of moderate intensity for a minimum of 30 minutes per day, five or more days per week) or vigorous physical activity (leisure activities of vigorous intensity for a minimum of 20 minutes per day, three or more days per week). Physical inactivity was considered to be <10 minutes per week of moderate or vigorous physical activity.

Productivity

Work loss days were assessed from the survey item "During the past four weeks, on how many days did you miss more than one-half day of work because you were not feeling well?" Presenteeism was measured by the six items in the Stanford presenteeism scale (Koopman et al. 2002, Turpin et al. 2004). The scale captures variation in attention to work, from substantial attention towards work to substantial distraction from work.

Health status

Health status was measured using the Physical Component Summary (PCS) and the Mental Component Summary (MCS) scales of the SF-12 (Ware et al. 1995, 1996). PCS-12 measures variation in physical functioning; MCS-12 measures variation in emotional functioning. Although we did not anticipate meaningful or significant change in these

Table 6.2   Sample size and effect estimates for achieving statistical significance (alpha of 0.05, 80 per cent power)

	Standard deviation	Sample size and effect estimates		
		12.5% of change in SD N=376	25% of change in SD N=96	50% of change in SD N=44
BMI	6.9	1.0	2.0	3.0
Percent fat intake	5.2	0.75	1.5	2.3
Daily fibre intake	5.7	0.83	1.7	2.5
Activation	17.7	2.6	5.1	7.7

measures in a 12-month period, we collected this information to be able to compare the study population with US national "norms" on these measures of health status.

STATISTICAL POWER

Using biometric and survey data from other KPGA research studies, we estimated that 50 participants would be sufficient to achieve 0.80 power to observe change in a measure over six months on the order of 1 standard deviation with an alpha of 0.05 (Table 6.2). For continuously distributed measures, we compared change in pre- and post-worksite wellness programme means using a paired t-test. For dichotomous measures, we compared change in pre- and post-worksite wellness characteristics using McNemar's test. We also examined the Pearson correlations of change in lifestyle and biometric measures to see if the correlations were statistically significant and in the direction consistent with the "common sense model" (Figure 6.1; e.g. positive correlations of decreased fat intake with decreased BMI and total cholesterol).

Discussion

In this chapter, we described the design, implementation and evaluation of a multifactorial worksite wellness programme intended to improve attitudes and behaviour towards a healthy lifestyle (patient activation, leisure physical activity, and dietary intake) and reduce prevalence of cardiovascular disease risk factors over a six-month period in a small employer group. Design, implementation and evaluation of the worksite wellness programme involved a partnership between the employer's leadership team and relevant professional staff from KPGA, a health maintenance organization with expertise in health education, chronic disease management, and health services research.

The Joint National Committee on Hypertension recommends a healthy lifestyle with regular leisure physical activity and dietary intake that is low in fat and high in fruit and vegetable and fibre intake as a means to achieve good blood pressure control and to reduce subsequent cardiovascular morbidity (Joint National Committee on Prevention 2003). Yet, many working age adults fail to adhere to these lifestyle recommendations. Among employed and insured adults in a recent Behavioral Risk Factor Surveillance

Survey, 49.0 per cent of 18–64 year old adults did not meet the Centers for Disease Control and Prevention recommended moderate or vigorous leisure physical activity levels and 77.1 per cent did not consume at least five daily servings of fruits and vegetables (Hughes et al. 2010). Almost one-third of US adults, both men and women, 40 to 59 years of age, are considered to be obese (Ford et al. 2010). And, large proportions of the US population have untreated or uncontrolled hyypertension (Egan et al. 2010).

Design, implementation and evaluation of this worksite wellness programme was based on clinical evidence and consensus as to the relevance of modification of lifestyle attitudes and behaviours for achieving cardiovascular disease risk factor reduction and to organizational change necessary for achieving effective modification of lifestyle in an employed population. The American College of Cardiology endorses worksite wellness programmes for reducing cardiovascular disease risks (Carnethon et al. 2009). Systematic reviews and meta-analyses of clinical trials indicate that improving lifestyle attitudes and behaviours reduces future cardiovascular disease risks (Dickinson et al. 2006, Fagard 2005, Neter et al. 2003, Kelley and Kelley 2000, Whelton et al. 2002).

Recent evaluations of worksite wellness programmes to reduce cardiovascular disease risks indicate that this goal can be achieved through programmes that improve lifestyle attitudes and behaviours (Aldana et al. 2002, Gemson et al. 2008, Merrill et al. 2009, Milani and Lavie 2009, Racette et al. 2009, Perez et al. 2009, Sternfeld et al. 2009, Williams et al. 2007). Although not reported here, the formal evaluation of this worksite wellness programme demonstrated statistically significant ($p \leq 0.05$) reductions in systolic and diastolic blood pressure and in percentage dietary fat intake over a six-month period (Roblin et al. 2011). While not statistically significant, the evaluation indicated the pathway through which reductions in blood pressure were attained was also through improvements in activation ($p=0.07$) and decreased physical inactivity ($p=0.08$).

In developing the worksite wellness programme, KPGA's staff followed various principles and guidelines that recommend specific components of worksite wellness programme that have been shown to improve the likelihood of programme success (Goetzel et al. 2007, Sorensen et al. 2011, Terry et al. 2008, Weiner et al. 2009). Working with the HMO primarily responsible for conducting the programme, the employer group's leadership created a culture to accept and promote the worksite wellness programme – including a wellness fair to inform employees about the programme and financial incentives (contribution flex spending account, on-site activities, and permitted time off to participate) to promote employee participation. The enrolment session included a health risk appraisal (HRA) with feedback to employees on request. Results of the HRA and baseline survey were used by KPGA staff to target and tailor programme activities towards reduction of cardiovascular risks, which were relatively prevalent in an otherwise healthy employee population. Commitment of the employer's leaders to the programme, combined with convenient access to programme activities and financial incentives, helped achieve employee engagement and a high rate of participation (60 of the 90 employees). A formal evaluation of the programme was conducted by health services researchers at KPGA. At a joint meeting of the employer leadership team and the KPGA team following conclusion of the programme, both parties concluded that the partnership which afforded the employer an opportunity for a worksite wellness programme by bringing the expertise of the HMO staff into programme design, implementation and evaluation was successful. Table 6.3 summarizes the key components of a successful worksite programme and design and implementation issues from the pilot cardiovascular risk reduction programme.

Table 6.3   Components of a successful worksite programme and design and implementation issues from the pilot cardiovascular risk reduction programme

Components of a successful programme	Notes from the pilot programme
Establish organizational commitment.	Frequent, regular meetings involving employer leadership team and KPGA team to develop framework regarding timing and content of the worksite wellness programme. Employer contributes financial incentives, facility space, time off for employee participation in programme activities.
Conduct a health risk appraisal (HRA).	Biometric assessment at baseline. Immediate feedback and private counselling by KPGA staff to interested employees.
Target and tailor programme activities to important employee health issues.	Biometric and survey assessment at baseline and tabulated and reported to employer leadership. Agreement between employer leadership and KGA team to tailor programme content with activities to reduce cardiovascular risk factors identified as prevalent in the baseline HRA and survey.
Achieve employee engagement and participation: access.	Initial 'health fair' to explain programme activities, introduce KPGA staff, and allow for employee questions and comments. Time off for employees to participate in selected activities. Activities scheduled "on site" when possible.
Achieve employee engagement and participation: incentives.	Monetary: employer contribution to employee retirement accounts for participation. Non-monetary: exercise DVD, pedometer and health diary provided to participants for home use.
Programme evaluation.	KPGA team brings in research staff after employer leadership requests a formal programme evaluation. KPGA research integrates behaviour model into programme evaluation design to inform survey data collection to complement biometric data collection. Collected biometric, behavioural and lifestyle data at baseline and six months later. Research provides evaluation summary (written report and in-person presentation) to KPGA team and employer leadership.

Table 6.4   Estimated worksite wellness programme costs per participant (amortized on 60 baseline participants)

	Employer	KPGA: implementation	KPGA: Evaluation
Staff for programme design	Unknown	Unknown	Unknown
Participant incentive	$100 (flex account contribution)	—	$50 (baseline and six-month AmEx gift checks @$25 each)
Staff time and supplies for biometric screening	Unknown: facilities and administration amortized cost	$130 (baseline and six-month follow-up)	—
Survey		—	$50 (baseline and six-month licence fees and printing)
Contract programme activities		$90 (exercise classes, massage therapy, healthy living seminars)	—
Staff for programme implementation and evaluation	Unknown	$100	$100
IRB fees	—	—	$25
Local travel	—	$10	$5
Supporting materials	—	$5 (pedometer, exercise DVD, lifestyle diary)	—
Total cost per participant	> $100	> $335	> $230

A comprehensive multifactorial worksite wellness programme, such as this pilot programme, can come at considerable cost to the employer, MCO partner and research group depending on the scope of activities and participation rate (Table 6.4). Because some costs are unknown and unaccounted (e.g. employer costs for amortized facility space and staff time for administration and participation), the costs that we attribute to this pilot worksite wellness programme undoubtedly underestimate the true costs per participant. Overall, the costs per participant in this six-month worksite wellness programme were at least $665 – $435 for employer and KPGA to conduct the programme and another $230 to evaluate the programme.

Conclusion

In summary, we have described the strategy and processes by which a worksite wellness programme (in this case, one intended to decrease cardiovascular risks) might be designed, implemented and evaluated. This pilot worksite wellness programme represented the partnership of a small professional services company and a mid-sized HMO in the Atlanta area. We believe that the pilot programme provides a potential model for developing similar relationships between HMOs – which typically have staff and resources for design and implementation of health education and chronic disease management programmes – with small and mid-sized employer groups which frequently have need for, but lack expertise and resources for, worksite wellness programmes. In addition, we described how a formal evaluation component might be added to the worksite wellness programme to determine if programme activities changed behaviour, attitudes and health in the intended and expected direction. While this worksite wellness programme was targeted towards, and achieved, changes in behaviour, attitudes and health risks related to cardiovascular disease risks, we believe that the framework that we have presented can be generalized to many aspects of the metabolic syndrome. Based on our experience, we encourage future partnerships between HMOs with established health promotion and disease management programmes and employers – particularly those with small (<100 employees) or dispersed employee populations – to develop partnerships improving the health of employed adults.

Disclosures

Funding for the evaluation of this worksite wellness programme was provided through discretionary funds of Kaiser Permanente Georgia's Research Department. Funds for costs of conduct of the worksite wellness programme were provided by the employer (who has asked to remain anonymous) and by Kaiser Permanente Georgia's Wellness and Health Promotion Department and Marketing Department. The authors have no industry or financial arrangements that might pose a conflict of interest with respect to conduct of this study.

Acknowledgements

Findings related to evaluation of the worksite wellness programme have been presented at the conference "Work, Stress, and Health," Orlando, FL, in May 2011 and "The HMO Research Network Annual Research Conference," Austin, TX, in April 2010. Roslin Nelson of Kaiser Permanente Georgia's Research Department assisted in data collection and data entry.

References

Aldana, S.G., Greenlaw, R., Diehl, H.A., Englert, H. and Jackson, R. 2002. Impact of the coronary health improvement project (CHIP) on several employee populations. *Journal of Occupational and Environmental Medicine*, 44, 831–9.

Anderson, L.M., Quinn, T.A., Glanz, K., Ramirez, G., Kahwati, L.C., Johnson, D.B., Buchanan, L.R., Archer, W.R., Chattopadhyay, S., Kalra, G.P. and Katz, D.L. 2009. The effectiveness of worksite nutrition and physical activity interventions for controlling employee overweight and obesity: A systematic review. *American Journal of Preventive Medicine*, 37, 340–57.

Baicker, K., Cutler, D. and Song, Z. 2010. Workplace wellness programs can generate savings. *Health Affairs*, 29, 304–11.

Benedict, M.A. and Arterburn, D. 2008. Worksite-based weight loss programs: A systematic review of recent literature. *American Journal of Health Promotion*, 22, 408–16.

Block, G., Gillespie, C., Rosenbaum, E.H. and Jenson, C. 2000. A rapid food screener to assess fat and fruit and vegetable intake. *American Journal of Preventive Medicine*, 18, 284–8.

Carnethon, M., Whitsel, L.P., Franklin, B.A., Kris-Etherton, P., Milani, R., Pratt, C.A. and Wagner, G.R. 2009. Worksite wellness programs for cardiovascular disease prevention: A policy statement from the American Heart Association. *Circulation*, 120, 1725–41.

Centers for Disease Control and Prevention. 2003. *Behavioral Risk Factor Surveillance Survey: State Questionnaire*. Retrieved on 6 June 2012 from www.cdc.gov/brfss/questionnaires/english.htm.

Centers for Disease Control and Prevention. 2005. Adult participation in recommended levels of physical activity – United States, 2001 and 2003. *Morbidity and Mortality Weekly Report*, 54, 1208–12.

Centers for Disease Control and Prevention. 2012. *Educational and Community-based Programs*. Retrieved on 6 June 2012 from http://www.healthypeople.gov/2020/topicsobjectives2020/overview.aspx?topicid=11.

Chan Osilla, K., van Busum, K., Schnyer, C., Wozar Larkin, J., Eibner, C. and Mattke, S. 2012. Systematic review of the impact of worksite wellness programs. *American Journal of Managed Care*, 18, e68–e81.

Conn, V.S., Hafdahl, A.R., Cooper, P.S., Brown, L.M. and Lusk, S.L. 2009. Meta-analysis of workplace physical activity interventions. *American Journal of Preventive Medicine*, 37, 330–339.

Dickinson, H.O., Mason, J.M., Nicolson, D.J., Campbell, F., Beyer, F.R., Cook, J.V., Williams, B. and Ford, G.A. 2006. Lifestyle interventions to reduce raised blood pressure: A systematic review of randomized controlled trials. *Journal of Hypertension*, 24, 215–33.

Egan, B.M., Zhao, Y. and Axon, R.N. 2010. US trends in prevalence, awareness, treatment, and control of hypertension, 1988–2008. *JAMA*, 303, 2043–50.

Engbers, L.J., van Poppel, M.N.M., Chin A Paw, M.J.M. and van Mechelen, W. 2005. Worksite health promotion programs with environmental changes: A systematic review. *American Journal of Preventive Medicine*, 29, 61–70.

Fagard, R.H. 2005. Effects of exercise, diet and their combination on blood pressure. *Journal of Human Hypertension*, 19, S20–S24.

Ford, E.S., Li, C., Zhao, G. and Tsai, J. 2010. Trends in obesity and abdominal obesity among adults in the United States from 1999–2008. *International Journal of Obesity (London)*, 35, 736–43.

Gemson, D.H., Commisso, R., Fuente, J., Newman, J. and Benson, S. 2008. Promoting weight loss and blood pressure control at work: Impact of an education and intervention program. *Journal of Occupational and Environmental Medicine*, 50, 272–81.

Goetzel, R.Z. and Ozminkowski, R.J. 2008. The health and cost benefits of work site health promotion programs. *Annual Review of Public Health*, 29, 303–23.

Goetzel, R.Z. and Pronk, N.P. 2010. Worksite health promotion: How much to we really know about what works? *American Journal of Preventive Medicine*, 38, S223–S225.

Goetzel, R.Z., Schechter, D., Ozminkowski, R.J., Marmet, P.F., Tabrizi, M.J. and Roemer, E.C. 2007. Promising practices in employer health and productivity management efforts: Findings from a benchmarking study. *Journal of Occupational and Environmental Medicine*, 49, 111–30.

Grosch, J.W., Alterman, T., Petersen, M.R. and Murphy, L.R. 1998. Worksite health promotion programs in the US: Factors associated with availability and participation. *American Journal of Health Promotion*, 13, 36–45.

Hibbard, J.H., Mahoney, E.R., Stockard, J. and Tusler, M. 2005. Development and testing of a short form of the patient activation measure. *Health Services Research*, 40, 1918–30.

Hibbard, J.H., Stockard, J., Mahoney, E.R. and Tusler, M. 2004. Development of the Patient Activation Measure (PAM): Conceptualizing and measuring activation in patients and consumers. *Health Services Research*, 39, 1005–26.

Hughes, M.C., Hannon, P.A., Harris, J.R. and Patrick, D.L. 2010. Health behaviors of employed and insured adults in the United States, 2004–2005. *American Journal of Health Promotion*, 24, 315–23.

Joint National Committee on Prevention. 2003. *Detection, Evaluation, and Treatment of High Blood Pressure*. Bethesda: National Institutes of Health.

Kelley, G.A. and Kelley, K.S. 2000. Progressive resistance exercise and resting blood pressure: A meta-analysis of randomized controlled trials. *Hypertension*, 35, 838–43.

Koh, H. and Sebelius, K. 2010. Promoting prevention through the Affordable Care Act. *New England Journal of Medicine*, 363, 1296–99.

Koopman, C., Pelletier, K.R., Murray, J.F., Sharda, C.E., Berger, M.L., Turpin, R.S., Hackleman, P., Gibson, P., Holmes, D.M. and Bendel, T. 2002. Stanford presenteeism scale: Health status and employee productivity. *Journal of Occupational and Environmental Medicine*, 44, 14–20.

Kuoppala, J., Lamminpaa, A. and Husman, P. 2008. Work health promotion, job well-being, and sickness absences – a systematic review and meta-analysis. *Journal of Occupational and Environmental Medicine*, 50, 1216–27.

Leventhal, J., Brissette, I. and Leventhal, E.A. 2003. The common-sense model of self-regulation of health and illness. In *The Self-regulation of Health and Illness Behavior*, edited by L.E. Cameron and H. Leventhal. London: Routledge, 42–65.

Linnan, L., Bowling, M., Childress, J., Lindsay, G., Blakley, C., Pronk, S., Wieker, S., and Royall, P. 2008. Results of the 2004 National Worksite Health Promotion Survey. *American Journal of Public Health*, 98, 1503–9.

McAndrew, L.M., Musumeci-Szabo, T.J., Mora, P.A., Lileikyte, L., Burns, E., Halm, E.A, Leventhal, E.A. and Leventhal, H.. 2008. Using the common sense model to design intervention for the prevention and management of chronic illness threats: From description to process. *British Journal of Health Psychology*, 13, 195–204.

Merrill, R.M., Aldana, S.G., Ellrodt, G., Orsi, R. and Grelle-Laramee, J. 2009. Efficacy of the Berkshire Health System cardiovascular risk reduction program. *Journal of Occupational and Environmental Medicine*, 51, 1024–31.

Meyer, D., Leventhal, H. and Gutman, M. 1985. Common-sense models of illness: The example of hypertension. *Health Psychology*, 4, 115–35.

Milani, R.V. and Lavie, C.J. 2009. Impact of worksite wellness intervention on cardiac risk factors and one-year health care costs. *American Journal of Cardiology*, 104, 1389–92.

Mohr, L.B. 1992. *Impact Analysis for Program Evaluation*. Newbury Park: Sage Publications.

Neter, J.E., Stam, B.E., Kok, F.J., Grobbee, D.E. and Geleijnse, J.M. 2003. Influence of weight reduction on blood pressure: A meta-analysis of randomized controlled trials. *Hypertension*, 42, 878–84.

Pelletier, K. 1996. A review of the clinical- and cost-effectiveness studies of comprehensive health promotion and disease management programs at the worksite: 1993–1995 update. *American Journal of Health Promotion*, 10, 380–88.

Pelletier, K. 2001. A review of the clinical- and cost-effectiveness studies of comprehensive health promotion and disease management programs at the worksite: 1998–2000 update. *American Journal of Health Promotion*, 16, 107–16.

Perez, A.P., Phillips, M.M., Cornell, C.E., Mays, G. and Adams, B. 2009. Promoting dietary change among state health employees in Arkansas through a worksite wellness program: The Healthy Employee Lifestyle Program (HELP). *Preventing Chronic Disease*, 6, A123.

Pronk, N.P. and Kottke, T.E. 2009. Physical activity promotion as a strategic corporate priority to improve worker health and business performance. *Preventive Medicine*, 49, 316–21.

Racette, S.B., Deusinger, S.S., Inman, C.L., Burlis, T.L., Highstein, G.R., Buskirk, T.D., Steger-May, K. and Peterson, L.R. 2009. Worksite opportunities for wellness (WOW): Effects on cardiovascular disease risk factors after 1 year. *Preventive Medicine*, 49, 108–14.

Roblin, D.W, Robinson, B.E and Benjamin, S.A. 2013. Evaluation of a worksite wellness program designed to reduce cardiovascular risks. *Journal of Ambulatory Care Management*, 36, In press.

Rossi, P.H., Freeman, H.E. and Lipsey, M.W. 1999. *Evaluation: A Aystematic Approach*. 6th edn. Thousand Oaks: Sage Publications.

Schmidt, H. 2012. Wellness incentives, equity, and the 5 groups problem. *American Journal of Public Health*, 102, 49–54.

Schmidt, H., Voigt, K. and Wikler, D. 2010. Carrots, sticks and health care reform – problems with wellness incentives. *New England Journal of Medicine*, 362, e3.

Seaverson, E.L., Grossmeier, J., Miller, T.M. and Anderson, D.R. 2009. The role of incentive design, incentive value, communications strategy, and worksite culture on health risk assessment participation. *American Journal of Health Promotion*, 23, 343–52.

Sorensen, G., Landsbergis, P., Hammer, L., Amick, B.C., Linnan, L., Yancey, A., Welch, L.S., Goetzel, R.Z., Flannery, K.M. and Pratt, C. 2011. Preventing chronic disease in the workplace: A workshop report and recommendations. *American Journal of Public Health*, 101, S196–S207.

Sternfeld, B., Block, C., Quesenberry, C.P., Block, T.J., Husson, G., Norris, J.C., Nelson, M. and Block, G. 2009. Improving diet and physical activity with ALIVE: A worksite randomized trial. *American Journal of Preventive Medicine*, 36, 475–83.

Terry, P.E., Seaverson, E.L.D., Grossmeier, J. and Anderson D.R. 2008. Association between nine quality components and superior worksite health management program results. *Journal of Occupational and Environmental Medicine*, 50, 633–41.

Turpin, R.S., Ozminkowski, R.J., Sharda, C.E., Collins, J.J., Berger, M.L., Billotti, G.M., Baase, C.M., Olson, M.J. and Nicholson, S. 2004. Reliability and validity of the Stanford Presenteeism Scale. *Journal of Occupational and Environmental Medicine*, 46, 1123–33.

Wagner, E.H., Austin, B.T. and von Korff, M. 1996. Organizing care for patients with chronic illness. *Milbank Quarterly*, 74, 511–44.

Ware, J.E., Kosinski, M. and Keller, S.D. 1995. SF-12: How to score the SF-12 physical and mental health summary measures. *Boston: The Health Institute*, New England Medical Center.

Ware, J.E., Kosinski, M. and Keller, S.D. 1996. A 12-item short-form health survey: Construction of scales and preliminary tests of reliability and validity. *Medical Care*, 34, 220–33.

Weiner, B.J., Lewis, M.A. and Linnan L.A. 2009. Using organization theory to understand the determinants of effective implementation of worksite health promotion programs. *Health Education Research*, 24, 292–305.

Whelton, S.P., Chin, A., Xin, X. and He, J. 2002. Effect of aerobic exercise on blood pressure: A meta-analysis of randomized, controlled trials. *Annals of Internal Medicine*, 136, 493–503.

Williams, A.E., Vogt, T.M., Stevens, V.J., Albright, C.A., Nigg, C.R., Meenan, R.T. and Finucane, M.L. 2007. Work, weight, and wellness: The 3W program: A worksite obesity prevention and intervention trial. *Obesity (Silver Spring)*, 15, 16S–26S.

Wilson, M.G., DeJoy, D.M., Jorgensen, C.M. and Crump, C.J. 1999. Health promotion programs in small worksites: Results of a national survey. *American Journal of Health Promotion*, 13, 358–65.

CHAPTER 7

Effects of Workplace-based Physical Exercise Interventions on Cost Associated with Sickness Absence and on Productivity

ULRICA VON THIELE SCHWARZ, HENNA HASSON
and PETRA LINDFORS

Abstract

Stating that physical exercise is related to improved health is hardly provocative. However, when it comes to implementing physical exercise at work, organizations are often hesitant to do so. This may follow from the scarcity of research linking health promoting interventions at work to organizational outputs such as productivity (Miller and Haslam 2009). This chapter aims to provide a research based framework for physical exercise interventions at work, present a case study of physical exercise in dentistry and outline a conceptual model specifying mechanisms linking the more well-known effects of physical exercise on individual outcomes to organizational outcomes. Specifically, the case is used to illustrate how physical exercise can influence productivity and organizational costs related to sickness absence.

Introduction: Physical Activity

Physical activity is one of the most common workplace health promotion initiatives. One reason for this may be that increasing levels of physical activity in the general population is one of the most powerful ways to improve public health (Marcus et al. 2006). It is well known that low levels of physical activity are related to several common mental and physical disorders such as heart disease, diabetes, hypertension, osteoporosis, cancer and depression, as well as emotional well-being (Colditz 1999, Friedenreich 2001, Galper et al. 2006, Stephenson et al. 2000, Sui et al. 2009). It is also clear that many individuals in Western societies are physically inactive. For instance, 38 per cent of women and 34 per cent of men living in Sweden, where the case presented in this chapter is set, fail to meet the general guidelines from the American College of Sports Medicine (ACSM) for physical activity (Swedish National Institute of Public Health 2008). This corresponds to a minimum of 30 minutes of daily physical activity of moderate intensity five days a week, or vigorous-intensity aerobic physical exercise for a minimum of 20 minutes on three days a week (Pate et al. 1995).

In this chapter, we will discuss both *physical exercise* and *physical activity*. Physical activity is defined as any bodily movement produced by skeletal muscles that result in energy expenditure. Physical exercise is a specific type of physical activity that involves exercise of greater intensity, over 60 per cent of maximal oxygen uptake, which causes rapid breathing and a substantial increase in heart rate (Haskell et al. 2007). Physical exercise is also distinguished by being planned, structured and involving repetitive bodily movement with the goal of enhancing or sustaining physical capacity (Caspersen and Merritt 1995, Pate et al. 1995). According to the ACSM, both physical activity of lower intensity and physical exercise are appropriate in meeting the guidelines for physical activity (ACSM and AMA 2007). However, only physical exercise of moderate and vigorous activity involves aerobic activity, which engages large muscular groups in dynamic muscular activity that affects the circulatory and metabolic systems (ACSM 1998, Kraemer et al. 2002). In the following parts of the chapter, we will use the term physical activity to denote physical activity of all intensities, that is, also including physical exercise. The term physical exercise will be used to refer to physical activity of medium to high intensity levels.

Workplace-based Interventions and Health Promotion

There are a number of different strategies for implementing physical activity in the workplace. Two important distinctions concern whom the intervention targets and to what extent the workplace is utilized as an instrument to initiate or maintain change. First, interventions can target all employees or focus on individuals with certain high-risks disorders. This chapter focuses on interventions that target all employees at the workplace. This means that we describe interventions that include both employees who are at risk for work-related ill health and employees who are not at risk, but not interventions that target specific diseases or problems and thus target specific individuals.[1] Second, interventions can focus mainly on individual change or on using the organizational context as a part of the intervention. Both these types of interventions are covered in this chapter. Interventions focusing on the individual are typically more common (Engbers et al. 2005, Yancey et al. 2004) and often include health check-ups, educational programmes, workplace exercise programmes or a combination of the above (Marshall 2004). Interventions that use the organizational context as part of the intervention often involve strategies such as the use of motivational prompts for the employees to be physically active, for example posting signs and creating walking trails (Yancey et al. 2004). Additional strategies involve using existing social structures for support (Campbell 2002, Cohen 1985) and incentive-based programmes (Cohen 1985, Marshall 2004). One

---

1   This means that the interventions described here can be categorized as either primary prevention or health promotion, but not as secondary/tertiary prevention or rehabilitation.

way of making use of the workplace for helping employees to initiate and maintain physical activity is to integrate physical activity into the workday, for example, in the case presented in this chapter, by providing time for physical activity during work hours. A recent review (Barr-Anderson et al. 2011) concluded that integrating short bouts of physical activity into the organizational routine during a workday has promising effects on physical activity levels but also on performance and clinical disease risk prevention. Moreover, physical activity interventions of this kind have been suggested to be more sustainable over time than individual-level interventions. Other studies have shown that allocating time for physical activity during work hours is efficient in attracting employees, reaching higher participation levels and affecting employee health positively (Eriksen et al. 2002, Pohjonen and Ranta 2001, Yancey et al. 2004).

EFFECTS OF WORKPLACE PROMOTION OF PHYSICAL ACTIVITY ON HEALTH

There is a general agreement on the beneficial effects of physical activity (Marcus et al. 2006). In the following section, we draw on this agreement but focus on what happens when physical activity interventions are conducted at a workplace. To provide a rationale for why workplace physical activity should lead to the positive effects that physical activity in general does, the first question that needs to be answered is whether workplace physical activity interventions actually lead to increased physical activity levels among employees. There is certainly evidence supporting that (Barr-Anderson et al. 2011, Chan et al. 2004, Conn et al. 2009, Dishman et al. 2009, Sternfeld et al. 2009, Titze et al. 2001). A recent meta-analysis and review concluded that although the effect sizes were moderate, the effects were still much better than what is typically found in other settings (e.g. communities and hospitals) (Abraham and Graham-Rowe 2009). However, low participation rates and a self-selection of participants are frequent problems associated with worksite physical activity interventions (Alexy 1991). Participation rates as low as 15 per cent to 30 per cent for white-collar workers are frequently reported (Conrad 1987). With some important exceptions (Eriksen et al. 2002, Pohjonen and Ranta 2001), participation rates may be even lower among blue-collar workers (Marshall 2004). Moreover, prior to the intervention, those participating tend to be younger, better educated, fitter and in better health than non-participants (Alexy 1991, Conrad 1987, Marshall 2004). In addition, it is important to keep in mind that workplace-based physical activity interventions target a population that in most cases would be considered healthy. The effect of physical activity on health in a fairly healthy population is likely to be smaller than effects in clinical groups. Consequently, effect sizes associated with workplace health interventions involving physical activities are often quite small (Dishman et al. 1998, Wilson et al. 1996). Another factor contributing to this is the fact that there are many possible determinants of the outcomes of workplace physical activity interventions (e.g. health) which makes the variance explained by each factor (such as physical exercise) small (Zapf et al. 1996). In sum, the challenge in workplace-based physical activity interventions often is to reach all employees.

After establishing that physical activity levels generally do increase following a workplace-based physical activity intervention, the next question is whether physical activity interventions at work have positive effects on health. Generally, the findings from earlier reviews and meta-analyses have been inconclusive (Marcus et al. 2006). This may, however, may be attributed to the small effect sizes. A recent meta-analysis shows positive effects on a number of health and work-related outcomes, including fitness, lipids, antropometric measures, diabetes risk, work attendance and job stress (Conn et al. 2009). In more detail, empirical studies have shown that workplace-based physical activity is related to increased physical fitness, improved perceived health status and prevention of early decline of work ability (Pohjonen and Ranta 2001). A recent large-scale cluster-randomized study also reported positive effects on systolic blood pressure and heart rate in the intervention group, despite that it failed to demonstrate an increase in self-rated physical activity (McEachan et al. 2011). Similarly, another study reported positive effects on total energy expenditure, cardiorespiratory fitness, percentage of body fat and blood cholesterol (Proper et al. 2003a). In addition, physical activity interventions have been related to a reduced risk of developing musculoskeletal problems (Proper et al. 2003b), less headache and neck pain (Sjogren et al. 2005) as well as to improved subjective physical well-being (Eriksen et al. 2002, Sjogren et al. 2006). In sum, it seems reasonable to assume that physical activity has merits as a work-related intervention that improves employee health.

WORKPLACE-BASED INTERVENTIONS, PRODUCTIVITY AND COSTS

Although healthy employees should be tempting in itself for an employer, it is important to translate employee health into outcomes that help build a business case for an organization, e.g. financial outcomes (Miller and Haslam 2009). However, economic evaluations have been described as an underdeveloped component within the Occupational Health and Safety (OHS) literature (Tompa et al. 2006). Many have also pointed out that the quality of the economic evaluations of worksite health interventions is generally poor (Pelletier 2005, Proper et al. 2004, Tompa et al. 2009, Uegaki et al. 2010, van Dongen et al. 2012). This lack of knowledge of how effects on employee health relate to organizational and financial output may be one of the reasons why organizations are still hesitant to implement such initiatives in practice. To date, relatively few studies have investigated the cost-effectiveness or budget impact of OHS interventions (Tompa et al. 2009, 2010). Three recent reviews of OHS and worksite health promotion (Pelletier 2005, Tompa et al. 2009), one of which focused solely on physical exercise interventions (Pronk 2009), conclude that there is some support for positive effects of different worksite health interventions on costs. Another review, focusing on combined physical exercise and dietary interventions, concludes that these interventions are more costly but at the same time more efficient in terms of health improvements than is treatment-as-usual (van Dongen et al. 2012). However, this review also underscores that the findings do not allow conclusions of cost-efficiency, as it is dependent on the willingness of organizations to pay for health effects. This illustrates a dilemma for worksite health interventions: the interventions do not only have to result in positive effects on employee health but also need to be associated with positive effects on organizational "business results" in order to make a business case.

Two of several outcomes of interest from an organizational perspective include productivity and sickness absence. There is general agreement on the fact that that ill health among employees is related to decreased productivity (Schultz et al. 2009, Schultz and Edington 2007, Stewart et al. 2003). Evidently, ill health decreases productivity when health problems result in sickness absence and make employees unable to attend work. It then seems reasonable to assume that improved health is related to reduced sickness absence. In line with this reasoning, workplace physical activity programmes have been related to reduced absenteeism (Aldana and Pronk 2001, Conn et al. 2009, Falkenberg 1987, Parks and Steelman 2008). Reduced absenteeism, in turn, can be translated into cost savings. When an employee is unable to work, someone else has to do the job, the job is delayed, or the job is not done at all (Berger et al. 2001, Uegaki et al. 2010). All these alternatives are related to costs. If an employee is replaced by someone else, the means and methods of replacement are related to different costs (overtime, staffing agency, temporary workers, etc). If the employee is not replaced, the costs are related to productivity losses (Pauly et al. 2002). Along with decreased productivity because of sickness absence, ill health may also hinder employees from performing at their best while at work, e.g. sickness presenteeism. Physical activity interventions may therefore affect productivity by also decreasing sickness presenteeism, i.e. reduce the negative impact of poor health. Besides potential preventive effects, programmes may also have beneficial effects and be related to productivity *gains* and financial *profit*. Consistent with this, employee health improvements have been related to increased self-rated productivity (Lenneman et al. 2011; Pelletier et al. 2004). Also, participation in multi-worksite health promotion programmes have been related to decreased sickness absence but also to higher self-rated performance (Mills et al. 2007). In sum, there is limited but promising support for productivity related effects of worksite health interventions in general and physical activity interventions in particular. In the following section, we will describe a case that presents a physical exercise intervention project and its effects on productivity and sickness absence costs. Then, we will discuss potential mechanisms in the relationship between workplace-based physical activity interventions, employee health and organizational output.

Mandatory Physical Exercise During Work Hours: A Case Report

SETTING

The study was set in a large public dental health care organization in Stockholm, Sweden and took place between 2004 and 2006. Dentistry is a setting with a challenging psychosocial work situation with high demands, low rewards as well as third-party constraint (Bejerot 1998, O'Shea 1984, Öberg et al. 1995). It has been described as the most stressful of the health care professions (Freeman et al. 1995). The physical work environment is also demanding. For example, the job requires fixed postures and repetitive movements for extended periods of time, which generates a high muscular load. The visual demands are also high (Bejerot 1998, Mileråd et al. 1991, O'Shea 1984, Åkesson et al. 1997, Öberg et al. 1995).

At the time, the organization had had a recent comprehensive work environment assessment. At each of the 51 general dental health practices (GDPs), workshops had been performed in order to engage employees in improving the work environment. The focus was on small-scale improvements that could be initiated locally. However, a recurring suggestion from the workplaces involved reduced work hours and/or physical exercise during work hours. The human resource (HR) department decided to test this as an intervention project that was to be investigated by a group of researchers at Stockholm University.

Of the organization's 51 workplaces, six were selected to participate by the HR department. The inclusion criteria were that the participating workplace had to have at least 25 employees and had been profitable the previous year. Also, both management and a majority of employees had to agree to participate. All eligible workplaces were ranked based on short-term sickness absence (>14 days), and the top three and the bottom three were invited to participate. All of the approached workplaces accepted to take part. The low and high sickness absence workplaces were matched on the number of employees, resulting in three pairs that were randomly allocated to one of three conditions: physical exercise, reduced work hours and referents. In all, 201 individuals were currently working at the six workplaces. All but two volunteered to participate in the scientific study and at the one-year follow-up, 177 of the 201 (88.2 per cent) were still employed and willing to participate. Of the employees, 91 per cent were women, and the mean age was 45.2 years. The largest occupational groups were dental nurses (48 per cent) followed by dentists (32 per cent).

INTERVENTIONS

In the physical exercise condition, 2.5 hours of weekly working hours were allocated to mandatory physical exercise during two different days. Participants were free to decide the kind of physical exercise as long as it was of medium to high intensity, corresponding to 60 per cent to 89 per cent of maximum heart rate. The reason for targeting an intensity level rather than a specific physical exercise type was to increase participants' ability to influence the intervention while still minimizing the variation in the exposure. It was also believed to increase the likelihood that the physical exercise was perceived as pleasant. This is important, given that fulfillment of a self-interest in itself may be related to positive effects while forced activities may consume regulatory resources and have negative effects that may counteract the positive effects of the activity (Baumeister et al. 1998). Medium to high intensity level was considered appropriate given the duration and frequency of the exercise and since it made it possible to study effects relating to cardiovascular capacity. All employees recorded in writing the type of activity and the duration of each exercise session and these written records were checked weekly by an assigned employee.

To be able to separate effects relating to physical exercise during work hours from effects relating to the fact that employees spent fewer hours on carrying out their work tasks than they usually do, a reduced working hours condition was formed. Similar to the physical exercise condition, the reduced working hours condition involved full-time weekly hours being reduced from 40 hours/week to 37.5 hours/week. However, instead of allocating the reduced time to physical exercise, the employees in the reduced working hours condition were free to choose themselves how to spend the time. To account for part-time work, the reduction of working hours in both intervention conditions was less than 2.5 hours so that, instead of similarity in absolute numbers, the reduction was similar in relative numbers. Although part-time work was common (55 per cent), the great majority of the part-timers still had either a reduction of 2.5 hours (46 per cent) or 2 hours (39 per cent) each week. All employees in the intervention groups retained their salaries. Moreover, although no additional personnel were hired, all workplaces were expected to deliver full services throughout the study period. This means that the organization expected an increase in productivity that would make it possible to maintain production levels from the previous year. At the two workplaces acting as referents, no intervention was carried out.

STUDY DESIGN

To investigate the effects of physical exercise over time and in comparison to the reduced working hours and reference conditions, multiple data types were collected. To assess health effects, biomarkers and self-ratings in questionnaires were obtained before any intervention took place (T1), six months (T2) and 12 months after the intervention (T3) started. Productivity was assessed through a combination of self-ratings, using the baseline and 12-month follow up, and objective production levels. The objective data were obtained on the workplace level. The number of patients during the intervention year was compared to the corresponding time period the previous year. For cost estimates of sickness absence, workplace level data were used and the intervention year compared to the previous year.

MEASURES OF SICKNESS ABSENCE

In order to investigate the effects of worksite health-promotion interventions on sickness absence costs, studies show that the calculation of direct costs can often be collected rather easily, i.e. from company records (Nyman et al. 2010, Ozminkowski et al. 2002, Tompa et al. 2009, Uegaki et al. 2010). Direct costs include costs of salary, rehabilitation and, in countries without a national health insurance, health care costs. Indirect costs, however, cannot be found in registries and have to be estimated. This is usually done using the human capital method (Brooks et al. 2010; Mattke et al. 2007) or the friction cost method (Berger et al. 2001). In the human capital method, the costs associated with sickness absence are assumed to correspond to the number of days lost multiplied by the daily wage (Brooks et al. 2010). In contrast to the human capital method, the friction cost method takes into account that a missing employee is likely to be replaced, limiting the costs to the period it takes to replace or substitute the missing employee (the friction period) (Brouwer and Koopmanschap 1998, Koopmanschap et al. 2005, Koopmanschap et al. 1995, Mattke et al. 2007). However, the friction method assumes that an employee can be replaced without any productivity loss. This assumption can be questioned. Therefore, in this study we not only collected data on the friction period but also carried out interviews to get information from managers concerning replacement methods and estimates of productivity levels and productivity losses associated with different means of replacement. This allowed taking into account the fact that a replacer may not be as productive as an absentee. Also, both the human capital method and the friction cost method rely on individual salary conversion methods to estimate the cost of absence (Brooks et al. 2010). However, most employers would probably like their employees to add more value than that corresponding to their salary. In line with this assumption, previous research has shown that absence costs can greatly exceed the costs of an absentee (Nicholson et al. 2006, Pauly et al. 2008). Consequently, estimates of productivity losses were based on difference between absentee and replacement in value-added-productivity.[2]

---

2   In this case, this was calculated as the incomes minus operating expenses divided by number of employees.

OUR RESULTS

In order to evaluate effects of workplace physical exercise on health, productivity and sickness absence costs, we first need to know that there is an actual increase in physical exercise. To test this, differences between baseline and follow-ups in self-ratings of physical activity (low intensity) and physical exercise (medium to high intensity) were investigated. The results showed that both physical activity and physical exercise levels increased in all groups. However, the increase was greater in the physical exercise intervention group, particularly for physical exercise. Consequently, it seems reasonable to assume that the intervention had been effective in increasing physical exercise levels (see von Thiele Schwarz et al. 2008 for more details). The next question is whether these changes were related to health improvements. The answer is yes, at least in terms of decreased levels of glucose and self-rated upper extremity disorders in the physical exercise group. Also, both the physical exercise and the reduced working hours conditions showed a decrease in subjective health complaints, compared to referents (von Thiele Schwarz et al. 2008).[3]

Based on the fact that physical exercise levels increased and there were health improvements, we continued to investigate changes in productivity. The results showed that, at the workplace level, the number of treated patients increased in all conditions during the intervention year (von Thiele Schwarz and Hasson 2011). The largest increase was found in the reduced working hour condition. However, the workplace level production results could only be validated by self-ratings in the physical exercise group. This group showed a concurrent significant increase in self-rated productivity, e.g. increased quantity of work, work-ability and decreased sickness absence. We conclude that in this case, a reduction in work-hours could be used for health promotion activities with sustained or even improved production levels. This, in turn, means that productivity was increased since the same, or a higher, production level was achieved using less resources.

In the next step, we looked at the costs for sickness absence (von Thiele Schwarz and Hasson 2012). The results showed that the number of sickness absence days decreased in all conditions (physical exercise condition: 11.4 per cent; reduced working hours: 4.9 per cent, and referents: 15.9 per cent). Interestingly, the changes in costs were not linear to changes in sickness absence days: while costs were substantially reduced in the physical exercise group (22 per cent) and somewhat in reduced working hours (5 per cent) condition, the costs increased among referents (10 per cent). We also showed that the indirect costs, that is, the costs associated with replacement and productivity loss, were about one-third of the total costs. This indicates that considering direct costs (e.g. salary and rehabilitation) only, may result in an underestimation of the total costs (and costs reductions). By combining the friction method with indirect cost estimates based on value-added productivity, it is possible to get a broader overview but also more details on potential costs associated with this type of organizational initiative. In sum, this case shows how mandatory physical exercise during work hours can pay off not only in terms of employee health effects, but also in terms of increased productivity and decreased costs for sickness absence. The next question is how these effects can be understood. This is discussed in the next part of the chapter, which outlines a model of potential mechanisms.

---

3   The health-related effects are based on data from the 91 per cent women in the sample. Only women were included since biomarkers were investigated and these differ between women and men.

Synthesis of the Mechanisms Explaining Physical Exercise Effects on Organizational Output

Few people will argue against the notion that physical exercise has positive health effects. Drawing on this, physical exercise can be assumed to have effects on sickness absence and the corresponding costs. Empirical data seem to support this notion, even though the number of empirical studies is small. Even fewer studies have tried to explain the mechanisms linking physical exercise during work hours to productivity. One exception is Falkenberg's (1987) theoretical model delineating the relationships between physical fitness, physical activity and employee fitness programmes. This model is used here as a starting point, but our model will focus specifically on the potential mechanisms explaining the effects of physical exercise during work hours on productivity and sickness absence costs. Considering the small number of empirical studies investigating the relationship between these factors directly, we draw on research investigating different parts of the relationship and link these together conceptually.

Before we elaborate further on the links between physical exercise and sickness absence and productivity, the relationship between productivity and sickness absence needs to be addressed. Simply put, sickness absence decreases productivity. This is a result of an employee not being at work and thus not carrying out any work. Also, if work carried out by different employees is interdependent, an absent individual may have implications beyond this individual's own productivity. This is particularly the case when an absent individual makes it difficult for others employees who are at work to perform as usual (Nicholson et al. 2006, Pauly et al. 2002). However, the relationship between productivity and sickness absence is further complicated by the fact that the true effect of sickness absence on productivity is dependent on the means of replacement and of the productivity level of the replacer. Yet, this additional complexity will be left out from the discussion since the replacement method is unlikely to be related to physical exercise. In line with this, our model discusses productivity in terms of employee performance while at work.

Table 7.1 outlines factors that may contribute to the understanding of how physical exercise during work hours relates to productivity and sickness absence. First, and as underscored by Falkenberg (1987), when physical activity is part of an organizational effort, factors can either be related to effects of physical activity on the individual level, or to effects relating to the fact that the intervention is workplace based, e.g. performed during work hours, as in this case. Second, effects of physical activity can be immediate or accumulated. Immediate effects are those short-term effects that follow directly from an exercise session. Accumulated effects, in contrast, include effects that come about over time as an individual exercises regularly and/or the organization sustains an exercise initiative over time. Third, the effects are categorized by whether they are likely to affect how the employees perform at work (productivity) and/or to sickness absence. We will now present each step of the model, starting with immediate effects relating to the individual.

Physical activity in general and physical exercise in particular have (at least) two immediate positive effects:[4] they are related to positive mood (Reed and Ones 2006)

---

4   Most of the studies have considered cardiovascular training of higher intensity. We will, therefore, use physical exercise when the studies explicitly refer to physical activity of higher intensity, but use physical activity otherwise.

Table 7.1   A conceptual model delineating how physical exercise during work hours can affect productivity and sickness absence at the employee and workplace levels

		Productivity at work	Sickness absence
Physical exercise during work hours	Employee — Immediate effects	1. Improved mood. 2. Improved cognitive functioning (reaction time, faster mental processes, memory storage and retrieval).	
	Employee — Accumulated effects	1. Improved health (mental and physical), and decreased sickness presenteeism. 2. Improved sleep. 3. Improved energy, decreased fatigue. 4. Improved cognitive vitality, improved cognitive functioning, particularly flexibility. 5. Sustained cognitive functioning with age. 6. Optimizing stress responsivity and faster recovery after stress. 7. More appropriate arousal levels.	1. Improved health (mental and physical). 2. Improved sleep. 3. Improved energy, decreased fatigue. 4. Increased emotional stability.
	Workplace-based — Immediate effects	1. Improved scheduling of work and resources at work, and a better balancing of work and non-work factors. 2. Reach employees not exercising. 3. Recovery – pause.	1. Reach employees not exercising.
	Workplace-based — Accumulated effects	1. Increased organizational commitment. 2. Improved work climate. 3. Improved cooperation. 4. Improved goodwill.	1. Increased organizational commitment. 2. Improved work climate.

and to improved cognitive functioning (Chang et al. 2012), including shorter reaction time (Kashihara et al. 2009) and faster mental processes, memory storage and retrieval (Lambourne and Tomporowski 2010). We propose that these effects, in turn, are related to performance at work (rather than to sickness absence), particularly when work involves complex cognitive tasks. It can be speculated that positive mood may affect productivity by increasing job motivation, but also by improving the social climate at the workplace. Since the acute effects of physical exercise are short-lasting, it may be particularly beneficial to schedule physical exercise during the morning hours or around lunch time, in order to have the positive, short-lasting effect taking place during work hours.

Over time, the effects of physical exercise accumulate. Improved fitness is a common denominator for many of these effects (Blumenthal et al. 1990). It is likely that physical exercise at intensity levels that boost fitness may be particularly beneficial. Improved mental and physical health, sleep, energy and emotional stability and decreased fatigue are factors that are likely to be related to on-the-job productivity but also to decreased sickness absence. The positive effect of physical exercise on mental and physical health may be the most evident mediator. Since the reason for sickness absence is generally assumed to be ill health and sickness, the solid evidence for the health benefits of physical exercise on health makes it likely that improvements in health lead to lower sickness absence levels. Improved mental and physical health is also likely to contribute to improved on-the-job productivity, since presenteeism, e.g. performing worse at work to due to illness, should decrease. In addition, physical exercise, through these mediating variables, is proposed to affect the frequency of sickness absence.

While ill health is suggested as the main reason for the number of days employees are absent from work, other factors may become more important in understanding the frequency of sickness absence (von Thiele et al. 2006). For example, poor sleep, fatigue and lack of energy may lead to short but frequent spells of absence and to decreased productivity while at work. Epidemiological studies have consistently linked habitual physical exercise to improved sleep (Youngstedt and Kline 2006), and to increased energy levels and decreased fatigue (Puetz 2006, Puetz et al. 2006). From a cost perspective, the distinction between short- and long-term absences is important since in some compensatory systems, the compensation levels differ between short-term and long-term absences. In the case described here, short-term absence is particularly expensive for the employer. This explains why there was no link between the overall decrease in number of sickness absence days and the decrease in costs in the reference group. Among referents, the total number of days decreased, but there was a concurrent increase in the number of short-term sickness absence days along with increased overall costs for sickness absence.

The acute effects of physical exercise on mood and cognitive function seem to have accumulated impacts as well. Interestingly, the acute effects on mood have been suggested to be turned into more sustained changes in personality in terms of increased emotional stability (Falkenberg 1987). The relationships between emotional stability and productivity and sickness absence remain largely unknown. However, one possible interpretation is that emotional stability improves well-being and positive functioning, which in turn has positive effects on organizational outcomes. Regular physical exercise also has positive effects on cognitive functioning. These effects include improved cognitive flexibility, and, to some extent, cognitive speed and attention (Masley et al. 2009). Moreover, physical exercise has been related to increased cognitive vitality, at least among older adults (Colcombe and Kramer 2003). Since low levels of physical exercise are related to decreased cognitive functioning, physical exercise may be related to, at least, sustained cognitive functioning (Singh-Manoux et al. 2005). These are factors that can be assumed important for productivity. Overall, physical exercise may have more widespread neurobiological effects, given that physical exercise has been suggested to affect brain plasticity in a way similar to that of learning and environmental enrichment (Cotman and Berchtold 2002). Physical exercise, particularly forms involving aerobic exercise, has also been related to decreased stress response to a stress test but also to faster recovery (Blumenthal et al. 1990), which is in line with the mechanism described by Falkenberg (1987). However, the effect of physical exercise on stress has been questioned (de Geus and Stubbe 2007), meaning that this particular mechanism remains unclear.

The individual effects of physical exercise described above can be beneficial at an organizational level regardless of whether physical exercise is part of an organizational initiative or is pursued by the individual employee. However, some effects are related to the fact that an initiative is workplace-based. In general, there is a lack of research investigating the direct effects of physical exercise on work-related factors. This means that this section is largely hypothetical. To start out broadly, it has been recognized that workplaces offering workplace-based health initiatives are more attractive for employees and improve organizational goodwill (European Network for Workplace Health 2005). This may for example be important from a recruitment perspective by making the employer more attractive. This may increase the recruitment base and the likelihood of finding the most suitable employees, which, in turn, may affect productivity. Also, there are potential effects on organizational commitment. Organizational commitment may increase if the employees feel that the organization provides something extra for the employees, and shows that the organization is concerned with employee well-being. Organizational commitment has been related to sickness absence both directly and through employee health (Karlsson 2010). This was suggested as an important mechanism already in 1989 by Falkenberg (1987), but has to our knowledge not been investigated empirically and remains speculative. Also, we propose that by making physical exercise mandatory and to schedule it during work hours, factors relating to cooperation and planning of work are affected. In order to get work done and to use resources (e.g. rooms) efficiently, the employees in the case presented in this chapter were interdependent and had to cooperate. This may contribute to an improved social work climate, which in turn may be related both to increased productivity and reduced sickness absence.

In sum, we propose a model that links physical exercise to organizational output, e.g. productivity and sickness absence costs by effects relating (a) to the employee and (b) to the organization. We also emphasize that organizations may accentuate the effect of a physical exercise intervention by considering that it may have both acute and accumulated effects. This has, for example, implications for the timing of the physical exercise during the day. Additionally, this has implications on the expected time lag of effects, that is, when certain outcomes can be evaluated.

PRACTICAL IMPLICATIONS AND RECOMMENDATIONS

A workplace-based physical exercise intervention has to be properly implemented in order for it to have the expected effects. This requires organizational efforts in terms of implementation strategies. For instance, this involves allowing time for employees to adapt and respond to the organizational initiative. In the case presented in this chapter the original idea came from the employees in the organization. Such a bottom-up approach may increase motivation. Furthermore, the participating workplaces were involved well in advance of the start of the intervention, i.e. about six months before the beginning. During this time, meetings were held allowing the employees to talk freely about their wishes, needs and worries. Care was taken to clearly communicate organizational expectations. Care was also taken to make it clear what was negotiable and what was not, e.g. that the employees could choose type of exercise but not intensity level. As for factors that employees could not influence, rationales were given in order to increase commitment and motivation to participate. Employees were also engaged in problem-solving relating to various issues, for example, scheduling exercise with minimum interference with production. Another important aspect involved including middle managers, since their engagement and motivation is central for implementation and maintenance of change. This was achieved by allowing managers to discuss with colleagues, by providing structural support from HR and, in this case, researchers and by educating about the benefits of the intervention. After the intervention started, the discussion was kept alive by having fellow employees following up on the documentation of the physical exercise, and by regularly discussing issues relating to the intervention at staff meetings. Finally, the fact that the physical exercise was scheduled and took place during work hours, and thereby was integrated into the workday, meant that the intervention made use of the organizational structure, which helped providing cues and feedback that may have facilitated implementing and sustaining the intervention.

Scheduling physical exercise into work hours may be particularly valuable for some employees. It is widely recognized that a limited part of the population spends time on regular physical exercise. Even though workplace-based health programmes often have difficulties attracting employees with a sedentary lifestyle, the possibility to schedule exercise during work hours increases the likelihood for engaging a greater proportion of the employees. It is by reaching those who do not exercise regularly that the organization will benefit the most, both in terms of productivity and sickness absence: these employees contribute to a large part of the sickness absence and to the presenteeism which means that there is more room for improvement in this group as compared to the already fit employees. Another reason for assuming that scheduling of physical exercise is particularly beneficial is that those who work full-time have been found to exercise less (Mein et al. 2005), and the most common reason for not exercising is "lack of time." Consequently, exercise during work hours may be considered as a means for organizations to ensure that a greater proportion of their employees get the possibility to experience the positive consequences of physical fitness, while at the same time boosting the effects on productivity and sickness absence.

Conclusion

For an individual, physical exercise has widespread and positive physical, emotional and cognitive effects regardless of whether the individual exercises during work hours or during leisure time. For an organization, having fit employees is associated with gainful effects on organizational outcomes such as productivity and sickness absence. This means that organizations benefit from having exercising employees. One way of influencing employee fitness levels is to make exercise part of work: that is, allowing employees to exercise during work hours. This way, employees get help to overcome a common obstacle to exercise, e.g. lack of time. Also, the organizational context provides support structures that help the individuals' initiative and maintain behavioural change (Bamberger and Sonnenstuhl 1996, Hays et al. 2000). Making use of work hours for physical exercise may further help overcome one of the common shortcomings of workplace-based physical exercise interventions, namely that they tend to attract those who are already physical active. Making exercise part of work also means that the organization can have the additional benefit of effects relating to the fact that the initiative is workplace based. Taken together, organizations can, by implementing physical exercise programmes, promote employee health and well-being while increasing productivity and decreasing sickness absence costs.

References

Abraham, C. and Graham-Rowe, E. 2009. Are worksite interventions effective in increasing physical activity? A systematic review and meta-analysis. *Health Psychology Review*, 3(1), 108–44.

Åkesson, I., Hansson, G.Å., Balogh, I., Moritz, U. and Skerfving, S. 1997. Quantifying work load in neck, shoulders and wrists in female dentists. *International Archives of Occupational and Environmental Health*, 69(6), 461–74.

Aldana, S.G. and Pronk, N.P. 2001. Health promotion programs, modifiable health risks, and employee absenteeism. *Journal of Occupational and Environmental Medicine*, 43(1), 36–46.

Alexy, B.B. 1991. Factors associated with participation or nonparticipation in a workplace wellness center. *Research in Nursing and Health*, 14(1), 33–40.

American College of Sports Medicine (ACSM). 1998. American College of Sports Medicine Position Stand. The recommended quantity and quality of exercise for developing and maintaining cardiorespiratory and muscular fitness, and flexibility in healthy adults. *Medicine and Science in Sports and Exercise*, 30(6), 975–91.

American College of Sports Medicine (ACSM) and the American Heart Association (AHA). 2007. Physical activity and public health: Updated recommendation for adults from the American College of Sports Medicine and the American Heart Association. *Circulation*, 116(9), 1081–93.

Bamberger, P. and Sonnenstuhl, W. 1996. Tailoring union-wide innovations to local conditions: The case of member assistance program implementation in the airline industry. *Labor Studies Journal*, 21(3), 19–39.

Barr-Anderson, D.J., AuYoung, M., Whitt-Glover, M.C., Glenn, B.A. and Yancey, A.K. 2011. Integration of short bouts of physical activity into organizational routine: A systematic review of the literature. *American Journal of Preventive Medicine*, 40(1), 76–93.

Baumeister, R.F., Bratslavsky, E., Muraven, M. and Tice, D.M. 1998. Ego depletion: Is the active self a limited resource? [Research Support, U.S. Gov't, P.H.S.]. *Journal of Personel Social Psychology*, 74(5), 1252–65.

Bejerot, E. 1998. Dentistry in Sweden – healthy work or ruthless efficiency? (Vol. 14): *Arbete och Hälsa* [Work and Health].

Berger, M.L., Murray, J.F., Xu, J. and Pauly, M. 2001. Alternative valuations of work loss and productivity. *Journal of Occupational and Environmental Medicine*, 43(1), 18–24.

Blumenthal, J.A., Fredrikson, M., Kuhn, C.M., Ulmer, R.L., Walsh-Riddle, M. and Appelbaum, M. 1990. Aerobic exercise reduces levels of cardiovascular and sympathoadrenal responses to mental stress in subjects without prior evidence of myocardial ischemia. *The American Journal of Cardiology*, 65(1), 93–8.

Brooks, A., Hagen, S.E., Sathyanarayanan, S., Schultz, A.B. and Edington, D.W. 2010. Presenteeism. *Journal of Occupational and Environmental Medicine*, 52(11), 1055–67.

Brouwer, W.B. and Koopmanschap, M.A. 1998. How to calculate indirect costs in economic evaluations. *PharmacoEconomics*, 13(5 Pt 1), 563–9.

Campbell, M. 2002. Effects of a tailored health promotion program for female blue-collar workers: Health works for women. *Preventive Medicine*, 34(3), 313–23.

Caspersen, C.J. and Merritt, R.K. 1995. Physical activity trends among 26 states, 1986–1990. *Medicine and Science in Sports and Exercise*, 27(5), 713–20.

Chan, C.B., Ryan, D.A.J. and Tudor-Locke, C. 2004. Health benefits of a pedometer-based physical activity intervention in sedentary workers. *Preventive Medicine*, 39(6), 1215–22.

Chang, Y.K., Labban, J.D., Gapin, J.I. and Etnier, J.L. 2012. The effects of acute exercise on cognitive performance: A meta-analysis. *Brain Research*, 1453, 87–101.

Cohen, W.S. 1985. Health promotion in the workplace. A prescription for good health. *The American Psychologist*, 40(2), 213–16.

Colcombe, S. and Kramer, A.F. 2003. Fitness effects on the cognitive function of older adults: A meta-analytic study. *Psychological Science*, 14(2), 125–30.

Colditz, G.A. 1999. Economic costs of obesity and inactivity. *Medicine and Science in Sports and Exercise*, 31(11 Suppl), S663–667.

Conn, V.S., Hafdahl, A.R., Cooper, P.S., Brown, L.M. and Lusk, S.L. 2009. Meta-analysis of workplace physical activity interventions. *American Journal of Preventive Medicine*, 37(4), 330–339.

Conrad, P. 1987. Who comes to work-site wellness programs? A preliminary review. *Journal of Occupational Medicine: Official Publication of the Industrial Medical Association*, 29(4), 317–20.

Cotman, C.W. and Berchtold, N.C. 2002. Exercise: A behavioral intervention to enhance brain health and plasticity. *Trends in Neurosciences*, 25(6), 295–301.

de Geus, E.J.C. and Stubbe, J.H. (2007). Aerobic exercise and stress reduction. In *Encyclopedia of Stress (Second Edition)*, edited by G. Fink, B.S. McEwen, R. de Kloet, R. Rubin, G. Chrousos, A. Steptoe, N. Rose, I. Craig and G. Feuerstein. New York: Academic Press, 73–8.

Dishman, R.K., DeJoy, D.M., Wilson, M.G. and Vandenberg, R.J. 2009. Move to improve: A randomized workplace trial to increase physical activity. *American Journal of Preventive Medicine*, 36(2), 133–41.

Dishman, R.K., Oldenburg, B., O'Neal, H. and Shephard, R.J. 1998. Worksite physical activity interventions. *American Journal of Preventive Medicine*, 15(4), 344–61.

Engbers, L.H., van Poppel, M.N.M., Chin A Paw, M.J.M. and van Mechelen, W. 2005. Worksite health promotion programs with environmental changes: A systematic review. *American Journal of Preventive Medicine*, 29(1), 61–70.

Eriksen, H.R., Ihlebaek, C., Mikkelsen, A., Gronningsaeter, H., Sandal, G.M. and Ursin, H. 2002. Improving subjective health at the worksite: a randomized controlled trial of stress management training, physical exercise and an integrated health programme. *Occupational Medicine (Oxford, England)*, 52(7), 383–91.

European Network for Workplace Health. 2005. The Luxembourg Declaration on Workplace Health Promotion in the European Union.

Falkenberg, L.E. 1987. Employee fitness programs: Their impact on the employee and the organization. *Academy of Management Review*, 12(3), 511–22.

Freeman, R., Main, J.R. and Burke, F.J. 1995. Occupational stress and dentistry: Theory and practice. Part II. Assessment and control. *British Dental Journal*, 178(6), 218–22.

Friedenreich, C.M. 2001. Physical activity and cancer prevention: From observational to intervention research. *Cancer Epidemiology, Biomarkers and Prevention: A Publication of the American Association for Cancer Research, Cosponsored by the American Society of Preventive Oncology*, 10(4), 287–301.

Galper, D., Trivedi, M.H., Barlow, C.E., Dunn, A.L., and Kampert, J.B. 2006. Inverse association between physical inactivity and mental health in men and women. *Medicine and Science in Sports and Exercise*, 38(1), 173–8.

Hays, C.E., Hays, S.P., DeVille, J.O., and Mulhall, P.F. 2000. Capacity for effectiveness: The relationship between coalition structure and community impact. *Evaluation and Program Planning*, 23(3), 373–9.

Karlsson, M.L. 2010. *Healthy Workplaces: Factors of Importance for Employee Health and Organizational Production*. Doctoral thesis, Karolinska Institutet, Stockholm.

Kashihara, K., Maruyama, T., Murota, M. and Nakahara, Y. 2009. Positive effects of acute and moderate physical exercise on cognitive function. *Journal of Physiological Anthropology*, 28(4), 155–64.

Koopmanschap, M.A., Rutten, F.F., van Ineveld, B.M. and van Roijen, L. 1995. The friction cost method for measuring indirect costs of disease. *Journal of Health Economics*, 14(2), 171–89.

Koopmanschap, M.A., Burdorf, A., Jacob, K., Meerding, W.J., Brouwer, W. and Severens, H. 2005. Measuring productivity changes in economic evaluation: Setting the research agenda. *PharmacoEconomics*, 23(1), 47–54.

Kraemer, W.J., Adams, K., Cafarelli, E., Dudley, G.A., Dooly, C., Feigenbaum, M.S., Fleck, S.J., Franklin, B., Fry, A.C., Hoffman, J.R., Newton, R.U., Potteiger, J., Stone, M.H., Ratamess, N.A., Triplett-McBride, T. and American College of Sports Medicine 2002. American College of Sports Medicine position stand. Progression models in resistance training for healthy adults. *Medicine and Science in Sports and Exercise*, 34(2), 364–80.

Lambourne, K. and Tomporowski, P. 2010. The effect of exercise-induced arousal on cognitive task performance: A meta-regression analysis. *Brain Research*, 1341(0), 12–24.

Lenneman, J., Schwartz, S., Giuseffi, D.L. and Wang, C. 2011. Productivity and health. *Journal of Occupational and Environmental Medicine*, 53(1), 55–61.

Marcus, B.H., Williams, D.M., Dubbert, P.M., Sallis, J.F., King, A.C., Yancey, A.K., Franklin, B.A., Buchner, D., Daniels, S.R. and Claytor, R.P. 2006. Physical activity intervention studies: What we know and what we need to know: A scientific statement from the American Heart Association Council on Nutrition, Physical Activity, and Metabolism (subcommittee on physical activity); Council on Cardiovascular Disease in the Young; and the Interdisciplinary Working Group on Quality of Care and Outcomes Research. *Circulation*, 114(24), 2739–52.

Marshall, A.L. 2004. Challenges and opportunities for promoting physical activity in the workplace. *Physical Activity*, 7(1, Supplement 1), 60–66.

Masley, S., Roetzheim, R. and Gualtieri, T. 2009. Aerobic exercise enhances cognitive flexibility. *Journal of Clinical Psychology in Medical Settings*, 16(2), 186–93.

Mattke, S., Balakrishnan, A., Bergamo, G. and Newberry, S.J. 2007. A review of methods to measure health-related productivity loss. *American Journal of Managed Care*, 13(4), 211–17.

McEachan, R., Lawton, R., Jackson, C., Conner, M., Meads, D. and West, R. 2011. Testing a workplace physical activity intervention: A cluster randomized controlled trial. *International Journal of Behavioral Nutrition and Physical Activity*, 8(1), 29.

Mein, G.K., Shipley, M.J., Hillsdon, M., Ellison, G.T. and Marmot, M.G. 2005. Work, retirement and physical activity: Cross-sectional analyses from the Whitehall II study. *European Journal of Public Health*, 15(3), 317–22.

Mileråd, E., Ericson, M.O., Nisell, R. and Kilbom, A. 1991. An electromyographic study of dental work. *Ergonomics*, 34(7), 953–62.

Miller, P. and Haslam, C. 2009. Why employers spend money on employee health: Interviews with occupational health and safety professionals from British Industry. *Safety Science*, 47(2), 163–9.

Mills, P.R., Kessler, R.C., Cooper, J. and Sullivan, S. 2007. Impact of a health promotion program on employee health risks and work productivity. *American Journal of Health Promotion*, 22(1), 45–53.

Nicholson, S., Pauly, M.V., Polsky, D., Sharda, C., Szrek, H. and Berger, M.L. 2006. Measuring the effects of work loss on productivity with team production. *Health Economics*, 15(2), 111–23.

Nyman, J.A., Barleen, N.A. and Abraham, J.M. 2010. The effectiveness of health promotion at the university of Minnesota: Expenditures, absenteeism, and participation in specific programs. *Journal of Occupational and Environmental Medicine*, 52(3), 269–80.

Öberg, T., Karsznia, A., Sandsjö, L., and Kadefors, R. 1995. Work load, fatigue, and pause patterns in clinical dental hygiene. *Journal of Dental Hygiene*, 69, 223–9.

O'Shea, R.M. 1984. Sources of dentists' stress. *Journal of American Dental Association*, 109(1), 48–51.

Ozminkowski, R.J., Ling, D., Goetzel, R.Z., Bruno, J.A., Rutter, K.R., Isaac, F. and Wang, S. 2002. Long-term impact of Johnson and Johnson's Health and Wellness Program on health care utilization and expenditures. *Journal of Occupational and Environmental Medicine*, 44(1), 21–9.

Parks, K.M. and Steelman, L.A. 2008. Organizational wellness programs: A meta-analysis. *Journal of Occupational Health Psychology*, 13(1), 58–68.

Pate, R.R., Pratt, M., Blair, S.N., Haskell, W.L., Macera, C.A., Bouchard, C., Buchner, D., Ettinger, W., Heath, G.W., King, A.C., Kriska, A., Leon, A., Marcus, B., Morris, J., Paffenberger, R., Patrick, K., Pollock, M., Rippe, J., Sallis, J. and Wilmore, J. 1995. Physical activity and public health: A recommendation from the Centers for Disease Control and Prevention and the American College of Sports Medicine. *JAMA*, 273(5), 402–7.

Pauly, M.V., Nicholson, S., Polsky, D., Berger, M.L. and Sharda, C. 2008. Valuing reductions in on-the-job illness: "Presenteeism" from managerial and economic perspectives. *Health Economics*, 17(4), 469–85.

Pauly, M.V., Nicholson, S., Xu, J., Polsky, D., Danzon, P.M., Murray, J.F. and Berger, M. L. 2002. A general model of the impact of absenteeism on employers and employees. *Health Economics*, 11(3), 221–31.

Pelletier, B., Boles, M. and Lynch, W. 2004. Change in health risks and work productivity over time. *Journal of Occupational and Environmental Medicine*, 46(7), 746–54.

Pelletier, K.R. 2005. A review and analysis of the clinical and cost-effectiveness studies of comprehensive health promotion and disease management programs at the worksite: Update VI. *Journal of Occupational and Environmental Medicine*, 47(10), 1051–8.

Pohjonen, T. and Ranta, R. 2001. Effects of worksite physical exercise intervention on physical fitness, perceived health status, and work ability among home care workers: Five-year follow-up. *Preventive Medicine*, 32(6), 465–75.

Pronk, N.P. 2009. Physical activity promotion in business and industry: Evidence, context, and recommendations for a national plan. *Journal of Physical Activity and Health*, 6 Suppl 2, S220–235.

Proper, K. I., Hildebrandt, V. H., Van der Beek, A., Twisk, J. and Van Mechelen, W. 2003a. Effect of individual counseling on physical activity fitness and health: A randomized controlled trial in a workplace setting. *American Journal of Preventive Medicine*, 24(3), 218–26.

Proper, K.I., de Bruyne, M.C., Hildebrandt, V.H., van der Beek, A., Meerding, W.J. and van Mechelen, W. 2004. Costs, benefits and effectiveness of worksite physical activity counseling from the employer's perspective. *Scandinavian Journal of Work, Environment and Health*, 30(1), 36–46.

Proper, K.I., Koning, M., van der Beek, A.J., Hildebrandt, V.H., Bosscher, R.J., and van Mechelen, W. 2003b. The effectiveness of worksite physical activity programs on physical activity, physical fitness, and health. *Clinical Journal of Sport Medicine*, 13(2), 106–17.

Puetz, T.W. 2006. Physical activity and feelings of energy and fatigue: Epidemiological evidence. [Review]. *Sports Medicine*, 36(9), 767–80.

Puetz, T.W., O'Connor, P.J. and Dishman, R.K. 2006. Effects of chronic exercise on feelings of energy and fatigue: A quantitative synthesis. [Meta-Analysis]. *Psychological Bulletin*, 132(6), 866–76.

Reed, J. and Ones, D.S. 2006. The effect of acute aerobic exercise on positive activated affect: A meta-analysis. *Psychology of Sport and Exercise*, 7(5), 477–514.

Schultz, A.B. and Edington, D.W. 2007. Employee health and presenteeism: A systematic review. *Journal of Occupational Rehabilitation*, 17(3), 547–79.

Schultz, A.B., Chin-Yu, C. and Edington, D.W. 2009. The cost and impact of health conditions on presenteeism to employers: A review of the literature. *Pharmacoeconomics*, 27(5), 365–78.

Singh-Manoux, A., Hillsdon, M., Brunner, E., and Marmot, M. 2005. Effects of physical activity on cognitive functioning in middle age: Evidence from the Whitehall II prospective cohort study. *American Journal of Public Health*, 95(12), 2252–8.

Sjogren, T., Nissinen, K.J., Jarvenpaa, S.K., Ojanen, M.T., Vanharanta, H. and Malkia, E.A. 2005. Effects of a workplace physical exercise intervention on the intensity of headache and neck and shoulder symptoms and upper extremity muscular strength of office workers: A cluster randomized controlled cross-over trial. *Pain*, 116(1–2), 119–28.

Sjogren, T., Nissinen, K.J., Jarvenpaa, S.K., Ojanen, M.T., Vanharanta, H. and Malkia, E.A. 2006. Effects of a physical exercise intervention on subjective physical well-being, psychosocial functioning and general well-being among office workers: A cluster randomized-controlled cross-over design. *Scandinavian Journal of Medicine and Science in Sports*, 16(6), 381–90.

Stephenson, J., Bauman, A., Armstrong, T., Smith, B. and Bellew, B. 2000. The costs of illness attributable to physical activity in Australia: A preliminary study. *Population Health Division Publications*. Canberra: Commonwealth of Australia.

Sternfeld, B., Block, C., Quesenberry Jr, C.P., Block, T.J., Husson, G., Norris, J.C. and Block, G. 2009. Improving diet and physical activity with ALIVE: A worksite randomized trial. *American Journal of Preventive Medicine*, 36(6), 475–83.

Stewart, W.F., Ricci, J.A., Chee, E. and Morganstein, D. 2003. Lost productive work time costs from health conditions in the United States: Results from the American Productivity Audit. *Journal of Occupational and Environmental Medicine*, 45(12), 1234–46.

Sui, X., Laditka, J.N., Church, T.S., Hardin, J.W., Chase, N., Davis, K. and Blair, S.N. 2009. Prospective study of cardiorespiratory fitness and depressive symptoms in women and men. *Journal of Psychiatric Research*, 43(5), 546–52.

Swedish National Institute of Public Health. 2008. Folkhälsoenkät 2007 – national data [Public Health survey 2007 – data from a national sample].

Titze, S., Martin, B.W., Seiler, R., Stronegger, W. and Marti, B. 2001. Effects of a lifestyle physical activity intervention on stages of change and energy expenditure in sedentary employees. *Psychology of Sport and Exercise*, 2(2), 103–16.

Tompa, E., Dolinschi, R. and de Oliveira, C. 2006. Practice and potential of economic evaluation of workplace-based interventions for occupational health and safety. *Journal of Occupational Rehabilitation*, 16(3), 375–400.

Tompa, E., Dolinschi, R., de Oliveira, C. and Irvin, E. 2009. A systematic review of occupational health and safety interventions with economic analyses. *Journal of Occupational and Environmental Medicine*, 51(9), 1004–23.

Tompa, E., Verbeek, J., van Tulder, M. and de Boer, A. 2010. Developing guidelines for good practice in the economic evaluation of occupational safety and health interventions. *Scandinavian Journal of Work, Environment and Health*, 36(4), 313–18.

Uegaki, K., Bruijne, M.C., Beek, A.J., Mechelen, W. and Tulder, M.W. 2010. Economic evaluations of occupational health interventions from a company's perspective: A systematic review of methods to estimate the cost of health-related productivity loss. *Journal of Occupational Rehabilitation*, 21(1), 90–99.

van Dongen, J.M., Proper, P., van Wier, M.F., van der Beek, A., Bongers, P.M., van Mechelen, W. and van Tulder, M. 2012. A systematic review of the cost-effectiveness of worksite physical activity and/or nutrition programs. *Scandinavian Journal of Work, Environment and Health*, 38(5), 393–408.

von Thiele Schwarz, U. and Hasson, H. 2011. Employee self-rated productivity and objective organizational production levels: Effects of worksite health interventions involving reduced work hours and physical exercise. *Journal of Occupational and Environmental Medicine*, 53(8), 838–44.

von Thiele Schwarz, U. and Hasson, H. 2012. Effects of worksite health interventions involving reduced work hours and physical exercise on sickness absence costs. *Journal of Occupational and Environmental Medicine*, 54(5), 583–44.

von Thiele Schwarz, U., Lindfors, P. and Lundberg, U. 2006. Evaluating different measures of sickness absence with respect to work characteristics. *Scandinavian Journal of Public Health*, 34(3), 247–53.

von Thiele Schwarz, U., Lindfors, P. and Lundberg, U. 2008. Health-related effects of worksite interventions involving physical exercise and reduced workhours. *Scandinavian Journal of Work Environment and Health*, 34(3), 179–88.

Wilson, M.G., Holman, P.B. and Hammock, A. 1996. A comprehensive review of the effects of worksite health promotion on health-related outcomes. *American Journal of Health Promotion*, 10(6), 429–35.

Yancey, A.K., Lewis, L.B., Sloane, D.C., Guinyard, J.J., Diamant, A.L., Nascimento, L.M. and McCarthy, W.J. 2004. Leading by example: A local health department-community collaboration to incorporate physical activity into organizational practice. *Journal of Public Health Management and Practice*, 10(2), 116–23.

Youngstedt, S.D. and Kline, C.E. 2006. Epidemiology of exercise and sleep. *Sleep and Biological Rhythms*, 4(3), 215–21.

Zapf, D., Dormann, C. and Frese, M. 1996. Longitudinal studies in organizational stress research: A review of the literature with reference to methodological issues. *Journal of Occupational Health Psychology*, 1(2), 145–69.

8 Services Vouchers: Tools to Prevent Stress and Enhance Well-being

NATHALIE RENAUDIN

Abstract

In the current context of high unemployment rates and stretched budgets, job insecurity is rising along with pressure on employees. This particular situation is combined with structural factors, such as the ageing population, or the increasing number of women in the workforce. More and more employees are facing stressful situations in their workplaces whilst also shouldering the responsibility of one or more dependents at home. Some companies provide voucher programmes for their employees, in the form of paper or cards, to help them to gain access to specific goods and services and to ease everyday pressures. This chapter will analyse the services voucher system, as well as its developments and impacts in the countries in which it operates. The study is based on the representative experience of countries with different social, economic, political and cultural contexts.

Introduction

Based on studies from international organizations (ILO, OECD, WHO, etc.), the European Union, or national institutes, voucher programmes have been designed and managed for the last 50 years to facilitate the implementation of social policy in order to meet the needs of companies as well as public authorities. Such solutions help promote health and well-being for both the general population and specific workforces. They constitute important prevention measures for companies, especially SMEs, who are looking for external solutions.

The first section of this chapter develops the background of the services voucher scheme, detailing the history and the mechanics of the system. Vouchers for meals during the working day were the first to be introduced and thus serve as a guide to the early development of the system. They are detailed in the second section, which also highlights the economic and social impacts such schemes have on all the involved stakeholders. The third section is dedicated to the services vouchers which have been adapted in the 1990s to improve work–life balance, with a specific focus on the childcare voucher in Great Britain and the Universal Employment Services Vouchers (Chèque Emploi Services Universel – CESU) in France. The fourth section covers environmental practices in companies, such as commuting, which has become an important source of stress and will detail interesting measures adopted in the United States (Commuter Check) and Brazil, Spain and Uruguay (transport vouchers). It will also examine some corporate social responsibility (CSR) initiatives, where vouchers scheme can play a key role, such as the "Ecocheque" (environmental or "green" vouchers) in Belgium. Last but not least, the fifth section will discuss holiday vouchers in Romania, as well as wellness vouchers in Sweden or Finland, which constitute important tools to build mentally healthy workplaces.

Origin of the Voucher System

Workplaces are important settings for health promotion and disease prevention. People need to be given the opportunity to make healthy choices in the workplace in order to reduce their exposure to risk. Further, the cost to employers of morbidity attributed to non communicable diseases is increasing rapidly. Workplaces should make possible healthy food choices and support and encourage physical activity (World Health Organization 2004).

Meal vouchers were introduced in Great Britain following the Second World War in order to offer an alternative to companies that could not provide employees with a canteen (especially SMEs) and to ensure that workers could take a lunch break, fundamental for their health and well-being.

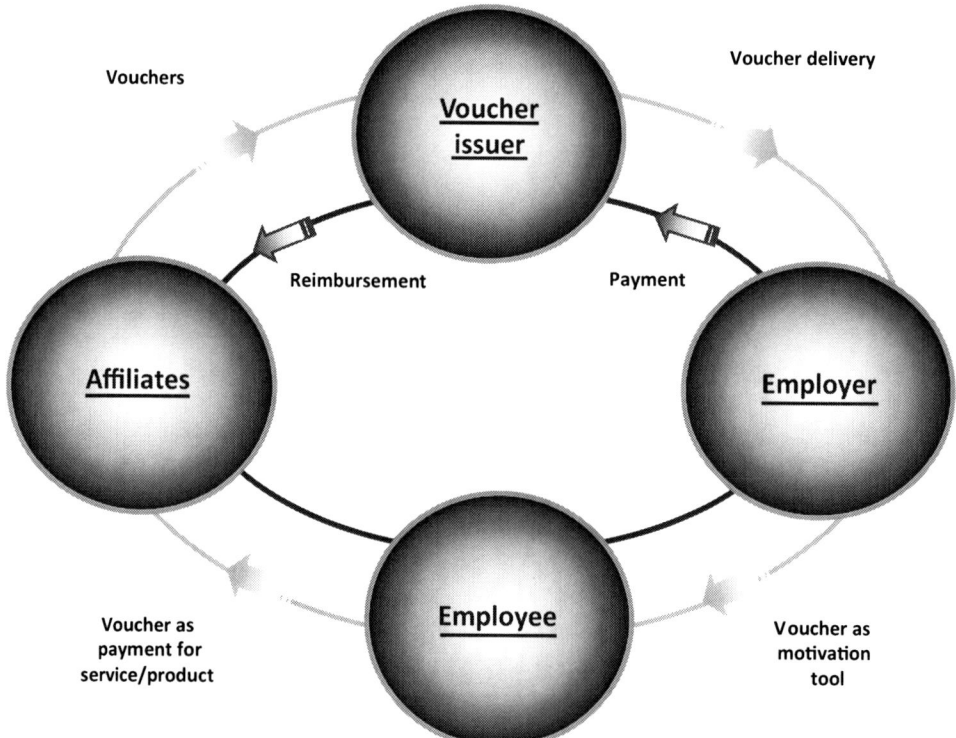

Figure 8.1   Vouchers scheme principle

The system operates on the basis of very simple principles thanks to comprehensive and favourable social laws, dedicated to employees' nutrition: companies buy vouchers from professional issuers. Then they distribute them to their staff. The user spends the vouchers at their face value in affiliated restaurants or food stores, which then are redeemed by the issuer. In this system, the professional issuer coordinates the overall organization of the system and therefore ensures the security and the reliability of the activity: administration, production, affiliation of service providers and communication with companies (see Figure 8.1).

Tax exemption is implemented by governments to encourage employers to give their employees access to a decent meal. The shortfall brought about by tax exemptions is largely compensated by the VAT and tax revenue generated by the increased activity and jobs creation. Indeed, by implementing preferential taxation for meal vouchers, the state is focusing its support on a dynamic sector, encouraging consumption. Concurrently, the government can be satisfied that the benefits offered cannot be misused or manipulated away from their intended purpose. "The setting up of vouchers enables employers to achieve a higher level of motivation and satisfaction in the workplace reduce absenteeism and obtain greater efficiency from their workers and increased productivity" (Senator of the Republic of Mexico, Maria de Los Angeles Moreno Uriegas, 2011).

The Importance of Nutrition for Health and Well-being

Christopher Wanjek, in his research *Food at Work*, for the International Labour Organization (ILO), places emphasis on the lack of attention to the issue of food in the workplace (Wanjek 2005). It is intended as a practical rather than a theoretical analysis, as many case studies are given as examples. In this way, this study shows what employers, workers and government can do to improve food at work, everywhere in the world, and why they have to do it.

THE LEGAL AND SOCIAL ORIGINS

The ILO has normative legislative recommendations that date back to as early as 1921 describing minimum worker relaxation time (Wanjek 2005). A more recent document, the 1971 *Expert Consultation on Worker's Feeding* (published jointly by the WHO and FAO) points out that workers' statutory rights should not be dependent on the size or industry of the company for which they work: "[Statutory obligations] should not be governed by the size of the unit, should embrace all types of industry in both urban and rural areas, and should be related to the needs of the workers and the industry."

Further, the FAO report (1971) entreats governments to adopt legislation that promotes this positive environment, suggesting:

> *that governments promulgate laws and regulations requiring the establishment of workers' feeding programmes with a view to improving the health, welfare and productivity of workers ... Such laws and regulations should have as their objective the adequate feeding of the worker and his family; should be designed to stimulate the establishment of appropriate food services ... and should recognize the economic limitation of the worker, the undertaking, the industry and the country.*

For all these reasons a number of structures and instruments have been set up at national level in order to provide different programmes that encourage employers to provide meals for their employees.

The introduction of the meal vouchers system within a company leads to better work conditions, nutrition and health for employees. It offers employees the opportunity to enjoy a real break at lunch time in a convivial and friendly place that they have chosen themselves. The food they get there is generally more diversified and the quality is better than packed lunches prepared at home in the morning, or the night before.

In *Food at Work*, Wanjek (2005) shows that good nutrition, like other vital occupational safety and health issues, is the foundation of workplace productivity and safety – priority concerns shared by unions, workers, employers and governments around the world. On the other hand, unhealthy foods can lead to obesity and chronic diseases, while macro- and micronutrient deficiencies can cause malnutrition. In both these instances, the effects are detrimental to the creation of a strong, well-equipped workforce.

The first action plan for food and nutrition issued by the European Regional Office of the WHO pointed out:

> *The prevalence of obesity affects up to 20–30% of adults, while the rate is rising among children. The result is an increased risk of cardiovascular diseases, some cancers and diabetes. It is estimated that obesity costs certain health services around 7% of their total healthcare budget. About one third of cardiovascular diseases, which represent the primary cause of death in the Region, is linked to an unbalanced diet, and 30 to 40% of cancers could be prevented by improving people's diet.*

IMPROVEMENT IN JOB SATISFACTION, WORK ATMOSPHERE AND EMPLOYEES' HEALTH

The most important feature of the vouchers is that they can be adopted by any company, regardless of size: private or public, isolated in an industrial area or located in city centres, with one location or multi-sited. They represent substitutes to canteens with more freedom of choice and use.

Vouchers enable individuals to apply for positions they would not have considered otherwise. Based on a study carried out by a French institute (IPSOS 2010), vouchers help employers to attract new employees even in small companies: more than half of employees state that their applications were motivated by the voucher scheme. It thus represents an advantage for SMEs who are able to attract high-skilled employees thanks to the vouchers. In the above-mentioned inquiry, 61 per cent of employers declared that the voucher supply was a major advantage in attracting candidates during theconsideration hiring process. In the same study, it is also shown that 64 per cent of the employees estimate that the vouchers enhance their well-being at work. Seemingly, according to a survey in Greece led by PricewaterhouseCoopers (2011), more than half of interviewed companies consider vouchers as part of their recruitment strategy, as they recognize that it is important for employees to have regular interaction with colleagues. Non-work opportunities to spend time with colleagues are likely to improve mental health and worker morale (Wanjek 2005).

The voucher system has also turned out to be a tool maintaining social peace in businesses due to its positive psychological effect on the beneficiaries. In this regard, 91.9 per cent of the employers in Spain declare that food vouchers help maintain a peaceful work atmosphere in their companies (Murillo 2011).

In Romania, where food represents a major expenditure in household budgets (43.3 per cent),[1] we could even say that food vouchers are considered to be a way of protecting part of the household income dedicated to food.

In Romania, the minimum monthly wage is €70 and the average monthly wage is between €130 and €150. Under such circumstances, food vouchers represent an increase in purchasing power of between 20% to 40%, over 90% of which is channelled to food purchases; this is highly significant in a country where the food budget frequently represents 50% of income. It is clear that in the case of the Czech Republic and Romania, vouchers do not just subsidize an employee's lunch; they make it possible for the employee/employee's family to purchase enough food (ICOSI 2004).

In the same vein, more than 88 per cent of the Bulgarian beneficiaries think that the voucher schemes affect their household food budget and consumption.

In Brazil, the introduction of meal vouchers was part of a wider political measure called PAT, which is the Brazilian acronym for Worker Food Programme.[2] Introduced in 1976, with more than seven million voucher users per day, this programme meets its objective successfully. While 46.1 per cent of the country's population were malnourished in 1975, by 1990 this had reduced to 30.7 per cent (Mazzon 2009). The improvement of the nutritional quality of the employees' meals also contributed to the change in sanitary conditions in the workplace (Mazzon 2001). The results of the programme represent the best guarantee of its success and illustrate that the system is sustainable (IBOPE 1996).

In Bulgaria, according to the survey led by the National Public Opinion Center (NAPOC) together with the Bulgarian parliament, in 2002, employees were not satisfied with the system which provided meals before the voucher scheme was introduced: there were very few external providers, there was a lack of restaurants and the quality of the meals was poor. The results were logical: a high proportion of employees did not eat diversified food (58 per cent), or good quality food (52 per cent). In a report, Doctor Lyudmila Ivanova (2002) underlined the importance of food in the workplace, which is an issue too often neglected (especially the consumption of vitamins and proteins). The introduction of the meal voucher was a good way of encouraging employees to eat a complete and balanced meal. As they increase the purchasing power, in the Czech Republic, voucher holders visit restaurants for lunch 53 per cent more often than non-holders (GFK 2009). In the same vein, and as Figure 8.2 shows, 70 per cent of meal voucher holders would not visit restaurants as often for lunch if the voucher scheme ended.

---

1   According to the National Statistics Institute of Romania, food and non-alcoholic beverages accounted for 43.3 per cent of the household budget in 2011.

2   The PAT is dedicated to employees, but complementary initiatives were developed for people in need. In 2003, Lula da Silva, the Brazilian president, implemented the Zero Hunger programme to feed every Brazilian in accordance with the Right to Food enshrined in the Constitution.

Food vouchers are a better policy compared to	Approved by	Disapproved by
Increasing minimum wage	48.28%	20.29%
Subsidized employment programs	34.48%	34.48%
Mandatory in-kind social benefits/assistance	27.59%	3.45%
Decrease in the social security contributions	55.17%	27.59%
Sectorial tax privileges	27.59%	6.9%

Figure 8.2   Food vouchers compared to other social policies (Bulgaria)
Source: "Food Voucher in Bulgaria- Economic and Fiscal Impact," Study led by KC2 and Industry Watch, June 2010, page 17

The introduction of meal vouchers helps governments to fight malnutrition problems affecting employees as well as their families. The consequences of these problems can be very serious for businesses, including the reduction of physical ability and increased tiredness, psychological fragility, absenteeism, lack of motivation and/or satisfaction at work.

The impacts on productivity

In his study, Christopher Wanjek (2005) also stresses the fact that good nutrition and high productivity are closely linked. He demonstrates that ensuring that workers have access to nutritious, safe and affordable food, an adequate meal break and decent conditions for

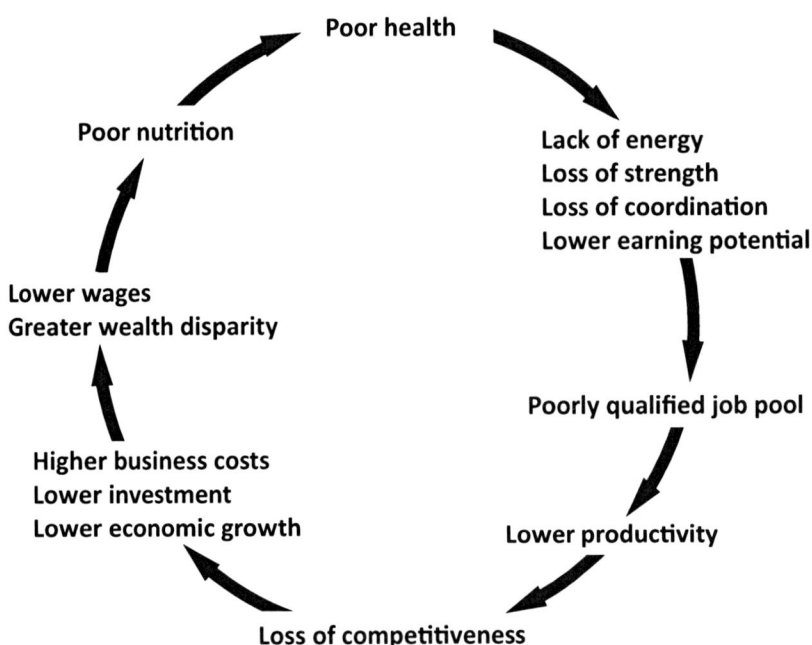

Figure 8.3   The cycle of poor nutrition and low national productivity
Source: Wanjek 2005; ILO 2005.

eating is not only socially important and economically viable but a profitable business practice too (see Figure 8.3).

Workers who have access to adequate nutrition can be up to 20 per cent more productive and less prone to accidents (Wanjek 2005). Employees' intellectual and physical abilities depend on the quality of the meal they enjoy during this break. Some 78 per cent of general medical practitioners in France consider that the reduction (or even removal) of the lunch break is even likely to increase the risk of work accidents (ICOSI 2004).

The influence of the Brazilian World Food Programme (PAT) has been such that productivity in Brazil rose 2.5 per cent a year between 1991 and 1998 on the back of improved worker health (Mazzon 2001). This productivity increase was similar in the United States during the same period. Workplace accidents are fewer, as are the number of days off work. The number of workplace accidents fell from 1.9 million in 1977 to 395,000, despite the doubling of the working population. The average number of medically-approved worker 'sick days' in Pernambuco, a state in the northeast region of Brazil, fell from 4.5 to 2.9 over the same period. In more than 30 years of the PAT's existence, the average number of workers benefiting from the scheme rose 8.7 per cent per year, while work-related accidents fell 3.7 per cent per year. Many factors other than food have helped to reduce the number of work-related accidents in Brazil, but the critical contribution of improving the food status of workers must be highlighted.

Nevertheless, workers do not always take a real lunch break. Several reasons could explain that phenomenon but the lack of time and the absence of incentives or facilities (vouchers or on-site canteens) are the major ones. The *Eurest Lunchtime Report 2000* also demonstrates that taking a midday break is a physiological need with direct consequences on workers' activity and productivity.

Whatever the underlying cause, skipping meals can have a profound effect on workers' ability to perform consistently with their potential. Workers who skipped breakfast score 15 per cent less on memory tests, and were prone to snacking later in the day. Some estimates put the cost of skipping breakfast in the billions for national economies – £17 billion ($26bn) in the case of the UK in 2007, the equivalent of 97 million lost working days (IPSOS Mori 2007; Druce 2007). In Mickelson's student study (1999), nearly a fifth of those who did not eat breakfast said it was because they had insufficient time. In the UK, a third of workers would rather get a few minutes of extra sleep than eat in the morning. In Morocco, (and, ostensibly, other practising Muslim countries) productivity noticeably falls during the annual month-long Ramadan as Siham Ali in Rabat and Hassan Benmehdi in Casablanca show (2009). They demonstrate that not eating impacts on a person's ability to sustain work.

Based on assumptions stating that for every extra 1 kcal of food, productivity increases by 2.27 per cent and for every US $1 invested in nutrition, there is a return of between $5 and $20 (WHO/FAO 2003), the Brazilian programme has contributed to the reduction of food insecurity and hunger by one-third between 2000–2002 and 2005–2007 (Oxfam 2010). In parallel, from 1975 to 2007, work-related accidents dropped by more than one-third while the economically active population more than doubled (Mazzon 2009).

Providing food to employees not only curbs work accidents, but it also has a significant impact on absenteeism. Such an impact is very important for companies' finances. Bearing in mind that employee absence accounts for the equivalent of 35 per cent of base payroll

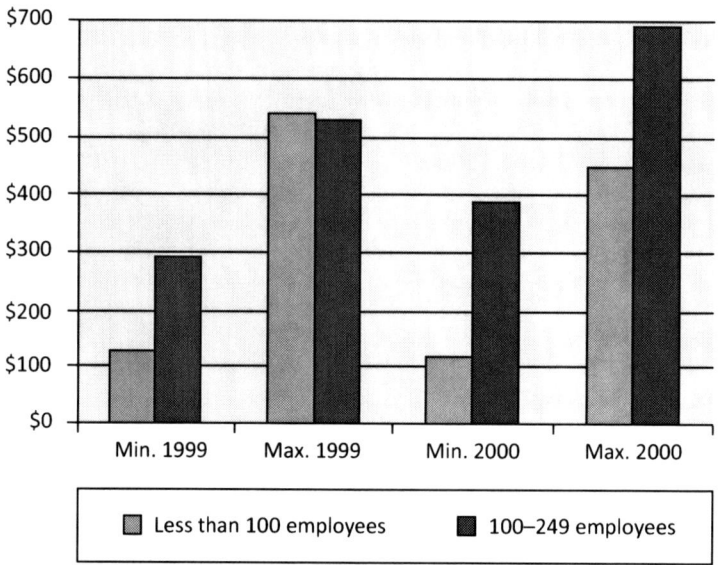

Figure 8.4   Annual cost of absenteeism to small businesses (per employee)
Source: Unscheduled Absence Survey – CCH incorporated October 2007.

(see Figure 8.4), it is worth noting that, in Spain, employers evaluate that absenteeism has fallen by 37.7 per cent since the setting up of the voucher schemes (Murillo 2011).

However, improving performance is not only a question of work but also a question of social relations at the workplace, as already underlined. Indeed, a high rate of absenteeism could contribute to the erosion of staff relations (ANACT 2009). Beyond limiting absenteeism and work accidents, the vouchers also maintain social harmony at work.

ECONOMIC BENEFITS: TURNING INFORMAL ECONOMY INTO FORMAL ACTIVITY, ENSURING A RECOLLECTION OF TAXES

The voucher system guarantees total transparency in the use and the circulation of the dedicated funds. It provides public authorities with a solution to increase transparency in a specific sector of the economy without adding administrative burden (for instance in managing social data).

With the use of invoices between voucher issuers and restaurants, the state offers itself a better visibility of the informal economy, which is very important in these sectors, and thus improves its rate of fiscal recovery: according to a 2007 Eurobarometer, the informal economy amounts to 20 per cent in Europe (Tudorel et al. 2010) – with an average of 18 per cent – and is much more prevalent in labour-intensive sectors such as catering and domestic services (Scheinder 2009). Total transparency in the use and the circulation of the dedicated funds is guaranteed by the nature of the scheme.

To target this additional consumption at the formal sector of the economy, this generates an improved collection of VAT throughout the production and distribution chain, as well as an increase in the amount of taxable income of all these industrial and business sectors, which compensate the effect of the tax incentive granted. The voucher system results in an economic benefit in every country that has set up the scheme. In

France (Ernst & Young 2006), the voucher system resulted in a benefit of €47 million, in Turkey, the scheme resulted in an economic surplus of €17.28 million (Istanbul Ekonomi 2010), in Bulgaria (2010) the voucher scheme amounted to a benefit of €15.5 million and, finally, the system resulted in a surplus of €128 million in Spain for the same period.

In the medium term, what the voucher system does achieve is an increase in consumption of foodstuffs produced mainly in the country, which leads to the creation of jobs in the production and food services industry. Indeed, in Bulgaria, every money unit (lev) spent in the food procurement scheme generated between 20 and 25 per cent more value added in the domestic economy (KC2 and Industry Watch 2010). In Turkey in 2006, every €42.40 spent in the meal voucher system contributed to an increase of €2.98 in public revenue. Finally, the meal voucher scheme created more than 21,000 direct jobs in the catering sector in Spain (Murillo 2011). On the reverse, some macroeconomic studies show that in case of withdrawal, the impacts on the state budget would be very significant.

Services vouchers guarantee and increase the purchasing power for basic needs, promote health and well-being, fight against the informal economy and consequently constitute a way to collect complementary taxes and to boost a specific sector of economy.

A Solution to Ease Work–Life Balance

Member States [are invited] to explore ways in which reconciliation policies might be further upgraded by encouraging employers to eliminate the tension between workers' commitment to their careers and to their families by making those two spheres of life mutually compatible (European Council, Employment, October 2011).

The lunch break has, over time, tended to shrink but complementary factors have also led to stressful situations, especially because employees' habits have changed, along with their needs. The increasing number of women in the workforce and an ageing population implies that more and more employees are shouldering the responsibility of one or more dependents and need solutions to balance their private and professional life. The working time needs to be more flexible. Additionally, access to quality, affordable services (childcare, elderly care, home services, etc.) is also an important determinant of citizens' employment opportunities and stress prevention. Nevertheless, existing policies, programmes or services are too often inadequate to meet these specific workers' needs.

The lack of affordable, convenient and good quality childcare is a problem that negatively affects both workers and employers around the world. Restricted access to work, lower earnings, lower productivity and higher absenteeism are just a few of the consequences of the lack of suitable childcare, all of which jeopardize families' income security and company success (see Figure 8.5).

If well designed and targeted, public measures supporting workplace initiatives can:

- increase resources available for childcare;
- help ensure that provision is responsive to the needs of working parents;
- encourage greater participation of women in the labour force;
- improve visibility, recognition and working conditions of caregivers.

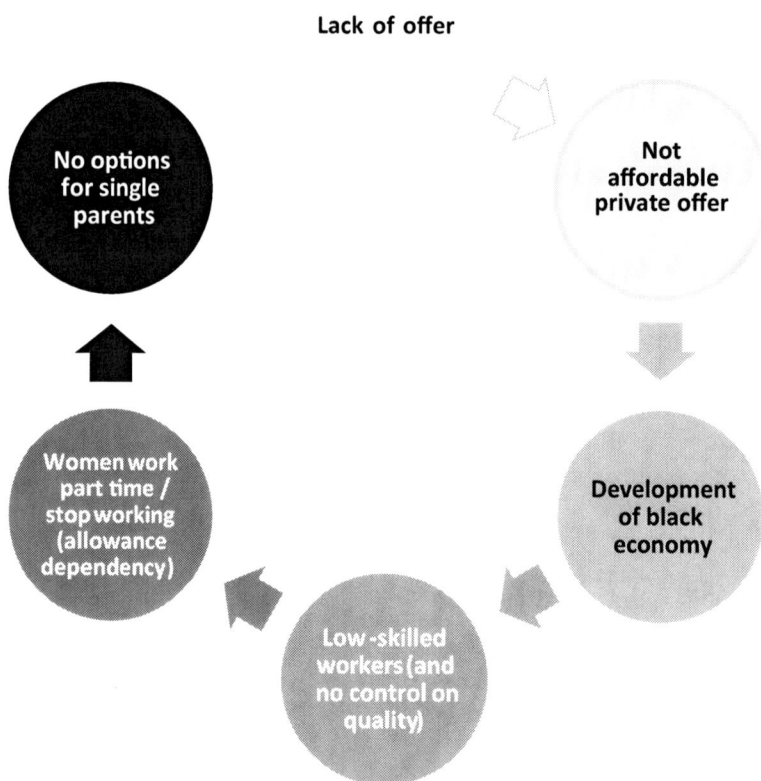

Figure 8.5 Work–life balance and childcare: The vicious circle

According to ILO research (Hein and Cassirer 2010), there are basically two types of measures at the level of the workplace that can help employees cope with their familial responsibilities: measures on the working time of employees so that they are able to carry out domestic tasks (for instance to fit in with school hours), and measures to access services provided by others, such as service vouchers.

Either for childcare or home services, this system is still based on shared responsibility between companies, employees, services providers and public authorities. "Successful partnerships often bring together actors that offer complementary financial, human, and technical contributions, but the sustainability of the entire initiative can be threatened when one partner must withdraw" (Hein and Cassirer 2010).

FOCUS ON THE CHILDCARE VOUCHER

In Europe, countries with government measures that support the costs of childcare tend to have higher rates of women's labour force participation and fertility, and lower gender inequality, as is particularly the case in the Nordic countries.

The childcare voucher system was adopted by the British government in 1989 in order to involve employers and have them share the employees' concerns about work–life balance. For many families, the cost of childcare still represents a barrier to returning to or remaining in employment. Parents in the UK have to spend a higher proportion of their income on childcare than any other country in the OECD. As a recent Aviva *Family Finances Report* underlines (2011), since the third quarter of 2010, 32,000 more women have been forced to leave work to look after their family due, in part, to the high cost of childcare. The figures demonstrate that childcare costs are one of the key motivations for choosing to stay at home and care directly for children instead of working.

Childcare vouchers can contribute to the cost of childcare while parents are at work. A subsidy of £55 per week can be allowed for childcare and the vouchers are exempt from tax and National Insurance. Parents with children up to the age of 15 can then choose freely the form of childcare from a wide range of services (in registered settings, ensuring that children receive high quality pre-school care: day nurseries, workplace nurseries, childminders, pre-school playgroups, nursery schools, out-of-school clubs and holiday clubs, as well as nannies and relatives). This ensures that once a child reaches school age, parents are able to fund after-school clubs to allow them to return or remain in employment.

Childcare vouchers also provide children with the best possible start in life. Both nurseries and childminders help the early development of skills that benefit children throughout their lives.

Some employers may place certain restrictions on the eligibility for childcare vouchers, for example only offering vouchers towards registered childcare or for parents with children under the age of five. Nevertheless, they are extremely popular with employers, as an integral part of recruitment and retention strategies, and the ability to use vouchers for various kinds of childcare makes them more useful to a larger group of parents and provides greater flexibility than some other types of support, such as workplace nurseries.

Among parents paying for childcare, the percentage receiving help from their employer more than tripled between 2004 and 2007 from 1 per cent to 3.4 per cent. Moreover, the childcare workforce increased 21 per cent to 275,000 between 1997 and 2001. Today, childcare vouchers are an employee benefit that helps approximately 450,000 individuals to afford good quality childcare. In the Netherlands also, the childcare sector has evolved since 1990 from a small sector with 8,000 employees into a mature sector employing over 60,000 employees in 2003.

FOCUS ON HOME SERVICES VOUCHERS

The creation of the Cheque Emploi Services Universel (CESU) in France was part of the development plan of care and support services launched by Minister Jean-Louis Borloo in 2005. The objective was to create 500,000 jobs in three years and to encourage the development of home services, with three specific goals:

- simplifying access to, and lowering the costs of, quality services;
- improving the income conditions of employees in this sector;
- giving the associations and the firms a favourable legislative context in order to promote their activities.

The French government also wanted to help workers to choose freely whether or not they wanted to take care of their relatives personally. As demonstrated, the costs are determinant in such choices. By providing allowances for both people at work and people at home, the state helped to base the choice on other grounds than financial ones.

The coverage of services is wider in France than in Great Britain as the CESU vouchers can be used for childcare in and outside the home, as well as for eldercare, care for the disabled, and domestic services. It is currently used for:

- home services: 41 per cent;
- gardening: 15.5 per cent;
- dependence (elderly or disabled people): 17.5 per cent;
- childcare (childminding, homework help, etc.): 12 per cent;
- others (computer assistance, etc.): 14 per cent.

The CESU scheme is tax exempted for both the employer and the employee: employers do not pay social security and have 25 per cent tax credit on their contribution (up to an annual ceiling of €1,830 per employee). Employees do not pay income tax on up to €1,830 annually.

Based on the impact study led by the National Association of Services for Individuals in 2010, 390,000 jobs were created in France between 2005 and 2010 in the home services sector (a 15 per cent increase) and 275,000 people were receiving CESU vouchers.

A similar system in Belgium dedicated to home services enabled the creation of 62,000 jobs and turned an informal economy into a formal one. The scheme has become very popular among workers and 5.5 per cent of the active population used the service vouchers after five years.

The Belgian service vouchers ("Titres-services") were introduced in 2003, with the objective of creating 25,000 jobs by 2007. Regarding service providers, 50 per cent among them were previously unemployed, 98 per cent are women and one third are single mothers. This is particularly relevant as the sector offers both flexibility and a job close to home. It also enables low-skilled people to (re-)enter regular employment. According to a survey conducted in 2005 on behalf of the Ministry of Employment, Labour and Social Dialogue, 62 per cent of workers employed in the service vouchers scheme hold a certificate of secondary education at most, and 28 per cent a certificate of lower primary education.

The National Employment Office (ONEM) annual report of 2011 stated that there were 2,708 Titre-services providers in 2011 (+5.1 per cent compared to 2010) and 136,915 workers in 2010 (+13.8 per cent compared to 2009). Moreover, there were 834,959 users of Titres-services vouchers in 2011, that is to say an increase of 9.8 per cent compared to 2010. According to Gosta Esping-Andersen (1999), 15 servicing jobs are created for every 100 additional women in work.

Public interventions implemented in Belgium, France and Sweden have shown a positive effect on the creation and/or growth of respectively 1,022, 8,300 and 13,500 SMEs.

Service vouchers represent a solution for Member States to implement efficient measures to promote work–life balance while encouraging employers to participate in conciliation policies. They help employers to tackle absenteeism in their company, to favour their employees' well-being and prevent stress. They are also a tool to enhance corporate social responsibility, especially in SMEs. Since workers spend an important part of their lives at work, companies' activities have repercussions on people's daily lives. Companies are thus socially responsible for helping their employees to improve their quality of life.

The Green Vouchers: Transport and Environment as Part of Employees' Daily Life

Time pressure can affect workers at several points in the day. Lengthening working hours are supplemented by long commutes to work – the average commute in Great Britain is over 45 minutes each way (BBC 2003) and the distance travelled is increasing (BBC 2006). In Spain, with these time constraints, the siesta is gradually becoming a thing of the past (Deschenaux 2008). Similar trends are apparent in the United States, with workers increasingly feeling as though the lunch 'hour' is a myth (Armour 2006).

Exempted from tax and social charges, commuting and mobility solutions are offered by organizations to facilitate and co-finance employees' commuting costs. Several solutions exist according to each specific national need: vouchers redeemable for transit passes, cards, tokens and other fares at public transportation agencies and all other transit agencies among the affiliates' network and also a portfolio of transportation passes (season or single card) from private or public transportation companies.

In the United States, commuting vouchers emerged in the 1970s, as part of a wider traffic mitigation strategy. First authorized in 1984 with a tax-free monthly maximum of $15, transit benefit legislation was expanded numerous times and now allows a monthly maximum of $230, equaling tax-free parking benefit (Oram et al. 2010). Employees can save approximately 40 per cent of taxes, which makes public transit less expensive and more appealing to commuters.

The legislation allows employer-paid subsidy plans, employee-paid pre-tax salary deductions, and combinations of the two. Thousands of employers of all types and sizes in the public, private and non-profit sectors use the programmes. The US government offers fully subsidized transit fares to all of its employees. In 2008, a similar tax-free provision was added for bicycle commuters, and the City of San Francisco adopted an ordinance requiring provision of transit benefits by all employers with 21 or more employees.

A study done by San Francisco's Bay Area Rapid Transit District (BART) showed that after 12 years of the programme, participation grew by 44 per cent annually, to reach 39 per cent of users receiving transit benefits.

In Brazil, in order to ensure the mobility of low income workers, a dedicated law was introduced in 1985 to offer transport vouchers. At that time, vouchers were optional. Two years later, it became compulsory for employers to pay for employees' commuting costs. Today, 50 per cent of commuters using public transport in Brazil use the vouchers (Louven 2006).

The objective of such legislation is precisely to encourage employees not to use cars. It is also a way to reduce traffic congestion, greenhouse gas emissions and energy use. For companies, the system is very simple to implement with no administrative burden and no necessity for parking provision. In the United States, employers save approximately 10 per cent of tax and have all the benefits that have already been mentioned (employees' equity, recruitment, retention, productivity, etc.).

The use of transport vouchers is a way for employers to fulfil their employees' expectations and guarantees the correct use of the funds. This section also serves to highlight the responsibilities employers have towards their employees – showing that the provision of benefits is not a new phenomenon – and that it deserves recognition as being an important part of companies' human resources and corporate social responsibility policies.

In some countries, governments and companies have wished to go beyond this and to use the vouchers as communication tools, to promote public policy messages. This was the case in Belgium, with the green vouchers, called Ecocheques.[3] They allow for the purchase of ecological products and services in an affiliated network, in six specific areas: energy economy, water economy, sustainable mobility, garbage, ecodesign (products and services with the European Ecolabel) and nature. Each employer can provide employees with €250 per year. The maximum value of a green voucher is €10 and they are valid for two years.

Ecocheques are powerful tools to boost sustainable consumption from the demand side and sustainable production from the supply side. They do not only increase the purchasing power of the beneficiaries in specific purposes, they also constitute efficient communication tools to raise awareness and inform about responsible consumption.

The appearance of concepts such as human capital, social responsibility (CSR) or sustainable development shows an increasing interest of the companies for everything that deals with human and social fields. Favourable tax measures for companies on the one hand, and the interest of employees in social and environmental development, encourage the employers to evaluate the different social benefits they can offer to their employees.

Wellness Programmes and Leisure

In order to increase the health and morale of the working population, the Swedish tax authority implemented a tax-exempt wellness allowance that employers can distribute to their employees up to SEK 7,000 (about €700) per year. The allowance includes sport activities, massages and many other services. The voucher system was developed together with access to a web portal that encloses Wellness System functions (such as reports regarding the use, health screening and a personal follow-up). The voucher system replaces the complicated system of recollection of receipts and makes the benefit visible. It raises awareness of, and promotes, physical activity.

In 2009, a sport and culture benefit was also introduced in Finland, with a tax-exempt amount of €400 per year and per employee. Beyond sport activities, employees can use the vouchers for cultural activities (museums, theatres, cinema, as well as for artistic activities) to enhance their well-being and leisure time.

In Romania, holiday vouchers were introduced as an emergency measure in 2009. In the midst of the economic crisis, the government wished to help as many people as possible to go on holiday. They wanted at the same time to increase the purchasing power of low income employees, to boost tourism activities in Romania, to fight against the informal economy (which accounts for a high percentage of transactions in the tourism and catering sectors) and to address the high level of unemployment (especially for SMEs).

---

3   Meal vouchers are also used as communication tools to promote balanced nutrition. A public–private consortium was designed and received funding from the European Commission to develop the FOOD programme. See www.food-programme.eu.

In Switzerland, France, Hungary and Italy, holiday vouchers provide access to an extensive network of tourism professionals. The Romanian legal framework is one of the simplest: only employers can finance the vouchers and it is the same system for all the employees whatever the size of the company and the employee wage. The amount is exempt from all taxes, up to six times the minimum wage (RON 600) for each employee. The network is composed of 10,000 hotels, 2,500 travel companies, leisure and cultural centres (massages centres, museums, etc.), all approved by the tourism ministry.

Wellness, cultural and holiday programmes are also directly linked to employees' health and wellbeing. Tax exemption is granted by governments in order to encourage employers to give their employees access to adapted solutions through vouchers and to improve their everyday lives. This makes them relevant for any economic, social and cultural contexts. One of the biggest challenges is to involve all the stakeholders and assist them in identifying roles and responsibilities.

Conclusion

Because workers can spend 45 hours a week or more at their workplaces – a third of their waking lives, it is a logical place to establish a well-being programme and to render an environment conducive to healthy choices on the part of workers, as a healthy workforce is a prerequisite for sustainable development and social well-being.

Based on a 50-year history, the voucher system has already proven to be very effective: social subsidies are granted with better fund control, because vouchers guarantee the destination of the allocation, and create jobs. They turn informal economies into formal economies, generating tax revenue for the state. They improve employees' quality of life with greater purchasing power. Eventually, they are also easy to manage whatever the size of the company. Organizations gain in productivity and well-being and address the problems of absenteeism and presenteeism.

Today, service vouchers are used by 45 million people in 39 countries (19 of which are EU Member States), and are part of their users' daily lives, offering them administratively simple, cost-effective solutions that are highly motivational. Because they have a positive impact in the economic as well as the social world, they are an important part of modern life, especially in the current context of economic crisis.

References

Ali, S. and Benmehdi, H. 2009. Month of lassitude impacts Moroccan productivity. *Magharebia*, 25 September.

APETDS/IPSOS. 2010. *Perception et usage des titres restaurant.*

Armour, S. 2006. Lunch break becomes briefer as "hour" shrinks. *USA Today*, 6 December. Average commute 'now 139 hours'. 2006. *BBC*, 4 August.

Aviva. 2011. *Family Finances Report 2011*. London: Aviva.

CCH. 2007. *Unscheduled Absence Survey.*

Council of the European Union.2011.*Conclusions of 3114th Employment, Social Policy, Health and Consumer Affairs Council meeting "Managing demographic challenges through better reconciliation of work and family life."* Brussels: Official Journal of the European Union.

Deschenaux, J. 2008. Less time for lunch: The siesta in Spain is disappearing under the pressures of international business and big-city commuting. *HR Magazine*, June.

Druce, C. 2007. Skipping breakfast and lunch not good for economy. Retrieved on 21 August 2013 from www.catererandhotelkeeper.co.uk/articles/22/11/2007/317453/skipping-breakfast-and-lunch-not-good-for-economy.htm.

Ernst & Young. 2006. *Analyse d'impact du Titre-Restaurant en France*.

Esping-Andersen, G. 1999. *Social Foundations of Postindustrial Economies*. Oxford: Oxford University Press.

Food and Agriculture Organization. 1971. *Joint Expert Consultation of FAO/WHO on Workers' Feeding*.

Hein, C. and Cassirer, N. 2010. *Workplace Solutions for Childcare*. Geneva: International Labour Organization.ICOSI. 2004. *Meal Vouchers: A Tool Serving the Interests of the Social Pact in Europe*. Gennevilliers: ICOSI/CODESI.

Instituto Brasileiro de Opinião Pública e Estatística (IBOPE). 1996. *Enquiry on Public Opinion about the Worker Food Program and the Vouchers use*.

IPSOS Mori. 2007. *BaxterStorey Workplace Productivity Survey*.

Ivanova, L. 2002. *Nutrition of Bulgarian Working Population*.

GfK. 2009. *Purchasing behaviour of meal voucher users*. Czech Republic

KC2 and Industry Watch. 2010. *Food Voucher in Bulgaria: Economic and Fiscal Impact*.

Kilpatrick, K. 2010. *Fighting Hunger in Brazil, Much Achieved, More to Do*. Retrieved on 21 August 2013 from http://www.oxfam.org/sites/www.oxfam.org/files/cs-fighting-hunger-brazil-090611-en.pdf

Louven, M. 2006. Vale-transporte: mudança prejudicará trabalhador. *Globo*, 18 February.

Lunch Ticket System and Public Revenue: Review. *Istanbul Ekonomi*, May 2010, 1–36.

Mazzon, A. 2001. *Worker Food Program, 25 Years of Contribution for the Development of Brazil*. Sao Paulo: Economics Department, Administration and Accountability, University of Sao Paulo.

Mazzon, A. 2009. *Proposed Model for a Worker's Food Programme in Argentina*. Sao Paulo: Burson Martsteller SA.

Mickelson, R. A. 1999. *The Breakfast Habits of Middle School Students*. The Graduate College, University of Wisconsin-Stout.

Murillo, J., Romani, J. 2011. *Analisis del impacto economico de los vales comida en España*. Barcelona: Parc Cientific, Universitat de Barcelona.

Oram, R., Baker, S. and Judd, D. 2010. Tax-free transit benefits at 30: Evolution of a free parking offset. *Journal of Public Transportation*, 13(2), 1–22.

PWC. 2011. *The market Survey Report on the use of food vouchers in Greece*.

Rousseau, T., Arezki, S., Bérard, D., Douillet, P., Gagné, B. and Lemettre, C. 2009. *L'absentéisme-Outils et méthodes pour agir*. Retrieved on 21 August 2013 from http://www.psppaca.fr/IMG/pdf/Absenteisme_-_outils_et_methodes.pdf

Scheinder, F. 2009. *The Shadow Economy in Europe: Using Payment System to Combat the Shadow Economy*. Retrieved on 21 August 2013 from www.visaeurope.com/en/about_us/idoc.ashx?docid=4d53b726-cd71-4ba5-a50b-735d11ca4075&version=-1.

Tudorel, A., Iluzia, A. and Hertiliu, C. 2010. *La tendance de l'économie souterraine au niveau d'un pays pendant la période de transition* (Cmnd. inria-00494713). Bucharest: 42èmes Journées de Statistique.

UK commute 'longest in Europe'.2003. *BBC*. 22 July.

Wanjek, C. 2005. *Food at Work: Workplace Solutions for Malnutrition, Obesity and Chronic Diseases*. Geneva: International Labour Organization.

World Health Organization. 2004. *Global Strategy on Diet, Physical Activity and Health*. Retrieved on 6 September 2013 from http://www.who.int/dietphysicalactivity/strategy/eb11344/strategy_english_web.pdf

3 Civility, Engagement and Participation

9

Civility, Respect and Engagement in the Workplace (CREW): Creating Organizational Environments that Work for All

KATERINE OSATUKE, MAUREEN CASH, LINDA W. BELTON and SUE R. DYRENFORTH

Abstract

This chapter explains and describes an intervention that promotes civil climate within organizations. CREW (Civility, Respect, and Engagement in the Workplace), an intervention designed within the US Veterans Health Administration (VHA), was used to improve organizational-level outcomes through bettering the culture of interactions between employees. *Civility* in CREW refers to workplace behaviours that express interpersonally valuing and being valued by others, and are based on a consciously cultivated awareness of one's interpersonal impact. The documented success of CREW in achieving the intended outcomes (Osatuke et al. 2009) resulted in its quick spread. The intervention model has been freely shared with interested organizations within and outside the United States, in government and private sector, in healthcare and other industries. This chapter explains the CREW approach, processes involved in implementing it within the largest healthcare system in the United States, summarizes its outcomes and sustained impact. To aid organizational leaders and consultants in evaluating their interest, we include illustrations of CREW concepts, tools and results. We also discuss organizational contexts, barriers encountered and working solutions. This chapter is based on summarizing the authors' experience of implementing CREW at more than 1,000 workplaces within the US Department of Veterans Affairs (VA), from 2005 until now.

Introduction

ORGANIZATIONAL BACKGROUND: HOW CREW WAS DESIGNED AND LAUNCHED IN THE VA

The Veterans Health Administration (VHA) is the largest branch of the US Department of Veterans Affairs (VA) and the biggest healthcare system in the country. It includes 152 primary, secondary and tertiary medical centres serving over eight million veterans from the Second World War through the Iraq and Afghanistan wars. The medical centre structure is extended by VA community living centres, domiciliaries (residential and treatment facilities), and more than 800 community based outpatient centres (CBOCs) which all provide basic care close to the veteran's home. VHA fulfils its mission of "caring for those who have borne the battle" (Lincoln 1865) through a comprehensive system of healthcare and social support, extensive research in veteran-related health issues, and leadership in developing and providing services in several specialized areas particularly relevant for veterans' healthcare (such as prosthetics, post-traumatic stress disorder, health needs of the homeless, telehealth, and emergency preparedness). Clinical, administrative and policy functions of VHA are implemented through a variety of national, regional and local programmes that collectively employ well over 300,000 employees. These employees, spread over all the 50 states in a variety of professional settings, represent a great diversity of occupations and demographics that is currently a part of the VHA workforce.

In the past 15 years, VHA has undergone a significant organizational transformation, by embracing cutting-edge strategies in patient-centred care, customer service, systems redesign, electronic medical records, other information technology, health informatics and healthcare quality. VHA healthcare services result in clinical outcomes and patient satisfaction ratings that surpass those of other hospital systems; indeed, VHA now provides "the best care anywhere" (Longman 2007, Congressional Budget Office 2009). As VHA was reaching these heights, its senior leaders also became aware of impending challenges in succession planning. Fiscal Year 2004 data suggested that a large proportion of experienced leaders would be retiring over the following 10 years. The National Leadership Board (NLB), then the governing body of VHA, began considering how to manage these losses and charged the Human Resources Committee (HRC), an entity responsible for planning employee resources and their technical aspects, with formulating strategies for training and developing VHA leaders. As the HRC examined this task, it quickly broadened to include a closer look at work environments and leaders' role in supporting high performance. Considering what leaders should do to support optimal workplace functioning suggested an important concept: *organizational health*. It was defined as a state of systemic well-being that nurtures success in complex and chaotic organizations like VHA. Organizational health is a two-way street: both the organization and the employee commit to each other. The HRC was interested in exploring organizational health in VHA system-wide, from both managers' and employees' perspectives.

One tool used to examine employees' perspective was the VHA All Employee Survey (AES), a voluntary, confidential census survey of workplace perceptions and job satisfaction. The AES asks employees to rate civility, conflict resolution, diversity acceptance, co-worker and supervisor support, respect and cooperation, and other domains relevant to organizational health. The AES began in 2001, was significantly restructured in 2004–2005, and since 2006 has been administered annually by VHA's National Center for Organization Development

(NCOD), with response rates typically well over 65 per cent. This participation, unusually high for voluntary organizational climate assessments, provides employee feedback at fine-grained organizational levels: AES results are available not only for medical centres but also for small clinics, services and workgroups. In 2004, these data allowed HRC to estimate baseline civility perceptions in VHA. The HRC also conducted over 400 in-person interviews with VHA employees around the country and examined longitudinal AES data in parallel with interview results from 2004, looking for areas in greatest need of attention to support organizational health. Both data sources were highly consistent. VHA employees described strong dedication to the mission of caring for veterans and also shared their deep desire for a civil, respectful workplace climate, which they did not consistently experience. Civility thus emerged as a key factor in VHA employee satisfaction and a potential focus for actions to improve VHA's work environment.

Having arrived at this conclusion, the HRC then needed to persuade the NLB that an intervention promoting civility would be of value to the organization. While many leaders saw civility as simply the right thing to do, others required evidence of return on investment as a condition for lending their support to an organizational initiative focused on civility. A business case (strong fiscal argument) for strategically focusing on civility within VHA appeared imperative. The HRC, with support from NCOD, summarized the information available regarding strong, consistent relations of employee satisfaction to a number of monetized organizational outcomes, such as employee time off work (sick leave usage), frequency of filed grievances and complaints, patient satisfaction ratings for quality of clinical care, productivity and organizational performance metrics for VHA facilities. These data illustrated relevance of civility to VHA workforce recruitment, employee engagement, retention, succession planning, ability of VHA organizations to carry out their mission of providing accessible, high quality patient care, and even to employee health outcomes. An HRC presentation made to the NLB in March 2005 (Belton 2005) summarized these findings. HRC asked VHA leaders to endorse the concept of CREW as a corporately-supported principle and allow designing a programme to promote it throughout the organization. NLB agreed CREW should become a system-wide initiative and recommended that, once created, it be piloted to a small number of VHA facilities prior to a system-wide implementation.

Developing the programme took approximately a year. The pilot in September 2005 included eight sites. Their CEOs all had a strong interest in CREW, lending support for employees' participation. Diverse characteristics of the sites (location, organizational structure, variety of outpatient and inpatient services and administrative and clinical units) allowed exploring viability and applicability of the initiative. The pilot was successful: the programme demonstrated the intended outcomes of improving workplace civility (Osatuke et al. 2009). Lessons from the pilot also shaped how CREW was implemented.

For example, early in the process, CREW developers faced the question of whether CREW should be mandatory (i.e. if civility positively impacts business outcomes as described, should not all employees be required to participate?). The initial plan was to initiate CREW at the top of the organizational hierarchy, then implement it across the country. A national rollout was to immediately follow the pilot: CREW was expected to be mandated at each facility. The pilot experience suggested, however, that the change process facilitated by CREW is most successful when people engage in it voluntarily within their own workgroups, rather than as entire facilities at a prescribed time. As in many organizations, the task of mandatory training in VHA is onerous and employees often approach it perfunctorily, with minimal engagement and little behavioural change. CREW participation needed to be seen as a benefit or a perk rather than a mandate to the participants. The CREW slogan became "Changing VHA culture one workgroup at a time," and promotion through attraction was deemed a key strategy in spreading the programme throughout VA. Voluntary participation became one of the few things about CREW that is mandatory. Sites choose whether they wish to implement CREW, and individual employees attend CREW meetings in their workgroup by choice only. However, this does not mean that staff can opt out of behaving in a civil manner. As one leader put it, attending a CREW meeting is optional; being civil and respectful in the workplace is not. Another lesson from the pilot was about the way that civil interactions within a work unit seem to spread beyond the single group. One of the pilot groups described it as "viral spread." When a nursing unit in that medical centre put CREW principles into action, other units (particularly the lab) noticed that their environment had become more pleasant, their staff more congenial, their interdepartmental encounters more relaxed and less contentious. The lab employees began asking "What are you doing differently on your unit?." When they heard about the CREW process, they asked the facility leaders how they could bring it to their own area. It became clear that word of mouth and CREW behaviours modelled by employee participants were the programme's best marketing tools. One important lesson from the pilot thus was that CREW should not be only a top-down initiative, but also bottom-up and side to side.

Foundations of the CREW Intervention Approach

CREW intervention aims at changing organizational culture. Its specific goal is to improve the work climate and organizational health, within one workgroup at a time, through supporting civil and respectful interactions on the job. A classic definition of civility in the workplace refers to interpersonal behaviours that demonstrate respect and "love of thy neighbour" (Anderson and Pearson 1999). Being polite, courteous and respectful towards each other are concepts that most of us likely learned in grade school. However, remembering to do these things can become more difficult when facing the many demands on our time and attention within the workplace environment. As the review of current literature in the process of developing the programme showed, most of the existing research and organizational interventions centre on incivility as it relates to organizational citizenship behaviours, disruptive behaviour, stress, aggression and interpersonal justice. In the CREW approach, a deliberate decision was made to focus on civility – a decision influenced by the Appreciative Inquiry framework (Cooperrider et al. 2003) and, more generally, by insights from the field of positive psychology (Seligman and Csikszentmihalyi 2000). These models emphasize positive outcomes achieved by looking for what is already working, building upon already existing strengths and assets, and defining change in terms of what is wanted (rather than what is *not* wanted). CREW interventions are based on first identifying, then recreating, successful, satisfying work experiences, building on teams' strengths, and finding opportunities for improvement. This served as an overarching philosophy for CREW workplace interventions. Concepts from a variety of models and theories were then incorporated in CREW as potentially useful tools and heuristics for elaborating more specific aspects of workplace interactions, relevant for the programme's focus on civility. These concepts served as additional language to help articulate and explain CREW's intervention approach and some of the interactional dynamics that it addresses.

The following examples illustrate how concepts from other models were incorporated into CREW. From Berne's (1964) transactional analysis came the idea that respectful conversations become more likely when the engaged parties approach each other as adults. Being mindful about interpersonal communications in this sense minimizes the impact of inherent power differentials, particularly in those workplace relationships that have a potential of shifting to a parent/child model (conversations between employees and supervisors, between individuals at different levels of experience, education, organizational tenure). As another example, Maslow's (1973) hierarchical model of needs suggests that meeting basic human needs of safety and social connection is prerequisite to achieving higher developmental goals. This notion helped underscore the importance of intentionally supporting these needs when designing CREW intervention process – e.g. by creating deliberate spaces and times for discussing and attending to these needs. As yet another example, fundamental attribution error (Ross 1977) refers to people's tendency to ascribe personality-based explanations to strangers' behaviour but situational explanations for their own behaviour or the behaviour of those they know well. For example, if a co-worker I only know by name speaks abruptly to me, I am more likely to judge him or her as rude, but if a co-worker with whom I have a personal relationship does the same, I am more likely to think he or she is simply having a bad day. In CREW, activities that allow workgroup members to get to know each other as people, not just as co-workers, reduce anonymity and humanize participants to each other, leading to fewer fundamental attribution errors and a greater cohesiveness.

Civility within the CREW initiative is defined as an essential behavioural expectation of all employees in an organization. Civil behaviours express interpersonal norms, or rules of engagement, for how co-workers relate to each other, to their customers and stakeholders. CREW interventions increase workplace civility by means of helping organizational members reconnect to the fundamental values of courtesy, politeness and consideration in their workplace. Since within CREW, civil behaviours are understood as situated in the context of a specific workplace, defining what civility means locally is an important task at the start of any CREW intervention. With the support of CREW facilitators, group meetings give members a chance to step away from their everyday routine, reflect upon the interpersonal process typical of their group, recommit to the basic values of respect and mutual consideration, and articulate specific ways to express these values in their daily job interactions. The co-workers are then able to take steps towards improving their workplace climate in very practical ways that express their refreshed understandings of these shared values.

Consistently civil *behaviours* at work are instrumental for creating healthy *relationships*. Unlike civility that refers to *organizationally* supported behaviours (behaviours expected as part of the job), respect, the next fundamental CREW concept, connects individuals at a *personal* level. Respect is an appreciation for the inherent worth of another person. It is born over time from active listening and understanding, cultural and personal sensitivity, and compassion. Facilitated CREW meetings provide opportunities to grow mutual respect, by creating a place for honest conversations that allow addressing all kinds of workplace topics, including conflict-laden or difficult ones, in an interpersonally safe and civil manner. In the CREW model, respectful relationships between co-workers are understood to create an atmosphere of trust and psychological safety at the workplace which, in turn, enables employee engagement.

Engagement, the third key concept of the CREW approach, is the degree of personal investment and passion for the mission of the organization, as expressed through employees' work performance. When employees are engaged, they feel empowered to take initiative and responsibility for addressing the shared organizational tasks. Specifically in VHA, engagement means that all staff members have the authority, training and organizational support needed to make decisions "on the spot" in the best interest of the veteran.

To summarize, at its most fundamental, the CREW intervention is about carving out time for guided conversations about the interpersonal climate at the work unit. For this to work, the facility's top managers and union leaders, workgroup managers, as well as workgroup members themselves, must be willing to participate in and remain committed to sustaining this process. One of the foundations of CREW is that civil, respectful and engaged team climate is rooted in honest conversations. For example, if I believe you have made a racially insensitive remark, I can respond with aggressive or passive-aggressive remarks – which perpetuate the problem and invite co-workers to choose sides, creating an incivility spiral (Anderson and Pearson 1999). I can also file a formal grievance – which can take months to wind its way through the system and result in substantial operational costs (Osatuke et al. 2010). Alternatively, we can open a dialogue about how your remark affected me, giving worth to each other's perspectives (respect) and thus partnering to resolve any unintended impact and preventing a toxic environment for all group members. Over time, the CREW group process creates the confidence and trust necessary to take a risk and choose the latter option. Honest, and at times difficult, conversations can resolve, mitigate and often pre-empt problems seated in interpersonal issues. It is intuitive that people who enjoy coming to work and feel valued and affirmed at the workplace are less likely to be absent, stressed, have grievances or complaints, or be retired on the job (cf. Crawford et al. 2010, Leiter and Bakker 2010). Choosing civil behaviour thus has ripple effects on operational costs, productivity, morale and employees' well-being on the job (Carameli et al. 2012, Leiter et al. 2011, Leiter 2012).

CREW Intervention Processes

CREW does not work overnight. Changes in workplace climate take time because they involve deeply held values and ingrained habits (Schein 2006). CREW intervenes into the domain of organizational life that, by and large, is based on participants' psychological processes. The mechanism of change that explains positive outcomes is that, for the intervention period, organizations commit to giving time, attention and support to having regular workgroup-level conversations about civility. This time and commitment, rather than any particular procedure or component of the intervention, constitutes the active ingredient that brings positive change in CREW. It follows that one key to its success is the continuous support and resources available to participating sites throughout the intervention.

The CREW initiative is structured similarly across all participating groups in that each site identifies a CREW coordinator to organize their local programme, workgroups to participate, and facilitators to run workgroup meetings. Trained facilitators meet regularly with groups participating in CREW for approximately six months to afford

Table 9.1   Civility scale

1.	In my workgroup, co-workers are treated with respect.
2.	Cooperation and teamwork exist in the workgroup.
3.	In my workgroup conflicts are resolved fairly.
4.	Co-workers take personal interest in each other.
5.	Co-workers can be relied upon when help is needed.
6.	Discrimination is not tolerated.
7.	Differences among individuals are respected and valued.
8.	Supervisors work well with employees of different backgrounds.

Notes: civility score is computed as an average of these eight item ratings by individual employees, on a scale from 1 (strongly disagree) to 5 (strongly agree). The same measure of civility is used within CREW and within annual VHA All Employee Survey (AES). This scale was first presented in Meterko et al. (2007) and later published within Osatuke et al. (2009). Its psychometric properties are extensively discussed in Meterko, Osatuke, Mohr, Warren and Benzer, Civility in the healthcare workplace: Psychometric properties of a workgroup climate measure (manuscript submitted for publication).

the group the needed time for joint reflection and actions towards creating a respectful, civil workplace. The facilitators come to the meetings prepared to facilitate discussions, encourage problem-solving, and conduct exercises and activities that help examine and improve participants' ways of relating to one another. These regular meetings provide both a forum for discussions and a place to practice new (civil) behaviours and ways of interacting. As part of the process at every participating site, workgroup members complete an eight-item civility survey, both before and after their participation in CREW, using a standard, psychometrically established measure of civility (see Table 9.1, Meterko et al. 2007, Osatuke et al. 2009). NCOD supports administering this assessment and processes participants' data, ensuring confidentiality. The pre- and post-survey summaries are provided as feedback and serve as discussion pieces for the group conversations about what is reflected in these scores and in their change from start to finish. These ratings also become part of the aggregated results collected by VHA's National Center for Organization Development (NCOD) that oversees the initiative nationally.

Beyond these standard elements of the programme, the content of CREW meetings differs substantially across sites. There is no manual for CREW, and no two CREW sites look exactly alike. The programme allows for a tailored fit to the needs and culture of the local workgroups, which is a key feature supporting its success. CREW uses a principle-driven rather than protocol-driven style that philosophically stems from the client-centred roots of organization development (Osatuke et al. 2009, 2012) and can be a drastic mismatch to some participants' anticipation of a checklist or recipe format ("tell us what to do, in what order, to create a civil environment"). While the content of CREW (its focus on civility) is standard, implementation of the initiative is not. Participants often adapt the provided materials, create new process-tracking tools, construct a local business case, try new techniques, and so forth. Shared understandings and experiences of civil behaviours that the group members accumulate through their participation in the programme all become ground for establishing their own unique interpretation of CREW. Through this process, they come to define their own civility-related values as the cultural norm within their group (see Osatuke et al. 2012, for a more detailed discussion).

Anyone can contact NCOD to obtain information on the CREW initiative (see example below). Marketing materials for the programme include an informational PowerPoint presentation, brochures, an FAQ sheet and a promotional DVD. Once a site has determined its interest in the programme, it must appoint a local CREW coordinator while NCOD assigns a companion for the site. A "readiness call" is then scheduled. It includes the director and the president(s) of the union(s) at the site, the newly appointed CREW coordinator, an NCOD CREW senior staff member, and the NCOD companion. It is critical that a site interested in CREW has the support of its management and union leadership, so the readiness call will not be conducted unless these leaders attend. The readiness call presents an opportunity to go over the site agreement, answer any questions, and begin establishing the working relationship between the site and NCOD. Once all questions have been answered and there is consensus that the site is ready, all parties on the call sign the site agreement and send it to NCOD. The site agreement lists the support that NCOD will provide to the site and guidelines for success based on previous experiences with the CREW initiative.

Example: How Medical Center X Decided to Join CREW

The nurse manager of the medical/surgical unit at the VHA Medical Center X read about CREW in the *Environment of Care* newsletter published by The Joint Commission. She contacted NCOD for more information and was sent the PowerPoint presentation and the FAQ sheet, and encouraged to call if she had any further questions or wanted to talk about possible participation. The nurse manager called, spoke to the NCOD logistic and operations coordinator for CREW, and stated she thought CREW would be great for her hospital. The NCOD coordinator recommended that she talk with her leaders to share the information about CREW. A few days later, the executive assistant to the director of Medical Center X called NCOD, asking to discuss their site enrolling in CREW. NCOD requested they make their union leaders aware of their interest, and work with them to appoint a CREW coordinator, to organize the initiative locally. The Medical Center X director discussed CREW with the union president and they named the Equal Employment Opportunities (EEO) manager as CREW coordinator. While the site was making these arrangements, NCOD assigned Medical Center X to a staff member within NCOD who would serve as their CREW companion. A phone call was then scheduled with the hospital director, the union president, the local CREW coordinator, the NCOD CREW logistics and operations coordinator, and the NCOD companion. All parties attended the call, went over the site agreement, and NCOD staff answered questions about various aspects of the initiative, highlighting the need to keep employee participation voluntary and discussing workgroup selection guidelines. The participants at the call also discussed how to implement CREW in a way that would be consistent with the local culture at the site, and maximize the chances for success. At the end of the call, the site agreement was signed, then scanned and sent to NCOD.

Once a site has signed for CREW, the rollout begins. Typically, it starts with a CREW coordinator and two or three trained facilitators working closely with NCOD, to plan how best to promote the initiative at their site, choose appropriate workgroups, and begin marketing the CREW concepts. Marketing strategies include town hall meetings held by facility leaders and "email blasts" to all employees at a site showcasing the benefits of CREW and sharing the opportunities for participation. Promotional events are common that introduce and explain CREW to the entire site, often in the format of an open fair with promotional materials, snacks and civility-related games and activities. NCOD recommends that new sites start with just one or two carefully selected workgroups and complete their process before engaging more groups. This allows the involved staff time to gain the experience, evaluate their initial forays and modify any processes if they choose. CREW interventions typically last for six months. Throughout this time, local staff members implementing CREW maintain regular, frequent contact with their NCOD companion for intensive support and guidance.

An important part of success is the workgroup selection – ideally, a collaborative process including the CREW coordinator, executive leaders and union leaders at the site. NCOD encourages the sites to initially choose relatively well functioning groups, with mid-level civility scores on the AES and no active investigations, grave disciplinary concerns, or filed complaints or union grievances. This guideline is particularly useful for inaugural CREW groups. An intervention that targets cultural change needs to earn a good reputation and gain momentum at a site, which happens if workgroups feel fortunate for a chance to participate and do not view the programme as an attempt to punish or *fix* troubled work units. Motivated workgroups begin CREW and typically reach some demonstrable success, whether statistically (comparisons of pre- to post-ratings of civility) or anecdotally (stories about the programme exchanged between employees at the facility). After a few workgroups have gone through the CREW process successfully, it is common for other units' supervisors to begin asking that their areas be included as well. As CREW participation expands, its impact on the site is better understood (and often used for marketing).

At sites that already have workgroups enrolled in CREW, CREW marketing includes advertising testimonials of participants (see Table 9.2) and having employees from participating groups and the CREW coordinator join staff meetings of units considering CREW, to share their experience. As an example of CREW's impact on the organizational culture, at one site active in CREW for 18 months, the number of EEO cases went down

Table 9.2   CREW testimonials from employee participants (also displayed on CREW webpage, NCOD Internet site)

The link between employee satisfaction, patient satisfaction and clinical outcomes is clear. CREW has provided the necessary tools for us to achieve unprecedented levels of civility, respect and engagement in the workplace, resulting in patient satisfaction. Rica Lewis-Payton, Director, Birmingham VA Medical Center, Birmingham, AL
Through CREW, employees were able to identify and change unappreciated behaviors in the workplace. Employees work together where they resisted before, and are cordial where they avoided one another before. Sincere commitment to CREW will make a difference where you work. Cliff Wrencher, CREW Facilitator and President of the American Federation of Government Employees, Central Alabama Veterans Health Care System

Table 9.2   Continued

My CREW experience as a facilitator has been fantastic. Employees engaged in open and honest dialogue, renewing lines of communication and gaining awareness of behaviors. They were able to see others' points of view and resolve some of their own issues.
Bettie Bookhart, CREW Facilitator, VBA Regional Office, St. Petersburg, FL

The routine reinforcement and progression used in CREW ensures sustained improvements. Every medical center should consider having CREW in place for both teams that need help and for already high-functioning teams.
Tom Mattice, Director, Richard L. Roudebush VA Medical Center, Indianapolis, IN

CREW gave us a better understanding of ourselves and our coworkers. CREW was able to lay the groundwork for building that relationship that enabled us to provide the highest level of patient care. I would recommend CREW in any workplace.
Michael Clayton, RN in Case Management, Birmingham VA Medical Center, Birmingham, AL

Everyone wants to do a good job! CREW helped everyone realize that all services have obstacles to face. Acknowledging these obstacles and working WITH each other to face challenges helps everyone feel better and ultimately helps the patient.
Irene Watson, RNP, Hospitalist, Providence VA Medical Center, Providence, RI

I was admittedly skeptical going in but we saw outstanding results. It taught us that each of us communicates differently and that many of our disagreements were actually due to lack of communication rather than actual issues.
Kyle Inhofe, Chief of Human Resources, Oklahoma VA Medical Center, Oklahoma City, OK

It was a good way to get to know your fellow employees, build team work and communication.
Jeanette Rowland, Prosthetics Representative, Dayton VA Medical Center, Dayton, OH

CREW has assisted the QM staff to learn about each other. By taking time to really know each other, conflict is decreasing and understanding is increasing.
Margaret A. Russell, Chief of Quality Management, Aleda E. Lutz VA Medical Center, Saginaw, MI

significantly facility-wide since CREW's implementation. Figure 9.1 shows more data from this facility.

NCOD provides intensive consultation and guidance to participating sites. A monthly national conference call is held for CREW coordinators and facilitators. It provides them with opportunities to network, share best practices and creative ideas, consult and strategize about ways to address local challenges. A SharePoint dedicated to CREW houses an extensive toolkit of resources (e.g. materials and activities that can be used during group meetings to facilitate discussions about civility). These are updated monthly and available to all local coordinators and facilitators through a link on the VA intranet website. The most important support, however, comes from an NCOD staff member assigned to each site, informally called a "companion" (the term emphasizes the relational and collaborative spirit of the initiative). Companions have frequent phone and email contact

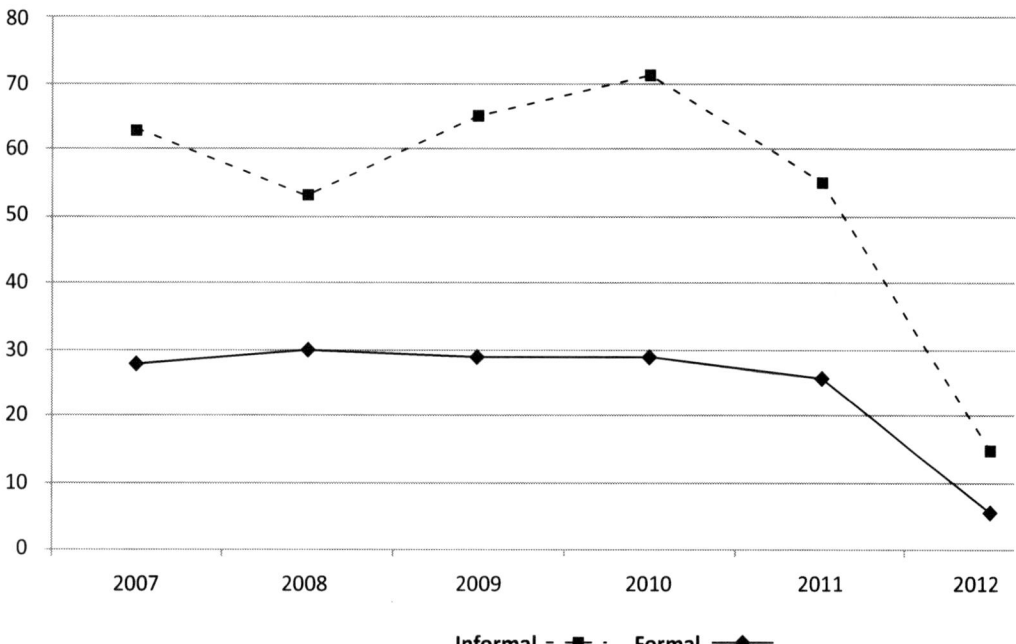

Figure 9.1   Decreased rates of informal and formal complaints to the VA Office of Resolution Management regarding Equal Employment Opportunities at a VA medical centre since it started participating in CREW

with CREW coordinators and facilitators, offering guidance and encouragement, sharing best practices, helping with marketing and promotion of CREW and with facilitators' and workgroups' selection, and assisting with obtaining resources in SharePoint and other formats that fit the site's needs. Companions also provide the surveys and data charts that facilitators then use to present pre-/post-intervention results to their workgroups. Each companion also works closely with local organizational leaders (e.g. medical centre directors) to support the intervention process at particular sites. Overall, the companions offer each site a link to NCOD and the CREW community nationwide.

The supervisor's support is important for the success of CREW. Once a list of potentially participating workgroups is generated, the site's coordinator approaches each group's supervisor to discuss the possibility of offering CREW to their employees. If the supervisor is agreeable and the workgroup elects to participate, the NCOD companion creates the pre-intervention survey link that the facilitators then send to the workgroup members, inviting them to rate workplace civility. When the survey closes, the NCOD companion creates a summary chart that serves as a discussion starter in an initial meeting. Workgroups typically meet two to four times per month, usually no more than an hour each time. The process tends to last for approximately six months.

Along with the appropriate workgroups, careful selection of CREW facilitators is crucial to success. NCOD provides training to CREW coordinators and facilitators several times per year. Beginning level materials for those completely new to the initiative and to group facilitation cover the basics of the CREW philosophy, the business case for civility, CREW's connection to other organizational health initiatives, and roles and responsibilities of all parties (management, union leaders, NCOD staff, coordinator, facilitators). Attendees learn basic facilitation skills needed to lead, guide and coach conversations about civility within workgroups. Attendees also practice these skills during the typically two-day training. More advanced materials for those who have completed the basic training cover more sophisticated facilitation techniques and include presentations on particular topics of interest for CREW and interactive discussions of common dilemmas and challenges.

CREW meetings tend to be structured around activities, interactional games or other tools that promote greater interpersonal understanding among group members. Facilitators typically open the CREW process by sharing with group participants the basic philosophy of the programme and its approach to interventions. The facilitators and group members then jointly customize the plan for CREW, fitting the needs of the specific workgroup, as well as reflecting the culture of their site and, more broadly, their community.

NCOD recommends that CREW meetings be co-facilitated. Having two facilitators in the room allows one to write discussion topics, comments and ideas on a flipchart while the other engages the group in discussion. As one facilitator leads the discussion, the other monitors participants' reactions and engagement in the process. If any participants leave the room unexpectedly, one facilitator can go after them to ensure their welfare while the other stays with the group.

As the end of the six-month intervention approaches, the NCOD companion creates a post-survey link for the facilitators to distribute electronically to the workgroup. As with the pre-intervention survey, the entire group is invited to rate their workplace climate. The NCOD companion creates a chart comparing the pre- and post-ratings; the facilitators are encouraged to share it with the workgroup and discuss what has changed, how, and what it means. NCOD also recommends that CREW coordinators share outcome data with their union and executive leaders, as feedback about CREW progress at their site.

Celebrating the accomplishments of each workgroup as it is winding down its CREW process is an important part of the programme. Like the inauguration of CREW at a site, it can take many forms, including a party with food and decorations, the site director or designee visiting the final meeting to thank workgroup members for their efforts, or some kind of trophy or recognition presented to the participating workgroup to display. For example, one site had a wooden oar (as a metaphor for the acronym *CREW*) made by their carpentry shop, on which they wrote the names of each workgroup that went through CREW, and displayed it in a prominent public place at the medical centre.

The national CREW initiative is structured at two levels. Level 1 sites are newly entering CREW; they need more intensive support. Level 2 sites have been in CREW for some time, have developed successful processes (e.g. an organized and effective workgroup selection, facilitator recruitment and marketing; a working plan for CREW maintenance and expansion) and are ready to be more autonomous. Such sites can choose to move to a more advanced level, using collaborative decision-making based on both the site's and NCOD's assessment of current progress. A site agreement is signed at each level of CREW.

CREW Intervention Outcomes

One intermediate outcome of the CREW process is a consensus reached by the workgroup members regarding the meaning of respect and ways to express it. For example, one

Human Resource CREW group decided to improve their communications by using more direct eye contact. All but one member worked at this skill. When a sufficient level of trust had been reached, that member spoke up at a CREW meeting saying that in his Native American culture, direct eye contact was a sign of disrespect. This example illustrates the role of shared assumptions about what behaviours signal respect and the importance of checking the implicit assumptions with co-workers in a multicultural workplace. CREW creates regular opportunities for such conversations. Once and only once there is a shared understanding of what civility is, what it means, and how it looks to the group members, are they in a good position to apply these understandings to action and take what they learned through CREW to the day-to-day work life.

One set of outcomes from the CREW process thus reflects better *interpersonal understanding* (not only cognitive, but also emotional: more empathy between co-workers). These gains are manifested through revised and improved assumptions about interpersonal meanings and processes that are part of employee interactions on the job. As a result of CREW meetings, understanding of co-workers' interpersonal behaviours becomes both more benign and more accurate, which invites more positive behaviour in response, creating a "civility spiral" (Osatuke et al. 2009, Leiter 2012). Pragmatically, interpersonal understanding improves relationships on the job, reduces stress, helps morale, and overall supports a more positive workplace climate – which contributes to collaboration around job-relevant tasks.

One CREW participant described his experience in the group as follows:

> Many of the CREW activities enabled us to understand different personality types and how their minds work; problem-solving types, and conflict styles. Once you understand all of this, it's easier to make an adjustment in how you react and interact with others. CREW is about getting groups together to work on problem solving and understanding that we don't all look at things the same way. No one way is better, more correct, or more important than another.

Sometimes employees are sceptical about CREW, but once they experience it, their view changes. The following two examples are unsolicited statements from CREW participants:

> My first thought about CREW was negative. [That it was just] another group established to identify problems and [do] nothing about them. I was pleasantly surprised that it was nothing I imagined. It was more about me and learning more about myself. Some of the benefits were learning about various personality styles and conflict styles, which taught me a great deal about myself and how others in my group were as well. Learning this information allowed me to understand why a certain person handled things differently than I do. As I learned this information, I began to pay more attention, and those experiences paid off.

> When I just joined CREW, I was not sure what to expect. As the weeks went on, the group became more talkative and shared more. I feel that communication did improve in the department as far as discussing issues and trying to come to some kind of agreement on how to handle them. In CREW, I learned about many different personality styles and how they have an impact on the work environment. Learning how to deal with different personality styles has a great impact on how others affect us. Many issues can't always be resolved, but it was very therapeutic to get together and discuss different issues; sometimes just starting the discussion can lead to a

resolution. I am happy to have been a part of CREW. It may take a while to resolve or even change the things we don't like; but all journeys start with baby steps.

Another set of CREW outcomes builds upon improved understandings and consists of *work process improvements*. We believe this second set of outcomes stems from a close connection between civility and *psychological safety*: employee perceptions of interpersonal risks and risk-taking consequences in their working environments (Edmondson 2004). Figure 9.2 shows data from VA's All Employee Survey illustrating this connection. In prior research, psychological safety has been connected to organizational effectiveness through its dramatic impact on people's ability to non-defensively discuss and correct work-related errors (Edmondson 1996). CREW seeks to establish a practice of routinely assessing interpersonal aspects of workplace interactions (as part of creating and maintaining the first set of outcomes). Once this practice becomes a norm within participating workgroups, obstacles to discussing potentially interpersonally risky work-related issues are removed. In other words, well-established interpersonal understandings based on shared values of civility and respect enable co-workers to directly address work-related challenges that otherwise might be too risky to face. When employees trust that in difficult conversations, their co-workers will remain civil and respectful (not hostile or blaming), work-related problems can be handled non-defensively, seeking mutually acceptable working solutions. That is, the large and positive impact of CREW on organizational outcomes likely operates through its

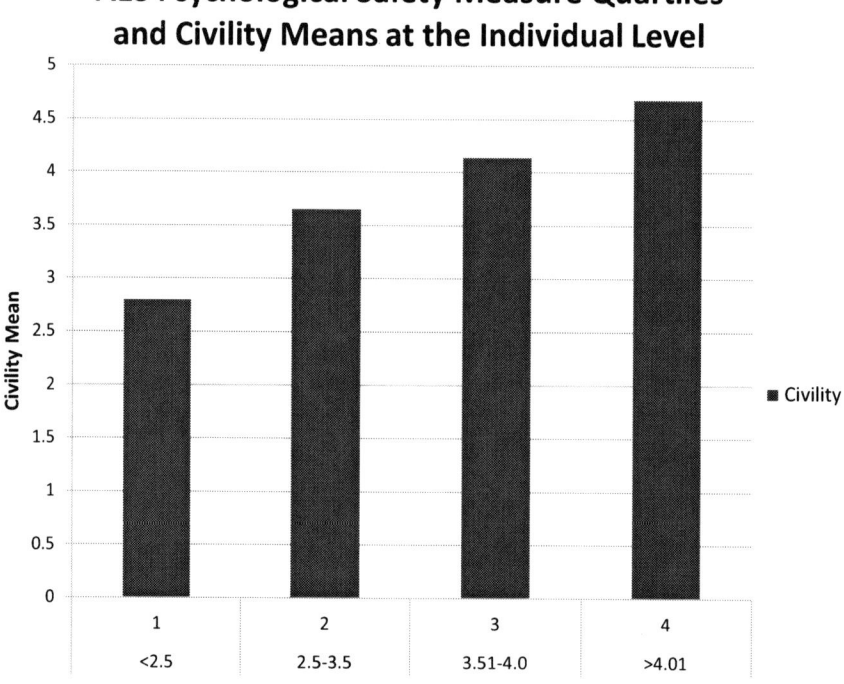

Figure 9.2 **Illustration of connection between psychological safety and civility**

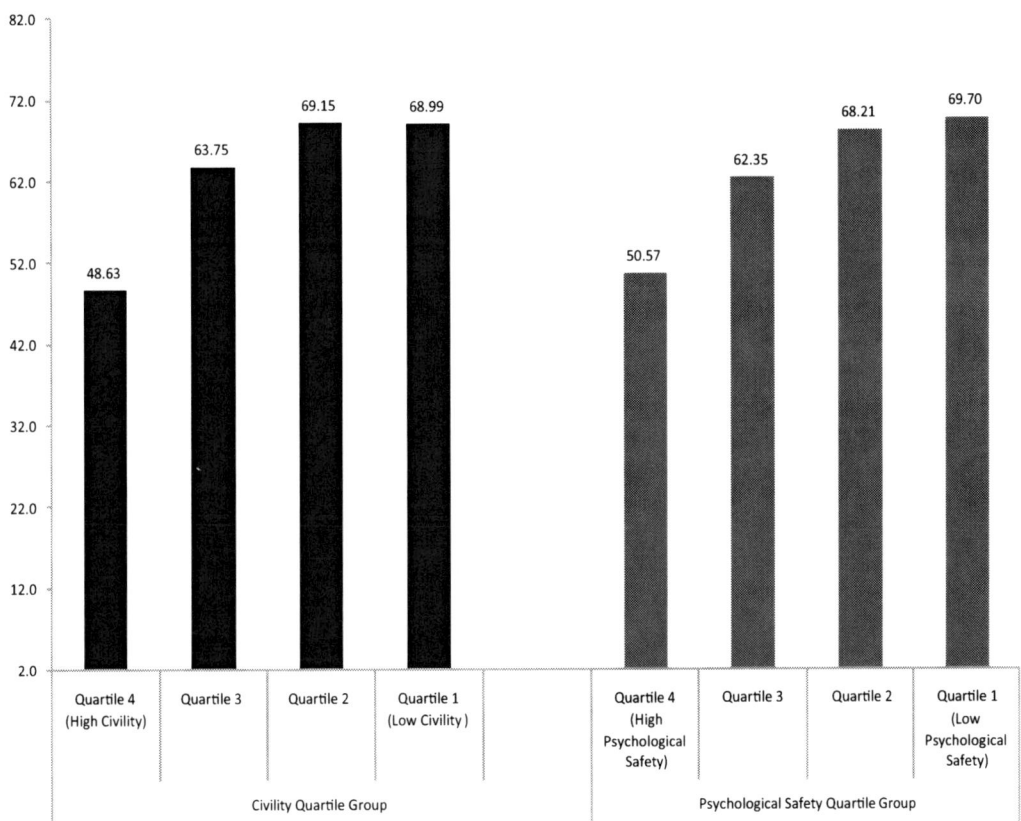

Figure 9.3   Relationship of civility and psychological safety to an important organizational outcome: Costs incurred from sick leave usage

beneficial impact on the psychological safety of the working environment (cf. Osatuke et al. 2012). Figure 9.3 provides an example of this impact.

The ability of CREW to enhance multiple employee and organizational outcomes has been statistically examined and documented elsewhere (e.g. Osatuke et al. 2009, Leiter et al. 2011, Leiter 2012). While this quantitative evidence is critical in documenting the impact of CREW, sharing participants' experiences of the CREW process and changes they notice in their workplace helps put the outcome data into human terms, making it easy to access from the participants' perspective. CREW outcomes typically noted in open-format qualitative comments to post-intervention CREW surveys include improved teamwork and trust, willingness to attribute positive motives to co-workers, acceptance of cultural differences, less need to resort to formal complaint processes, real-time problem resolution, increased self-confidence and self-respect, and a sense of personal connection to the mission of VHA. Another powerful evidence of CREW's impact is participants' stories, typically shared during monthly CREW calls when the intervention at a site "takes off." For example, at one medical centre, a timid Nutrition and Food Service worker expressed her enthusiasm for CREW by stating that in her 30 years on the job, it was the first time she felt empowered to express her thoughts in a group. At another site where a multidisciplinary patient care unit participated in CREW, a housekeeper was able to gently prompt a surgeon to wash her hands between patients. The group members saw it as expressing their new interpersonal norms for honest communication, established through their CREW meeting discussions and reflecting the heightened trust and psychological safety between group members regardless of their organizational status.

Since the time CREW first began as a pilot initiative with eight sites involved, CREW has been implemented in an estimated 1,000 workgroups throughout VA with nearly 600 workgroups having completed a full CREW process. At the time this chapter was written, nearly 10,000 employees have been touched by CREW. Participating workgroups have included clinical (dental, mental health, medical and surgical, emergency room, lab and pathology, primary care, chaplain services, etc.) and administrative functions (fiscal, HR, medical records, call centre, police, dietary, housekeeping, information and technology, engineering, etc.). Participating groups have included frontline employees working 24-hour shifts as well as more traditional day shifts, executive teams, union and management groups, inpatient groups, outpatient groups, interdisciplinary groups, and more.

Discussion

A programme like CREW is not without its challenges. Perhaps the most daunting one is logistical (implementing the initiative across a huge and diverse organization). Certain obstacles are encountered on a fairly regular basis. Time constraints are often at the top of the list. VHA employees are already challenged to "do more with less" (perform multiple roles with insufficient resources) at work. CREW coordinator or facilitator duties are ancillary to the job requirements, thus requiring time and energy beyond the normal work hours. It is critical that staff in these roles embrace them enthusiastically and have full support from facility leaders (e.g. training opportunities, "petty cash" funds to cover materials for the meetings, recognition activities and other small expenditures, time to fulfil their programme-related duties, and regular access to leaders for feedback about the programme).

Good group facilitation is also paramount in a successful CREW endeavour. Recruiting skilful facilitators can be a problem; coordinators and facilitators need to acquire skills and sharpen them by training. Sites use various strategies to identify good facilitators (e.g. self-selection, competitive selection, direct appointment). Some sites recognize the challenges of the role with a small bonus; a few carve out protected work time for CREW facilitation.

The unit supervisor's place in the CREW process is a recurring theme. In the perfect scenario, supervisors invite CREW into the workgroup to help a well-functioning team function even better, offer time and space for CREW groups to meet, and back-up staffing if necessary, provide a psychologically safe environment for honest conversations about workplace climate, and participate in CREW meetings in a non-hierarchical manner (as team members, not as supervisors). In a less than ideal scenario, supervisors may be informed of CREW by the facility leaders, feel threatened, resist, or even create obstacles to the CREW process. Workgroups often worry about including the supervisor in CREW meetings, fearing retribution for speaking openly and honestly. This is a valid concern; it can be addressed initially by meeting without the supervisor, to establish the process and mutual trust in the workgroup. The goal, however, is always to integrate the supervisor in the CREW process as soon as feasible. CREW is not intended as a method of disciplining employees for inappropriate (e.g. uncivil) behaviour, or a substitute for good supervision; attempts to use the programme in this manner typically bring poor outcomes.

A strong predictor of success of CREW is its support by the local CEO. For the intervention to succeed, staff at a given site needs to see that their CEO does not consider civility as a frill, but as an essential component to the success of the organization. The endorsement must be strong, consistent and visible, whether delivered by the CEO briefly joining a CREW meeting, providing resources, publicizing CREW achievements, or modelling CREW behaviours. As the ultimate endorsement, several medical centre CEOs described CREW as not just a programme, training or event, but as the way that business is done within their organizations.

Workplace leaders have much to gain from supporting CREW. Savvy leaders recognize that relationship issues, if unaddressed, inevitably become performance and public relations issues. As one medical centre director put it, "The link between employee satisfaction, patient satisfaction and clinical outcomes is clear. CREW has provided the necessary tools for us to achieve unprecedented levels of civility, respect and engagement in the workplace, resulting in patient satisfaction." This statement is consistent with statistical data showing associations of high workplace civility to improved customer satisfaction and loyalty, employee engagement, innovation, turnover, safety, quality and productivity (e.g. Osatuke et al. 2009, Leiter et al. 2011, Leiter 2012). Realistically, however, not all workplace leaders embrace and model civility and respect. The question then arises: "Can I practice CREW if my supervisor does not support it?." Our answer is a resounding *yes*. Members of the CREW community are leaders in their own right. Whether one's sphere of influence on the job encompasses 200 people or 20 or 2, in the CREW philosophy, any individual is able to model how to improve the workplace in a powerful way, whatever their place on the organizational chart. Two examples below were provided by CREW coordinators from two different facilities.

EXAMPLE 1

In a Food Service at a VHA medical center, small thefts began occurring from one employee's locker; a mug was stolen, then some candy, then a jar of honey. The workgroup, traditionally unsociable and suspicious of one another, decided to put CREW into action by replacing the stolen items. This reflected a new group dynamic that had developed in the group meetings. As this dynamic was brought into daily work life (by the group replacing what was stolen), the thefts stopped and were never an issue again.

EXAMPLE 2

The following story is from a CREW coordinator at another site:

One of the best examples of CREW for me was a personal experience I recently had while sitting in the hospital at the bedside of my dying mother. Granted, it was not in a VA hospital, but it really doesn't matter; it happened because of my involvement in CREW. I am the CREW coordinator at my station. As a result, I found myself looking at my own family crisis from the other side, as a customer needing services in the hospital. I watched things from this perspective with a new set of eyes and, although my mother ultimately succumbed to her illness and died, my take from this experience was profound. I observed varying degrees of civility between staff members and family as they interacted around providing care to my mother. One food service worker, Eddy, and one nursing assistant, Ava, took an approach that really dignified my mother's needs and in doing so, demonstrated the CREW philosophy that I think we should all take note of. My mother couldn't eat much, but Eddy knew that orange sherbet was her favorite, so she always had orange sherbet, and Eddy took the time to encourage a few bites before it melted. Ava came to assist my mother when she became incontinent and said to my family: "why don't you guys take a break and go get some coffee while I make your mother more comfortable." Other nursing assistants have stated in the same situation (and I quote): "please step out of the room so I can clean up this mess." My mother could hear every word and she was mortified and uncomfortable, as were we, the family. By their attitude, Eddy and Ava made a difference in the climate of care, not only for my mother, but also to us, the family. Yet we could observe them being treated with flippant disrespect, impatience and disregard by the registered nurses, medical residents and physicians. I found it disheartening, so I conducted a little CREW experiment; I posted a thank-you note addressed to Eddy and Ava on my mother's hospital room door. The results were amazing. Eddy and Ava were immediately treated significantly better by the hospital staff, and we saw that the behavior of all staff entering my mother's room became more thoughtful. All employees also appeared more willing to help each other out – it seemed as if they took more pride in their work of caring for my mother. The atmosphere changed; we now felt they viewed Mother and us as people in crisis – not as a smelly mess to clean up, a tray to deliver, or blood to draw. This attitude makes all the difference in the world when caring for those who are suffering ... and it did not cost us anything more than a kind word and maybe a piece of paper and a marker.

CREW participants are occasionally confused about how CREW fits with other Veterans Health Administration organizational health initiatives such as Systems Redesign (focus on improving work processes), Patient-Centred Care (bettering veterans' experiences of the healthcare they receive at the VA), and Diversity and Inclusion (approaching work style and individual perspective differences as a resource that can benefit the organization). While leaders and employees find value in many of these initiatives, their sheer volume may be beleaguering. This may suggest that engaging in CREW would be unnecessary because "we are already doing something else." Fortunately, CREW does not have to be separated from other organizational initiatives. As an analogy, a physician treating a cardiac patient will apply an array of interventions: medication, diet, exercise, smoking cessation. They are ordered simultaneously, not sequentially, because they complement each other, with an ultimate goal of a healthy cardiovascular system. Metaphorically, this is how CREW works. VA sites may apply an array of interventions, all targeting organizational health as their goal. A healthy organization is a silo-free zone where all workplace climate elements are viewed together, not in isolation. Viewing civility as woven into the fabric of the organizational life brings exponential benefits to organizations; we, like others (Leiter et al. 2011), believe this conclusion extends beyond VA.

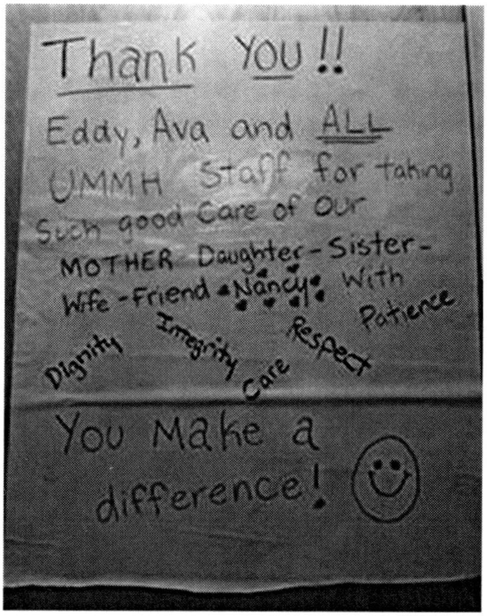

Figure 9.4   The thank you note from the second example

Conclusion

CREW has been successfully implemented in VA workgroups representing almost all professional disciplines within healthcare organizations (clinical, administrative, education, support services, security, trades, and others). However, these groups are not all equal in interest and engagement. There is generally a "hook-in" aspect of the CREW process or outcome that catches the interest of this particular group. For example, administrators may come on board because of the business case and high return on a low-cost investment. Physicians and nurses typically support CREW because of the impact of patient safety and quality of care. Support services often subscribe to CREW for the opportunity of honestly discussing workplace issues – a less common prospect on lower rungs of the hierarchical ladder. Finding the hook connects investment in CREW with concerns and priorities of employees in specific groups.

Over time, the CREW programme has had to respond to larger VA priorities as well. VHA is a government agency; the Congress of the United States is its Board of Directors. Inevitably, there are political and policy implications, including a competition for time, attention and scarce resources. The time and patience required to accomplish major culture change can be at odds with the timeframes of elected and appointed officials. Leadership changes related to the election cycle, retirement or mobility demand repeated contacts, marketing and communication, to keep the decision-makers aware of CREW and its advantages. With this said, CREW has a well-established reputation in VA as an intervention that works. The Joint Commission (TJC), a non-profit independent accrediting body for US healthcare agencies, recently recognized CREW as a national best practice in employee safety and well-being. Each VA hospital participates in TJC survey visits every three years in order to maintain accreditation as a safe and well-performing healthcare organization. CREW's recognition as a best practice at the healthcare workplace further validates its significant impact on patient care.

As CREW enters its seventh year, it becomes increasingly apparent that organizations of all kinds are clamouring for a more civil environment for both business and altruistic reasons. VHA's CREW initiative has been privileged to be the first of its kind. Nevertheless, its applications to private sector healthcare organizations (e.g. Canada's Enhancing Workplace Communities initiative – Leiter 2012) and a growing interest from other US government agencies (e.g. US Airforce), university settings (e.g. Xavier University, OH) and private sector groups (e.g. QUEST Diversity Initiatives, New Zealand Health Authority, BC Biomedical Laboratories, Johnson and Johnson) clearly suggest that the principles and process of CREW are inherently generalizable to a variety of work environments. CREW data have been published in peer-reviewed scientific journals and the programme has been profiled in management, human resource and diversity publications. CREW is a low-cost, low-tech, effective intervention that, given its custom-tailored delivery approach, can be organically grown and adapted to meet the needs of various organizational environments.

References

Anderson, L.M. and Pearson, C.M. 1999. Tit for tat? The spiraling effect of incivility in the workplace. *Academy of Management Review*, 24, 452–71.

Belton, L. 2005. *VHA Employee Satisfaction and Improved Organizational Outcomes*. Human Resource Committee report presented to the VHA National Leadership Board meeting, March 2005, Washington, DC.

Berne, E. 1964. *Games People Play*. New York, NY: Ballantine Books.

Carameli, K., Brown, A., Furst-Holloway, S., Cominsky, C., Moore, S., Carle, A., Howe, S., Osatuke, K. and Dyrenforth, S. 2012. *Monetizing Civility in the U.S. Department of Veterans Affairs*. Podium paper presented at the Academy Health annual research meeting in Orlando, FL in June 2012.

Congressional Budget Office. 2009. *Quality Initiatives Undertaken by the Veterans Health Administration*. Washington, DC: Congress of the United States, Congressional Budget Office. Retrieved from http://www.cbo.gov/publication/20928.

Cooperrider, D.L., Whitney, D. and Stavros, J.M. 2003. *Appreciative Inquiry Handbook*. Bedford Heights: Lakeshore Publishers.

Crawford, E.R., LePine, J.A. and Rich, B.L. 2010. Linking job demands and resources to employee engagement and burnout: A theoretical extension and meta-analytic test. *Journal of Applied Psychology*, 95, 834–48.

Edmondson, A. 1996. Learning from mistakes is easier said than done: Group and organizational influences on the detection and correction of human error. *Journal of Applied Behavioral Science*, 32, 2–28.

Edmondson, A. 2004. Psychological safety, trust, and learning in organizations: A group-level lens. In *Trust and Distrust in Organizations: Dilemmas and Approaches*, edited by R.M. Kramer and K.S. Cook. New York: Russell Sage Foundation, 239–72.

Leiter, M.P. and Bakker, A.B. 2010. Work engagement: Introduction. In *Work Engagement: A Handbook of Essential Theory and Research*, edited by A.B. Bakker and M.P. Leiter. New York, NY: Psychology Press, 1–9.

Leiter, M.P., Laschinger, H.K.S., Day, A. and Gilin Oore, D. 2011. The impact of civility interventions on employee social behavior, distress, and attitudes. *Journal of Applied Psychology*, 96, 1258–74.

Leiter, M. 2012. Forms of workplace mistreatment. In *Analyzing and Theorizing the Dynamics of the Workplace Incivility Crisis*, edited by M. Leiter. Amsterdam: Springer, 17–30.

Lincoln, A. 1865. Second inaugural address delivered 4 March 1865 in Washington, DC.

Longman, Philip 2007. *Best Care Anywhere: Why VA Health Care is Better than Yours*. Sausalito, CA: PoliPoint Press.

Maslow, A.H. 1973. *Dominance, Self-esteem, Self-actualization: Germinal Papers of A.H. Maslow*. New York, NY: Brooks/Cole.

Meterko, M., Osatuke, K., Mohr, D., Warren, N. and Dyrenforth, S. 2007. Civility: The development and psychometric assessment of a survey measure. In *Measuring and Assessing Workplace Civility: Do "Nice" Organizations Finish First?* moderated by M. Nagy. Panel presented at the 67th annual meeting of the Academy of Management, Philadelphia, 8 August 2007.

Osatuke, K., Moore, S.C. and Dyrenforth, S.R. 2012. Civility, Respect, and Engagement in the Workplace (CREW): An intervention promoting positive organizational culture. In *Analyzing and Theorizing the Dynamics of the Workplace Incivility Crisis*, edited by M. Leiter. Amsterdam: Springer, 55–68.

Osatuke, K., McNamara, B., Fishman, J. and Dyrenforth, S. 2010. Workplace civility perceptions relate to equal employment opportunity complaint rates. Paper presented at the 116th Annual national convention by the American Psychological Association in San Diego on 12 August 2010.

Osatuke, K., Moore, S.C., Ward, C., Dyrenforth, S.R. and Belton, L. 2009. Civility, Respect, Engagement in the Workforce (CREW): Nationwide organization development intervention at Veterans Health Administration. *The Journal of Applied Behavioral Science*, 45(3), 384–410.

Ross, L. 1977. The intuitive psychologist and his shortcomings: Distortions in the attribution process. In *Advances in Experimental Social Psychology*, edited by L. Berkowitz. New York, NY: Academic Press, 173–220

Schein, E. 2006. Culture assessment as an OD intervention. In *The NTL Handbook of Organization Development and Change: Principles, Practices, and Perspectives*, edited by B.B. Jones and M. Brassel. San Francisco, CA: Jossey-Bass, 456–65.

Seligman, M. and Csikszentmihalyi, M. 2000. Positive psychology: An introduction. *American Psychologist*, 55(1), 5–1.

10 CREW as a Work Engagement Intervention

MICHAEL P. LEITER

Abstract

Civility, Respect, and Engagement in the Workplace (CREW) is a process for improving the level of civility among members of workgroups. Elsewhere in this volume, Osatuke et al. (Chapter 9) describe the development of CREW within the Veterans Health Administration. That chapter also describes the CREW process in detail. This chapter reflects upon the qualities of CREW that contribute to its effectiveness as an organizational intervention.

Introduction

Intervention studies are relatively uncommon in research on worklife issues (Hammer et al. 2011, Macik-Frey et al. 2007, Scharf et al. 2008). One reason for this gap is that intervention studies are difficult to implement. They require a longer time line than required by cross-sectional or even longitudinal surveys. In addition to the logistics of designing, distributing, collecting and analysing surveys, intervention studies require the more challenging task of designing an intervention with the potential to make a difference in the psychological life of participants.

A second challenge for intervention research in work psychology arises from the working relationship of researchers with participating organizations. Applied research requires willing partners from private or public sector organizations. Although leaders within these settings may be seriously interested in the research issues and recognize a potential benefit from their participation, they may hesitate to commit to a long-term project. Research participation demands time and other critical resources from people in the settings. It is important that they are fully informed of the research objectives and procedures. Leaders must be aware of the procedures for gathering and sharing information in order to assure their cooperation in maintaining participant anonymity or confidentiality. Although research may be one dimension of the organization's mandate, it is unlikely to be their primary mission. Hospitals exist to provide health care; automobile manufacturers exist to build vehicles. Their participation in activities to generate new knowledge about worklife will invariably be a secondary consideration. When the organization encounters a serious crisis, that secondary commitment may fade away.

A third and fundamental issue is one of control. An organizational development (OD) intervention changes a management practice. Control over management practices is a core prerogative for executives and managers within participating organization. They do not readily relinquish those prerogatives to others. Successful partnerships for intervention research require an active and mutually supportive working relationship of researchers and managers to develop interventions that make sense to both parties. Building these relationships takes time and a convergence of interests. Maintaining these relationships requires an ongoing experience of activities that are successful from the perspectives of both parties. With the vagaries of career development and organizational events, the specific participants in these partnerships may change over time as individuals move on to other positions or retire. An ongoing and productive partnership is a rare and valuable thing.

EVALUATING INTERVENTIONS

Kompier et al. (2000) analysed workplace interventions and identified several process variables that contributed to the success of the interventions. Their work acknowledged the challenges inherent in conducting and evaluating organizational interventions. It also recognized the necessity of conducting such work. Organizations undertake major interventions in response to crises or simply as part of the process of evolutionary change and development. Without rigorous evaluation of interventions, organizational leaders risk investing time, energy and credibility into activities that lack a reasonable potential to make a lasting or meaningful contribution to the life of the organization. Decisions to participate in a programme may be more linked to current fads or interpersonal influence than objective criteria for programme efficacy. In light of these considerations, it is important to use intervention designs that assure that researchers and practitioners make the most of any opportunity to evaluate any programme to which they give serious consideration.

Kompier et al. (2000) considered effective programmes to be well managed overall. Interventions were more successful in sustaining employee well-being when:

1. organizations used a stepwise and systematic approach;
2. there was a clear structure (tasks, responsibilities);
3. consultants or researchers used a participative approach;
4. management and representatives of employees cooperated;
5. employees were recognized as "experts";
6. the responsibility of management was emphasized;
7. when monitoring and intervention were combined;
8. there was proper risk assessment using adequate instruments;
9. assessment for the company as a whole but also at the level of departments and job positions; and
10. clear facts and figures were used to present the programme to top management.

These 10 qualities suggest a well-considered and thoroughly managed intervention project. Bakker et al. (2013) reflected on their relevance to interventions designed to increase work engagement. The complete implementation of all 10 dimensions would produce a project that anticipates the challenges inherent in coordinating action across complex organizations and the scepticism employees have towards innovations in their workplace. Successful organizational change requires more than a good idea or valid insight into the way people experience their worklife. It requires as well a capacity to instill a new set of practices into day-to-day worklife.

Intervention Qualities of CREW

The following section considers the CREW intervention in regard to the 10 qualities from Kompier et al. (2000). The review considers CREW as it is implemented within the Veterans Health Administration as well as its structure within the research project within Canadian hospitals reported in Leiter et al. (2011). Because CREW's development is deeply rooted in an organizational development perspective, it provides a useful demonstration of these criteria to managing a complex intervention.

A STEPWISE AND SYSTEMATIC APPROACH

CREW is systematic but not completely stepwise. The process is systematic in that it follows an overall structure. It begins with persuasive communications. The external consultants meet with various constituencies including senior executives, managers responsible for participating units, and members of those participating units. Those meetings provide information about the CREW process, its rationale, and the evidence of CREW's impact in previous implementations. Next, the consultants conduct a baseline assessment and train people in the participating organizations who will play a leadership role in CREW. These gatherings are repeated at the halfway point and the end of the CREW process. Soon after the initial training gathering, groups begin their six months of CREW sessions. Towards the end of this series of meetings, they complete the assessment survey.

On this level of operation, the CREW process follows a stepwise approach. It departs from that ideal with regard to the content of the weekly sessions. The specific activities within these sessions are chosen in response to issues identified in the baseline assessment and to topics raised by participants in the sessions. The activities for the sessions are selected from a comprehensive toolkit of activities, discussion topics, and advice on group facilitation. Groups regularly create their own unique activities that fit their situations or challenges precisely.

The concept underlying CREW's system is that changes in social behaviour come about when people deliberately choose to participate in change. The marketing and training have an explicit objective of giving participants a sense of control. With a clear explanation of the process, people are more likely to feel part of the process rather than merely recipients. CREW also recognizes that groups vary considerably in the types of problems they experience and the approaches that best suit their efforts at improvement. One size does not fit all. Rather than a defined step-by-step process within the CREW sessions, the approach empowers leaders to adapt CREW's resources to the specific challenges for each group.

As noted by Osatuke et al. (2013) the flexibility that makes CREW responsive to local values and practices for showing respect creates a tension with establishing a strict invariable sequence or manualization of the CREW process. Instead, the *CREW Toolkit* that serves as a primary resource to facilitators contains a variety of discussion topics and exercises. Leaders select activities from this resource or generate new activities following their overall format. These selections reflect indications from the workgroup's assessment that a specific issue, such as inter-professional cooperation, presents an immediate challenge for the workgroup.

CLEAR STRUCTURE

CREW's structure is evident in its roles, its community gatherings, and its meetings. When operating as a system-wide intervention, as in the Veterans Hospital Administration (Osatuke et al. 2009 in press), CREW uses three primary roles. *Facilitators* lead the regular CREW sessions. Facilitators are usually employees of the participating organization, but not a member of the unit on which they are facilitating CREW. They supplement previous experience in training and facilitation with specific training on leading CREW sessions. The structure of the Facilitator role reinforces CREW as a structured process. Successful Facilitators are capable members of organizations with a history of effective group leadership and teaching. The fundamental activity within CREW sessions is talking about social interactions among colleagues. Even when relationships are going well, this may be a difficult topic of conversation. When relationships are going badly, it becomes an even more challenging topic. Focusing these conversations while assuring that people interact in a straightforward and sensitive fashion is a sophisticated ability that would be beyond the scope of a first-time Facilitator. However, it is an essential resource for CREW to have an effect.

The CREW approach does not assume that employees within participating organizations have these skills already developed. The initial training is intended to help Facilitators anticipate the specific challenges within the CREW process so that they can respond effectively. The training continues throughout the CREW process through ongoing contact between Facilitators and their mentors (Companions) and monthly conference calls that include Facilitators from all of the organizations implementing CREW at a given time. These conference calls, as well as the peer interactions that occur at the three CREW gatherings, bring a quality of organizational learning into CREW. The process recognizes that an important dimension of the Facilitator role is learning from and sharing knowledge with their counterparts in other settings.

Coordinators, members of the participating organization's senior management, provide moral and logistic support to the Facilitators while maintaining ongoing communication with the external consultants. A fundamental element of the Coordinator role is to legitimize CREW facilitation as an organizational activity. By designating a senior manager to act as Coordinator regarding CREW activities, the organization is explicitly accepting CREW facilitation as an official organizational activity that is part of the Facilitators' job responsibilities. Otherwise, the Facilitators could only rely on their personal persuasiveness to fulfil their role. They would also be vulnerable to complaints from individuals or unions who may question the legitimacy of their work.

Coordinators also attend to the logistics of the process. They maintain communication with the external consultants, keep senior management informed, and arrange for funds to cover expenses that arise from sessions or from Facilitators' participation in CREW gatherings. CREW is a complex organizational undertaking and it requires someone to take responsibility for its smooth operation.

Companions mentor the Facilitators through structured training sessions at the beginning of the programme followed by ongoing contact through individual or group telephone contact throughout the CREW process. Their role as mentors builds upon their training and experience. This background includes knowledge of social sciences, especially in relation to group process. This background may be in psychology, sociology, social work, nursing or management. Personal experience working in organizational settings, such as hospitals, is an asset for anticipating issues that can arise regarding management structures, office politics or formal labour relations. Effective Coordinators have a genuine interest in the social dynamics of workgroups. They elicit derailed reports of sessions from the Facilitators that allow them to reflect sympathetically on the challenges and successes that occur while implementing CREW. Their role is to help others succeed in their work. In light of the complex and novel elements of the Facilitator role, Companions provide an essential dimension to the CREW process.

CREW derives its structure as well from three community gatherings. The Orientation gathering occurs at the beginning of the CREW process, attended by the Facilitators, Coordinators and Companions. First line managers of the participating units as well as other leaders from the participating organization are invited to attend. One objective of the Orientation meeting is to inform all stakeholders about CREW, encouraging them to initiate further communications throughout their organizations about the programme. In addition, leaders use didactic and role-playing formats to train Facilitators on leading CREW sessions and Coordinators in their roles in providing support. The Orientation meeting builds enthusiasm for the goal of improving civility. To augment the impact of the CREW process, it builds a community of participants across organizations focusing on the shared objective of building civility at work. The experience of sharing perspectives and values for constructive collegial relationships with their counterparts in other organizations raises the CREW initiative from an organizational project to a larger context. Broadening the scope in this way could encourage greater dedication and involvement from participants.

The Midpoint gathering, true to its name, occurs around the three-month point in the six-month series of CREW meetings. The Midpoint brings together the Facilitators and Coordinators from the participating organizations. With a minimum of didactic sessions, the meeting provides a format for participants to share experiences. Across the participants will be Facilitators whose groups are making minimal progress towards serious conversations and others whose groups are enthusiastically initiating creative ideas for improving their workplace cultures. The larger community dynamic established in the Orientation meeting functions as an active resource for supporting the work of Facilitators and Coordinators.

The Celebration gathering at the end of the six-month series of CREW meetings makes the assessment process a shared experience. Leaders present the overall results of surveys to the group, contrasting indicators from before and after the CREW process. CREW Companions meet with the Facilitators and Coordinator from each organization to review their groups' profile of survey results. The Celebration gathering reviews ideas for maintaining gains and to continue progress after completing the CREW meetings. This meeting segues into an Orientation gathering for the units that are undertaking CREW for the following phase.

The CREW meetings have a clear structure of regular meetings that are open to willing participants. Although people who may greatly benefit from CREW participation may refuse to attend, the CREW perspective is to avoid pressuring people to attend. The target of intervention is improving the overall level of civility and respect throughout the workgroup. It is hoped that people who fail to attend sessions will benefit from their interactions with those who do attend. A standard format is for a 30-minute meeting weekly, but this practice varies with some groups holding 60-minute meetings every two weeks and others having 10-minute huddles a few times weekly. The objective is to fit the meeting schedule into the workflow of the unit to assure its implementation with the least disruption to the unit's productivity.

PARTICIPATIVE APPROACH

The model of change inherent in CREW is entwined with a participative approach. Social behaviour is a deeply personal experience. The emotional qualities of relationships have the potential for increasing resistance to change. By enlisting employees as active champions in change through their participation in CREW sessions, change becomes more likely.

Two of the CREW roles are organizational members. CREW does not rely on outside consultants to lead the groups but develops that capacity for this task within members of the organization. It is hoped that developing this capacity will allow the participating organizations to enrich their knowledge of workplace civility. The Facilitators' and Coordinators' ongoing membership in their organization can help not only to sustain the specific gains of CREW but help the organization to address broad concerns of collegiality. Facilitators and Coordinators become the means through which the organizational learning occurring during the CREW process becomes established in the organizational memory.

The flexible structure of the CREW process favours a participative approach. By foregoing a strict step-by-step procedure, CREW meetings create opportunities for participants to decide on issues to emphasize and values to support. Participants have responsibilities beyond attending and complying. The process relies upon participants to reflect on the activities critically and to devise ways of applying lessons from the meetings to their day-to-day worklife.

A participative approach encourages people to develop a richer understanding of themselves as part of a social context (Varela and Shear 1999). The sense of agency and efficacy that are integral to a participative approach encourage initiative in the change process (Senge and Scharmer 2001). An active role in a change process prompts people to think through its implications more thoroughly and to consider its broader implications. Senge and Scharmer (2001) make the case for an active role in change increasing the likelihood that participants will recognize their own blind spots. Active participation has an additional contribution to sustaining the gains of interventions (Eckman et al. 2012). Although CREW is primarily oriented towards the group processes within workgroups, a deeper level of insight from individual participants has the potential to both accelerate the change process and to sustain gains subsequently.

COOPERATION BETWEEN MANAGEMENT AND EMPLOYEES

CREW is a bottom-up initiative in its specific activities, but it relies upon executive level support in its overall approach. CREW programmes only proceed with an explicit commitment to civility and respect from senior management. Ideally, this commitment goes beyond leaders voicing their support for CREW to include executives being diligent about the civility inherent in their personal interactions at work. The commitment is also enhanced by including references to civility and respect in policy announcements and newsletter contributions from leaders.

The scope of CREW's activities makes such cooperation essential throughout the process. The training for Facilitators, logistic arrangements with Coordinators, and meeting time for participating employees divert a considerable amount of employee time. This level of activity can only occur with the active consent and explicit support of senior management. CREW involves direct expenditures as well in the travel costs associated with gatherings. Outside of the VHA, organizations contract with external consultants to provide the project management, assessment and mentoring functions of CREW.

Management becomes involved in the selection of Facilitators and providing the Facilitators with support throughout their involvement. The relationship of Coordinators with Facilitators works in two directions in that it keeps management informed of the overall direction of the CREW process and gives meaningful backup to the Facilitators' activities.

The most extensive cooperation between management and employees occurs when the executive team participates in CREW. In one of the organizations with which my group worked, the executive team of a continuing care facility decided that they would do CREW in the first wave. By doing so, they demonstrated their sincere commitment to the project, gave it a positive status, and acknowledged that everyone has the potential to improve their working relationships.

EMPLOYEES RECOGNIZED AS EXPERTS

The active roles given to employees as Facilitators and Coordinators recognizes and further develops their expertise. The flexibility built into CREW supports their expertise by presenting decisions about what activities to do and when to do them. CREW activities acknowledge participants as the final arbiters of what constitutes civility and incivility within their workplace. Some groups may prefer a direct and boisterous camaraderie while others prefer quiet companionship. By giving the latitude to determine the values of their workplace, CREW recognizes employees as experts on their work lives. The objective of CREW is to assist them in realizing those values in a shared and open manner.

This perspective stands in direct contrast to civility as etiquette. Rather than a universal or culture-wide set of rules or protocols that define polite behaviour, the CREW definition of civility rests on intrinsic qualities. Behaving with civility is to attend to others, to acknowledge their presence, to appreciate their contribution and to accommodate them within one's community. That quality of inclusiveness brings civility within the culture of the workgroup. The implicit understanding among co-workers and their immediate supervisors regarding the proper ways of showing respect and consideration become the standards against which behaviour is evaluated. This perspective does not mean that anything goes. Instead, it means that the final arbiters of civility within a workgroup are those members of the workgroup. In that way the CREW process acknowledges employees' expertise on what constitutes civility and respect within their workgroup.

As mentioned previously, the active role of employees within the CREW process is fundamental to its success. This quality implicitly acknowledges employees' expertise because it is their active participation that puts the process into effect. The final expertise is not with external consultants or with senior managers, but with the front-line employees who attend the sessions and enact the exercises from the *Toolkit*.

MANAGEMENT RESPONSIBILITY EMPHASIZED

Management responsibility is emphasized throughout the process. This responsibility begins with senior management requesting that CREW occur within the organization. Outside of the VHA system, senior management develops a contract with an external consultant to provide the project management, assessment and mentoring required for CREW. The *Coordinator* role establishes a clear attachment of the process to senior management. The *Facilitators* report to the *Coordinator* regarding their responsibilities in CREW. This link permits ongoing communication flow and supports the *Facilitators* by legitimizing their CREW activities as a job responsibility integrated into the organization's authority structure. The *Coordinator* is responsible for maintaining communication with the consultant and with senior management colleagues.

Through its commitment to CREW, management demonstrates that civility and respect among employees is a value in action, not simply an espoused value in the language of Argyris and Schön (1974). By implementing a structured system to assure workplace civility, senior management is going taking the issue seriously. Rather than relying on the good intentions of employees to interact with one another, management is implementing and explicitly supporting a process that monitors the level of civility in workgroups and responds accordingly when these levels are out of a comfortable range. Within the VHA system, the workgroup level of civility is an integral part of its annual employee survey report. The managers' performance review includes reference to the civility scores for their work units. They are expected to develop an action plan that may include CREW initiatives as part of their response to sub-optimal civility levels.

The combination of high level monitoring and intervention management brings upper management into the CREW process. By having a clear and consequential involvement at that level they maintain the process's relevance while the specific implementation occurs on the workgroup level.

INTEGRATION OF MONITORING WITH INTERVENTION

CREW includes assessments at the beginning and end of the process. The initial assessment serves as a diagnosis by comparing the unit response to established norms. All members of units hoping to participate in CREW are invited to complete an online survey. The results are organized by unit and presented as a profile. The Companions work with Facilitators and Coordinators to interpret the survey. The high points and low points on the responses may identify focus points for CREW activities. The VHA process (Osatuke et al. 2009) relies on an eight-item civility scale for their profile. The Canadian group (Leiter et al. 2011) supplements that scale with various other measures including items from the Maslach Burnout Inventory General Scale (MBI – GS; Maslach et al. 1996), Areas of Worklife Scale (Leiter and Maslach 2004), the Workplace Incivility Scale (Cortina et al. 2001), and the Rudeness Rationales Scale (Leiter et al. 2010). That group is testing out other measures, such as the Straightforward Incivility Scale (Leiter 2012). This collection of measures produces a rich profile of a workgroup's social context.

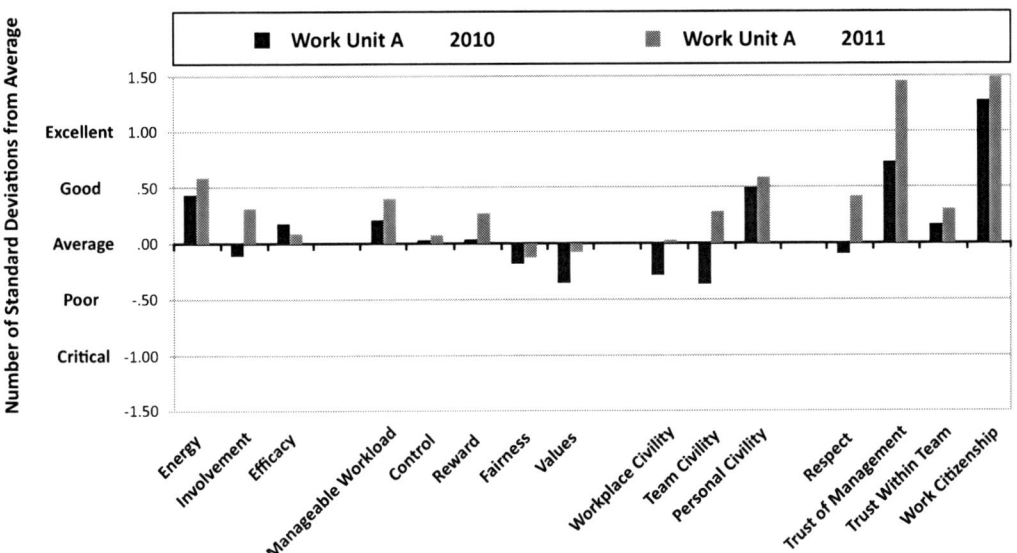

Figure 10.1 Profile of civility scores before and after CREW

The second assessment occurs towards the end of the CREW process. The results are summarized in a profile that presents a graphic depiction of the results before and after an intervention. The sample profiles in Figure 10.1 displays data from a health care unit within a Canadian facility before and after CREW. The midline of this profile indicates the normative value for each measure. The bars represent standardized scores in the positive (above the midline) or negative (below the midline) direction in standard deviation units. This unit's scores at Time 1 were negative on only a few measures – involvement (the reversed cynicism scale of the MBI – GS), fairness, value congruence (between individual and organizational values), workplace civility, team civility and respect. Most of these constructs directly reference civility while fairness and value congruence have implications for civility. In that way, the Time 1 profile indicated a unit for which CREW was an appropriate intervention because its challenges related to civility rather than other clearly defined workplace challenges.

The Time 2 scores indicate an overall improvement in the scores with those on workplace civility, team civility and respect moving from the negative to the positive area of the profile. The changes in the profile indicate that employee responses have moved in the preferred direction. They also indicate that the changes are incremental rather than revolutionary. Changes are most evident on constructs directly related to the quality of social interaction: CREW addresses a discrete range of workplace problems.

This structuring of assessment within the CREW process makes evaluation an integral part of the intervention. Implementing CREW requires organizations to evaluate the process and to report the results to members of the organization. The process thereby meets standards identified in Kompier et al. (2000) in that the survey provides a proper risk assessment using established, validated instruments, the assessment considers the overall organization as well as specific work units, and it generates standardized scores that inform management of the urgency for addressing civility and the effectiveness of the intervention.

PROPER RISK ASSESSMENT USING ADEQUATE INSTRUMENTS AND ASSESSMENT AT MULTIPLE LEVELS

The assessment approach used for CREW interventions uses surveys that have been developed for the specific purpose of evaluating the social context of work and employees' psychological connection with their worklife. All of the surveys have been evaluated by multiple researchers who have confirmed the factor structure and the reliability of the measures. The VHA developed the CREW Civility Scale (Osatuke et al. 2009) as an element of their All Employee Survey in which hundreds of thousands of employees participate annually. The employee survey data provides a normative reference point against which they can compare the results on the CREW Civility Scale before and after each CREW implementation. The broader range of measures used in CREW outside of the VHA arises from an interest in the broad implications of improved civility and greater latitude in the choice of survey instruments. The responses to these surveys are organized on the level of the participating workgroups and compared to the overall levels for the organization overall.

SUMMARY

This reflection on the qualities of CREW in relation to Kompier et al. (2000) demonstrate that the approach captures important elements of effective intervention design. CREW does not occur as an external process that descends upon the organization, but a complex activity to which a wide range of constituencies contribute. The diverse range of roles within the CREW process allows for distinct contributions from various parties. Full participation from various parties need not produce redundant activity. The intervention produces plenty of activity for all concerned.

Most importantly, the CREW process gives the participants a sense of ownership in the development. The high degree of customization inherent in the process calls upon a sense of agency from everyone. No one can simply follow a rearranged script. The process works best when each person attends to the distinct qualities of the present situation, listens attentively to their fellow participants, and generates a creative response. To the extent that these qualities are present in an intervention, people have a sense of control and empowerment as they undertake a change process. Those feelings will diminish the likelihood of active resistance to change.

A central point of the evaluation of CREW by Leiter et al. (2011) was that increases in civility mediated improvements in other measures, including cynicism, job satisfaction and organizational commitment. This finding confirmed a core construct that civility on the micro level of day-to-day interactions shape employees' psychological connections with work.

CREW and Work Engagement

Engagement as measured by the UWES was included in the survey used in Leiter et al. (2011) but its improvement in the CREW groups did not attain significance. In contrast, improvement was evident in the cynicism subscale of the MBI–GS. An additional analysis is presented here that examines the relationship of civility, incivility and engagement view on the relationship of civility with engagement as measured by the UWES.

Table 10.1   Correlations of engagement with social constructs

	Mean	SD	2	3	4	5	6	7
1. Engagement 2009	4.59	1.12	0.63	-0.11	-0.17	-0.17	0.32	0.32
2. Engagement 2008	4.62	1.08		-0.22	-0.18	-0.19	0.37	0.36
3. Supervisor incivility 2008	0.50	0.73			0.21	0.26	-0.32	-0.24
4. Co-worker incivility 2008	0.73	0.77				0.56	-0.27	-0.53
5. Instigated incivility 2008	0.49	0.44					-0.17	-0.41
6. Respect 2008	3.43	0.76						0.45
7. Civility 2008	3.74	0.67						

Note: N=446, all correlations $p<0.01$.

Table 10.2   Multiple regression on engagement 2009

Predictor	β	T	Significance	
Engagement 2008	0.587	14.88	>0.001	
Civility 2008	0.108	2.74	0.001	
				Partial
Supervisor incivility 2008	0.048	1.26	0.209	0.060
Co-worker incivility 2008	-0.005	-0.10	0.917	-0.005
Instigated incivility 2008	-0.016	-0.40	0.690	-0.019
Respect 2008	0.071	1.67	0.095	0.079

Table 10.1 displays the correlations of engagement at Time 1 and Time 2 with the social variables at Time 1. The correlations of respect and civility with engagement are nearly identical. These correlations are stronger than those of engagement with the incivility measures.

A multiple regression tested whether Time 1 social variables improved the prediction of Time 2 engagement beyond the contribution of Time 1 engagement. The analysis found that Time 1 civility made a significant contribution to the regression equation (see Table 10.2) increasing R^2 by .01 ($F_{(1,441)}=7.530$, $p=0.007$) resulting in an adjusted $R^2 = .399$ ($F_{(2,441)}=147.82$, $p<0.001$).

As noted, the correlations of civility and respect with engagement were similar as were their first-order partial correlations with Time 2 engagement after entering Time 1 engagement. The status of civility as the predictor is somewhat arbitrary as civility's relationship was not significantly stronger than that of respect and either would have made a significant contribution to the prediction. In contrast, none of the first-order partial correlations of the incivility constructs were significantly related Time 2 engagement after accounting for Time 1 engagement.

This pattern follows the prediction of the Job Demand/Resources (JD/R) Model that associates the construct of work engagement more closely with resources than with demands. The positive correlations of resources with the UWES definition of engagement are generally stronger than the negative correlations of demands with engagement.

The increment in explained variance in Time 2 engagement provided by Time 1 civility over Time 1 engagement demonstrates the relevance of workplace civility to work engagement. This finding is consistent with the proposition that civil relationships help people to access the important workplace resources that are available only through other people. Colleagues at work are a potentially rich source of knowledge, expertise, emotional support and practical assistance. Civil relationships make that resource more readily available; incivility creates barriers in the free flow of resources among people. The regression pattern indicates that greater access to such resources through civil relationship will have a positive relationship with subsequent levels of engagement. Although Leiter et al. (2011) did not find a significant improvement in the UWES construct of engagement following CREW, the constructs have clear relevance to one another.

Conclusion

This chapter considered the CREW intervention to increase workplace civility in relation to a framework outlined by Kompier et al. (2000) as indicators of successful organizational interventions. This reflection provides some indication of CREW's success in both the research context as reported by Osatuke et al. (2009) and Leiter et al. (2011). Within a research context these are complex intervention projects requiring active participation of organizational leaders and employees in collaboration with research teams. Their success with regard to a wide range of measures is consistent with CREW's faithful alignment with those principles.

In one regard CREW departs from the Kompier et al. (2000) conditions: it departs from a strict step-by-step procedure on the level of the agendas for the weekly meetings. To some degree this departure allows CREW to align more faithfully with other conditions. The flexibility inherent in the meeting agendas allows the participants greater control over the process. To support this level of flexibility, the CREW process develops Facilitators to a high level of proficiency and supports that training through ongoing mentoring, including peer mentoring.

The CREW process is far from perfect but it shows promise. Its compatibility with the conditions described here gives it a solid foundation for its impact. As the process develops within the VHA and is applied in a diverse range of settings, it will continue to evolve. For example, it may be that a different, perhaps simpler, version of CREW would be effective for relatively well functioning groups while a more intensive version would be necessary for workgroups with a toxic social environment. By explicitly referencing the qualities reviewed here, practitioners can assure that they maintain the core elements contributing to CREW's effectiveness.

References

Argyris, C. and Schön, D.A. 1974. *Theory in Practice: Increasing Professional Effectiveness*. San Francisco, CA: Jossey-Bass.

Bakker, A.B., Oerlemans, W.G.M. and Ten Brummelhuis, L.L. 2013. Becoming fully engaged in the workplace: What individuals and organizations can do to foster work engagement. In C. Cooper and R. Burke (eds) *The Fulfilling Workplace: The Organization's Role in Achieving Individual and Organizational Health*. London: Psychology Press, 55–70.

Cortina, L.M., Magley, V.J., Williams, J.H. and Langhout, R.D. 2001. Incivility in the workplace: Incidence and impact. *Journal of Occupational Health Psychology*, 6, 64–80.

Eckman, K., Blickenderfer, M. and Henry, S. 2012. *Native Shoreland Buffer Incentives Project: Final Report*. Water Resources Center, University of Minnesota, Minneapolis.

Hammer, L.B., Kossek, E.E., Anger, W.K., Bodner, T. and Zimmerman, K.L. 2011. Clarifying work–family intervention processes: The roles of work–family conflict and family-supportive supervisor behaviors. *Journal of Applied Psychology*, 96, 134–50.

Kompier, M.A.J., Cooper, C.L. and Geurts, S.A.E. 2000. A multiple case study approach to work stress prevention in Europe. *European Journal of Work and Organizational Psychology*, 9, 371–400.

Leiter, M.P. 2012. *The Straightforward Incivility Scale*. Working Document. Centre for Organizational Research and Development, Acadia University, Wolfville, NS, Canada.Leiter, M.P. and Maslach, C. 2004. Areas of worklife: A structured approach to organizational predictors of job burnout. In *Research in Occupational Stress and Well Being: Vol. 3. Emotional and Physiological Processes and Positive Intervention Strategies*, edited by P. Perrewé and D.C. Ganster. Oxford: JAI Press/Elsevier, 91–134.

Leiter, M.P., Laschinger, H.K.S., Day, A. and Gilin-Oore, D. 2010. *Rudeness Rationales: Whatever Were They Thinking?* Presentation at the Annual Conference of the Academy of Management. Montreal, Canada. Awarded Best Paper in Health Care Management.

Leiter, M.P., Laschinger, H.K.S., Day, A. and Gilin-Oore, D. 2011. The impact of civility interventions on employee social behavior, distress, and attitudes. *Journal of Applied Psychology*, 96, 1258–74.

Macik-Frey, M., Quick, J.E. and Nelson, D.L. 2007. Advances in occupational health: From a stressful beginning to a positive future. *Journal of Management*, 33, 809–40.

Maslach, C., Jackson, S.E. and Leiter, M.P. 1996. *Maslach Burnout Inventory Manual*, 3rd edn. Palo Alto: Consulting Psychologists Press.

Osatuke, K., Moore, S.C., and Dyrenforth, S.R. 2013. Civility, Respect, and Engagement in the Workplace (CREW): An intervention promoting positive organizational culture. In *Analyzing and Theorizing the Dynamics of the Workplace Incivility Crisis*, edited by M.P. Leiter. Amsterdam: Springer, 55–68.

Osatuke, K., Mohr, D., Ward, C., Moore, S.C., Dyrenforth, S. and Belton, L. 2009. Civility, Respect, Engagement in the Workforce (CREW): Nationwide organization development intervention at Veterans Health Administration. *Journal of Applied Behavioral Science*, 45, 384–410.

Scharf, T., Chapman, L., Collins, J., Limanowski, J., Heaney, C. and Goldenhar, L.M. 2008. Intervention effectiveness evaluation criteria: Promoting competitions and raising the bar. *Journal of Occupational Health Psychology*, 13, 1–9.

Senge, P.M. and Scharmer, C.O. 2001. Community action research. In *Handbook of Action Research*, edited by P. Reason and H. Bradbury. Thousand Oaks: Sage Publications, 195–206.

Varela, F. and Shear, J. 1999. First-person accounts: Why, what, and how. In *The View from Within: First-Person Approaches to the Study of Consciousness*, edited by F. Varela and J. Shear. Thorverton: Imprint Academic, 1–14.

11 *Participative Climate as a Key for Creating Healthy Workplaces*

KEIKO SAKAKIBARA SEKI, HIRONO ISHIKAWA and YOSHIHIKO YAMAZAKI

Abstract

An increasing number of studies have shown the importance of participative climate in order to manage growing demands in matters of change and workloads in contemporary organizations. By reflecting on the case of five major Japanese pharmaceutical companies undergoing recent mergers and/or acquisitions (M&A), the authors explore the association between the presence of participative climate on one hand, and health and work motivation on the other, in order to clarify how this work climate helps create healthy workplaces. The analysis helps build an improved understanding of the role and the repercussions of participative climate and provides implications for future research and practice.

Introduction

Participative climate can be defined as a work environment in which an employee has open communication with his supervisor and colleagues, and opportunities to actively participate in decision-making based on sufficient and timely information (Likert 1967). In this study, we explored associations with health and work motivation in order to clarify whether participative climate helps create healthy workplaces.

Stress and health of working people has been a critical issue worldwide. Japan is no exception. According to the National Survey of Health Conditions of Working People (Ministry of Health, Labour and Welfare 2008), 58 per cent of participants reported that they feel strong anxiety and distress in their job as well as in their working life. In another survey, 79.4 per cent of participants reported they feel fatigued after regular work (JILPT 2008). In response to this situation, the Japanese government has strengthened its policy to promote employee health, encouraging organizations to work positively on improving employees' mental and physical health (Ministry of Health, Labour and Welfare 2007).

Researchers and organizations have been discussing how to create healthy organizations, which are characterized by employees' health and organizational effectiveness (Cooper and Cartwright 1994, Grawitch et al. 2007). In these discussions, employee participation has been addressed as a critical factor for creating a healthy organization that should be enhanced in the workplace (Cooper and Cartwright 1994, Grawitch et al. 2006, House and Cottington 1986). In policy development, the World Health Organization (WHO) points out that involving workers and their representatives is one of the key features of healthy workplaces (Burton 2010). The American Psychological Association (APA) also positions employee involvement, synonymous with employee participation, as one of the practices of a Psychologically Healthy Workplace, that fosters employee health and well-being as well as enhancing organizational performance and productivity (APA 2012). As the theoretical model of APA's Psychologically Healthy Workplace Program, Grawitch et al. (2006) proposed the PATH (Practices for Achievement for Total Health) model. They identified five categories of indicators of healthy workplace practices that included employee involvement, and suggested that healthy workplace practices are those that enhance employees' well-being such as physical and mental health, motivation, commitment and job satisfaction (Grawitch et al. 2006).

The relationship between employee participation and well-being has been studied in Europe and the United States in fields such as occupational health, communication, management and leadership. In the field of occupational health, the participatory intervention approach has been frequently used to reduce occupational stress and improve employee health (van der Klink et al. 2001). The participatory intervention approach involves employees, management and researchers, who jointly participate in the problem-solving process (Heaney et al. 1993).

As in Western countries, participatory intervention studies are becoming increasingly common in Japan as well. Kobayashi et al. (2008) investigated the effects of participatory intervention on reducing job stressors and psychological distress of employees of a manufacturing company. They found that among women, who might have less opportunities than men to participate in decision-making in daily work due to their primary supportive roles, results were favourable in terms of job stressors such as supervisor and co-worker support, depression and job satisfaction in the intervention group versus the control group (Kobayashi et al. 2008). Tsutsumi et al. (2009) explored the effects of participatory intervention on workplace improvement not only in mental health but also in job performance among blue-collar employees of a manufacturing company (Tsutsumi et al. 2009). Their study found that mental health deteriorated in the control group while it did not change in the intervention group and job performance improved in the intervention group but decreased in the control group. This suggested that the participatory intervention approach contributes to both employees' health and productivity.

Though many studies have confirmed the effectiveness of participatory intervention on reducing work-related job stress, Mikkelsen et al. (2000) pointed out that in order to maintain the effects of participatory intervention, it is necessary for its integration into work routines (Mikkelsen, Saksvik, and Landsbergis 2000). It was suggested that if employee participation takes root in the culture/climate of a workplace environment, employees will spontaneously participate in decision-making processes and be able to solve problems that cause job stress.

Participative culture/climate has been studied mainly in the areas of management and communication. A study of 807 blue-collar employees in a manufacturing industry in the United States and China found that a higher participative climate in an organization is correlated with lower turnover intentions and withdrawal from work, as well as higher satisfaction with co-workers, supervisors and work even during restructuring

(Probst 2005). Another study of state employees in the United States demonstrated that participative work environments predict employees' work attitudes, such as organizational commitment and intrinsic job satisfaction (Tesluk et al. 1999).

A study examining the relationship between participative culture and the organizational commitment of staff of assisted living facilities found that a favourable perception of organizational culture was associated with a higher level of organizational commitment (Sikorska-Simmons 2005). Other than the association with individual behaviour, two studies examined the relationship of perceived participative culture to organizational performance, including the research performance of biological and chemical scientists in university research departments (Ryan and Hurley 2007), and residents' satisfaction with assisted living facilities (Sikorska-Simmons 2006).

Looking at studies in the field of leadership, there is a growing number of literature focusing on the effects of leadership styles on employee well-being (Kelloway and Barling 2010). It is reported that the quality of leadership style has been linked to various outcomes including psychological well-being (Arnold et al. 2007) and employee stress (Offermann and Hellmann 1996). For example, Nyberg et al. (2009) investigated the association of managerial leadership with ischemic heart disease (IHD) among employees in Sweden and found that higher leadership scores were associated with lower incidence of IHD. In their study, they define favourable leadership style as "the manager's consideration of the individual employee, provision of clarity in goals and role expectations, supplying information and feedback, ability to carry out changes at work successfully, and promotion of employee participation and control" (Nyberg et al. 2009). This definition can be categorized into participative/democratic leadership style where the leader shares the decision-making with group members by promoting the interests of the group members. It is similar to the definition of participative climate. Participative leadership and participative climate are however different in terms of what they focus on. In concrete terms, while leadership focuses on social influence enacted by individuals who hold formal leadership roles in organizations (Kelloway and Barling 2010), workplace climate focuses on employees' perception of their work environment created by interactions among members including management, supervisors and co-workers.

From the reviews mentioned above, we could not find any studies examining the relations of participative climate/culture to employees' health and work motivation. Therefore, we set the purpose of this study to examine whether participative climate contributes to the creation of healthy workplaces in which employees are healthy and motivated.

It seems increasingly difficult to create healthy workplaces in today's tough work environment. In the current difficult economic times, organizations tend to restructure and downsize their operation, which leads to an increase in perceived job insecurity (Worrall and Cooper 1997). In their review, Sparks et al. (2001) summarized how job insecurity puts negative impact on employees' health (Sparks et al. 2001). Downsizing also affects employees' work motivation, the source of organizational performance and effectiveness (Locke and Latham 1990). Kawai (2011) reported negative correlations between workforce reductions and employee motivation, and attributed it to job insecurity, the loss of friends and trust within a downsizing organization. It suggested that pursuing organizational effectiveness might be challenging because of difficulties to sustain employees' work motivation (Kawai 2011). If participative climate has a positive impact on health and work motivation, we would be able to recommend enhancing participative climate at work as a solution for creating healthy workplaces even in these difficult circumstances.

In expounding the hypotheses of this study, we referred to the Job Demands-Resource (JD-R) model (Bakker and Demerouti 2007, Demerouti et al. 2001). We cite a brief introduction of the JD-R model below (Bakker and Demerouti 2007).

The main assumption of the JD-R model is that every occupation has its own specific risk factors associated with job-related stress. These factors can be classified into two categories: job demands and job resources. Job demands are those physical, psychological, social or organizational aspects of the job that require sustained physical and/or psychological effort or skills and are therefore associated with a certain physiological and/or psychological cost. Job resources are those physical, psychological, social or organizational aspects of a job that:

1. are functional in achieving work goals;
2. reduce work demands and the associated physiological and psychological cost; and/or
3. stimulate personal growth, learning and development.

Thus, job resources not only deal with job demands but are also associated with motivation (Bakker and Demerouti 2007: 312).

Based on this theory, we considered participative climate as a job resource and expected a favourable perception of participative climate to be positively related to higher work motivation. Although the JD-R model does not mention the direct effects of job resources on health, we also expected a favourable perception of participative climate to have a positive relationship with better health since previous studies reported that a favourable workplace atmosphere is associated with better employee health (Ylipaavalniemi et al. 2005). In addition, we expected participative climate to modify the impact of work demand on health since the JD-R model hypothesizes that job resources buffer the negative effects of job demand on health.

Therefore, the hypotheses of this study are as follows:

Hypotheses 1: A favourable perception of participative climate is positively related to better health.

Hypotheses 2: A favourable perception of participative climate is related to higher work motivation.

Hypotheses 3: Participative climate modifies the impact of work demand on health.

Methods

PARTICIPANTS

Participants were white-collar staff workers working at five major Japanese pharmaceutical companies. Employees of pharmaceutical companies work in a stressful environment with rapid technical innovation, a dynamic, high-pressure global business climate (Cooper et al. 1996). We also assumed participants were under stress due to the recent mergers and/or acquisitions (M&A) of their companies. To be competitive in the global market, the pharmaceutical industry is constantly engaged in M&A activity and companies participating in this study had recently undergone M&A. A previous study identified poor communication as a potential source of dissatisfaction in the workplace, especially under dynamic conditions (Cartwright et al. 2000). Therefore, we considered participative climate, as characterized by open communication, sufficient and timely information, and opportunities to participate in decision-making, to be important and suitable to be examined in the current setting of pharmaceutical companies.

In February 2009, we conducted a self-administered questionnaire survey on work-related characteristics, health, work motivation and demographic information.

MEASURES

Participative climate

In this study, we developed a 20-item scale of participative climate. Items came from three sources, including the Organizational Culture Scale (OCS) (Glaser et al. 1987), our discussions with experts in the field of occupational health research, and discussions with white-collar employees who participated in our preliminary survey.

Here, we would like to clarify the difference between organizational culture and climate, and then explain why we referred to a scale that measures organizational culture when we develop a scale to measure organizational climate.

Denison (1996) mentioned that organizational culture and organizational climate have not been clearly distinguished from one another. He explained the difference between these concepts: culture refers to "an evolved context which is rooted in history, collectively held, and sufficiently complex to resist many attempts at direct manipulation." On the other hand, climate refers to a situation and its link to the thoughts, feelings and behaviours of organizational members. Thus, it is temporal, subjective and often subject to direct manipulation by people with power and influence (Denison 1996).

OCS is an instrument to measure employees' perception of organizational culture to ascertain its degree of participation and involvement. According to Denison's explanation of culture and climate mentioned above, items of OCS describe the situation in workplaces and they were actually used to measure changes in the work environment through intervention. Therefore, we considered that OCS can be used as a reference for developing measurements of organizational climate.

OCS consists of six subscales, a structure we maintained when developing items in Japanese. We translated and modified the original 31 items because some statements did not seem to relate to the reality of Japanese corporations. Then we conducted a preliminary survey of 37 white-collar employees to determine whether scale items are relevant to Japanese white-collar employees. Following the preliminary survey, we discussed the items with six participants of the preliminary survey. Based on an analysis of pre-survey data and opinions from the discussion including overlap between items, we chose 20 items for this study. Finally, we asked representatives of labour unions whether these items are relevant to employees in the pharmaceutical industry.

In our main survey, we performed confirmatory factor analysis to verify the structural validity of the scale. A model with six subscales provided an acceptable fit to the data when considering CFI=0.97, RMSEA=0.05, GFI=0.95. Cronbach's alpha was 0.82. Items of participative climate are shown in Appendix 1. The scale for alternative answers was from 1, "does not apply" to 5, "applies." Higher scores show that an employee's psychological work climate is more participative.

It is generally accepted that psychological climate is a property of the individual and that the individual is the appropriate level of analysis when outcomes, such as those of job satisfaction, employee motivation and psychological well-being, are on an individual level (Parker et al. 2003). Since the outcomes of this study were also individual level, we used participative climate as an individual-level variable.

Work demand and work control

In this study, we developed an eight-item questionnaire for work demand and a two-item questionnaire for work control to uncover the specific risk factors associated with job-related stress in the pharmaceutical industry. Items were developed by referring to the Job Content Questionnaire (Kawakami and Fujigaki 1996) and interviews with 10 representatives of labour unions. Examples of items of work demand (eight items) include, "It is impossible to finish work without working overtime" and "My work always requires new skills and knowledge," and, of work control (two items), "I can decide how to proceed with my own work" and "I can decide my own work schedule." In the main survey, we conducted an exploratory factor analysis, using the maximum likelihood method. Factors with eigenvalues of more than 1.0 were extracted and the Promax rotation method was used to obtain factor structure. Three factors including quantitative work demand, qualitative work demand and work control were extracted. Since there was no need to analyse quantitative and qualitative demand separately, we aggregated two of them as work demand. The result of exploratory factor analysis is shown in Appendix 2. The scale for alternative answers was from 1, "does not apply" to 4, "applies." Higher scores show that an employee has more work demand and more work control. Cronbach's alphas for work demand and work control were 0.77 and 0.76 respectively.

Health

We used fatigue and depression to examine the health condition of participants. As we mentioned above, fatigue and depression are serious health problems among Japanese working people. Ylipaavalniemi et al. (2005) found that poor team climate predicts subsequent doctor-diagnosed depression among Finnish hospital personnel (Ylipaavalniemi et al. 2005). Fukui et al. (2004) explored the association of organizational climate and psychological distress among 819 employees in Japan and found that the CES-D score of employees with a favourable perception of organizational climate was lower than those with a less favourable perception (Fukui et al. 2004). Meanwhile, a study in the Netherlands confirmed the close association between psychological distress and fatigue (Bültmann et al. 2002). Therefore, in this study we used depression and fatigue to assess health conditions related to employees' psychological climate.

For measuring fatigue, we used 18 items from the Cumulative Fatigue Symptoms Index (Yamazaki and Asakura 1999), examples of which include "My head feels heavy" and "I am often distracted" (0, "no" and 1, "yes"). Higher scores show that an employee has more fatigue. Cronbach's alpha was 0.85.

As for depression, we used a Japanese translation of the short version of the Center for Epidemiologic Studies Depression (CES-D) scale (Kinoshita 2001). The scale of alternative answers was from 0, "not at all" to 3, "almost everyday." Higher scores show that an employee is more depressed. Cronbach's alpha was 0.81.

Work motivation

Work motivation is a form of internal psychological energy (Pinder 2008) considered to be a source of organizational performance and effectiveness (Locke and Latham 1990). We measured work motivation using the MSQ (Motivation Status Quo) scale, which was developed based on data from a sample of over 100,000 subjects in Japan (JTB 1998). MSQ questionnaires examine to what extent individuals are interested in their current work and desire to continue to work. The measure has five items, examples of which include "I feel desire for my present work" and "I would like to continue my current work for a long time" (1 "This does not describe me" to 4, "This describes me"). Higher scores show that an employee has higher motivation. Cronbach's alpha was 0.87.

Control variables

Age, gender, marital status, occupation (engineer, sales/marketing, corporate staff), presence of chronic disease and company (company code 1 to 5) were included as control variables in regression equations.

Analysis

We performed hierarchical multiple linear regression analysis to examine the hypotheses. At first, we used work demand and work control as independent variables, and fatigue, depression and motivation as dependent variables to explore the effects of work demand and work control on fatigue, depression and work motivation (Model 1). Then we added the total score of participative climate as independent variables to examine H1 and H2 (Model 2). We also explored how the effects of work demand and work control on dependent variables would change when adding participative climate into the model. Finally, to examine H3, we added work demand as an interaction variable and participative climate as an independent variable to explore its buffering effects on fatigue and depression (Model 3). Control variables were also put into each model. We conducted additional analysis to explore the effects of each subscale of participative climate on dependent variables. In order to avoid multicollinearity, we put each subscale one by one in Model 2. The statistic packages SPSS 17.0J and AMOS 5.0 were used for analysis.

Results

A total of 680 questionnaires were distributed to participants through labour unions, and 650 were returned (97 per cent). Of these, 625 returned questionnaires without missing answers were included for analysis (the effective response rate was 92 per cent).

Table 11.1 shows respondent characteristics. Of the respondents, 75 per cent were male, and the mean age was 33.0 years. Most of the participants had an undergraduate degree or higher (95 per cent). Occupations of the respondents were engineer (47 per cent), sales and marketing (40 per cent) and clerical work (13 per cent). As for working hours, 62 per cent of respondents worked more than 50 hours per week, beyond which regular work becomes unhealthy (Messenger 2004).

Table 11.1 Demographic characteristics of participants (N=625)

	n	%	Mean	SD
Age			33.0	5.7
Gender				
Male	468	75		
Female	157	25		
Marital status				
Married	382	61		
Not married	243	39		
Education				
University/graduate school	593	95		
Less than college	32	5		
Occupation				
Engineers	291	47		
Sales/marketing	254	40		
Corporate staff	80	13		
Hours of work/per week				
>50 hours	238	38		
<=50 hours	387	62		

Table 11.2 shows means and standard deviations (SDs) of work demand and work control items. Among the work demand items, scores of qualitative work demand items such as "My work always requires new skills and knowledge" and "My work always requires creativity and innovation" were relatively high (3.4 and 3.3, respectively).

Table 11.3 shows the range, number of items, mean scores, SDs and mean per item of each subscale of participative climate. Comparing the mean score per item of each subscale, we found that the scores for "teamwork," "sufficient information flow about the organization and the job" and "supervisors' management" were relatively high.

Table 11.2 Means and SDs of work demand and work control items

Work Demand	Mean		SD
1. My work is busy.	3.1	±	0.69
2. Clients as well as internal/external work partners are demanding.	2.9	±	1.03
3. My work always requires creativity and innovation.	3.3	±	0.74
4. My work always requires new skills and knowledge.	3.4	±	0.70
5. My workload is heavy.	3.1	±	0.73
6. I am required to achieve agressive targets.	3.3	±	0.67
7. I don't have time to breathe as i always face looming work deadlines.	2.7	±	0.78
8. It is impossible to finish work without overtime.	2.9	±	0.89
Work Control			
9. I can decide how to proceed with my own work.	3.1	±	0.72
10. I can decide my own work schedule.	3.1	±	0.79

Table 11.3   Range, number of items, means, SDs and means per items of the subscale of participative climate (N=625)

	Range	Number of item	Mean		SD	Mean per item
Teamwork	3–15	3	11.6	±	2.2	3.9
Organizational trust and fairness	3–15	3	10.3	±	2.3	3.4
Sufficient information flow about the organization and the job	3–15	3	10.7	±	1.9	3.6
Valuing employees' opinions	3–15	3	10.3	±	2.3	3.4
Supervisors' mamagement	3–15	4	14.2	±	3.1	3.5
Efficient meetings	3–15	4	12.8	±	2.8	3.2

Table 11.4 presents range, means, SDs and intercorrelations between key variables. Participative climate was significantly and negatively correlated with depression and fatigue, and significantly and positively correlated with work motivation.

Table 11.5 shows the results of multivariate correlation of work demand, work control and participative climate on health and work motivation. Participative climate was negatively related to fatigue and depression, and positively related to work motivation. This means the higher participative climate is, the better employees' health and work motivation are, thus supporting H1 and H2. As for the effects of other independent variables in Model 1, work demand was positively related to all three dependent variables. Work control was positively related to work motivation and negatively related to fatigue and depression.

After adding participative climate in Model 2, the effects of work control on three dependent variables decreased or diminished. As shown in Model 3, participative climate

Table 11.4   Range, means, SDs and correlations of the variables used in this study (N=625)

	Range	Mean	SD	1	2	3	4	5
Work demand	8–32	24.7	3.9	0.13**				
Work control	2–8	6.3	1.4					
Participative climate	20–100	69.9	10.8	0.05	0.24***			
Fatigue	0–18	5.7	4.2	0.15***	-0.15***	-0.29***		
Depression	0–36	4.3	4.7	0.10*	-0.11**	-0.30***	0.67***	
Work motivation	5–20	15.2	3.1	0.07	0.14***	0.45***	-0.34***	-0.35***

Note: *p<0.05; **p<0.01; ***p<0.001.

Table 11.5   Standardized regression coefficients of work factors on fatigue, depression and work motivation (N=625)

Dependent Variables	Fatigue			Depression			Work Motivation	
	Model 1	Model 2	Model 3	Model 1	Model 2	Model 3	Model 1	Model 2
Work demand	0.19***	0.19**	0.19***	0.12**	0.13**	0.13**	0.11**	0.11***
Work control	-0.17***	-0.10*	-0.10*	-0.12**	-0.05	-0.05	0.17***	0.03
Participative climate		-0.27***	-0.26***		-0.30***	-0.27***		0.45***
Work demand X participative climate			0.04			0.10**		
Adjusted R²	0.14	0.20	0.20	0.10	0.17	0.18	0.17	0.35

Note: Adjusted for age, gender, marital status, presence of chronic disease, occupation and company.
*p<0.05; **p<0.01; ***p<0.001.

had the strongest association with fatigue, depression and work motivation among all independent variables.

With respect to the moderating effect of participative climate on health, the interaction variables of work demand and participative climate had a significantly positive association with depression but not with fatigue. Thus, H3 was partially supported.

As for the additional results of analysing the relationships of each subscale to fatigue, depression and work motivation, all of the items had significant and negative relationships with fatigue(-0.16– -0.21) and depression (-0.14– -0.23), and significant and positive relationships with work motivation (0.29–0.39).

Discussion

Our primary finding was that if employees' psychological climate is participative, employees can be healthy and motivated (H1 and H2 were supported). Since work motivation is a source of performance (Locke and Latham 1990), enhancing participative climate in the workplace may contribute not only to employees' health but also to organizational effectiveness by increasing employees' work motivation. Therefore, as we expected, we confirmed that participative climate can be a key in creating healthy workplaces. A previous study analysing correlates of work environmental factors to employees' perception of their workplace as healthy reported that good communication and supportiveness in the workplace was the strongest predictor of a healthy work environment (Lowe et al. 2003). The results of our study are consistent with these findings.

When examining H1 and H2, we found that the association of participative climate to employees' health and work motivation were stronger than those of work demand and work control to employees' health and work motivation. These findings were in line with the results of a previous study that revealed that favourable organizational climate predicts depression to a larger extent than work demands and work control (Ylipaavalniemi et al. 2005).

Our second finding was that participative climate buffers the negative effect of work demand on depression but not fatigue (H3 was partially supported). This suggests that the buffering hypothesis of the JD-R model was partially supported. A possible explanation for this result is that the psychological burden caused by work demand is modified through communication and cooperation with supervisor and co-workers, though the physical burden is determined by the workload itself and not modified by social factors at the workplace.

Our finding that associations of work control with fatigue, depression and work motivation were diminished or weakened when we added participative climate to the model suggests that participative climate may increase work control. Spreitzer (1996) found that employees whose workplaces have a participative climate perceive themselves to have a higher level of empowerment than employees in non-participative climates. Empowerment enables people to control things that affect them (Spreitzer 1996, Zimmerman and Rappaport 1988). Therefore, we assumed employees working in the participative climate feel empowered and have more control over their work.

With additional analyses, we found that each dimension of participative climate independently contributes to improving employees' health and work motivation. On the other hand, correlations of each subscale were from 0.29 to 0.65, all of which were significant. This suggests that by working on improving one dimension of participative climate, other dimensions are indirectly improved and then employees' health and work motivation are improved; which is worth noting from a practical standpoint., An unexpected finding was the positive relationship between work demand and work motivation. There have been few studies which deal with work demand and work motivation at the same time (Van Yperen and Hagedoorn 2003). The JD-R model also does not mention an association with these variables. Thus, we did not expect work demand to be positively related to work motivation. We may be able to explain this finding by employing the concept of challenging stressors and hindrance stressors (Lepine et al. 2005).

Hindrance stressors are those stressors restricting one's ability to accomplish one's work or one's development, such as role conflict, role overload and role ambiguity, while challenging stressors are those promoting one's growth and achievement, including a large workload, time pressure and responsibility (Lepine et al. 2005). The items of work demand used in this study were quantity of work (such as workload and time pressure) and quality of work (such as necessary knowledge, skills and creativity). These may belong to the category of challenging stressors described above. Therefore in this study we assume work demand is positively related to work motivation.

As for the items related to work demand, the scores for items measuring qualitative work demand, such as "My work always requires creativity and innovation" and "My work always requires new skills and knowledge," were relatively high. We considered these results reflect the characteristic of jobs in the pharmaceutical industry. Since employees of pharmaceutical companies work by using medical and other scientific knowledge while working alongside medical professionals, it is a constant challenge for them to maintain up-to-date knowledge and skills. A study on employees in the medical IT industry who also need to stay up-to-date on the latest IT technology and medical knowledge to discuss with doctors showed similar results (Yamazaki et al. 2011).

To enhance participative climate, management should work on improving organizational communication since open communication with sufficient information flow forms the core of participative climate.

Companies that participated in this study were actively improving communication within the organization. As discussed above, they had recently undergone M&A and may have been cognizant of the importance of communication. In interviews, labour union representatives mentioned that labour unions actively gather opinions from union members and exchange opinions with management to improve the workplace. Not only labour unions but also companies understand the importance of communication with employees and have adopted various policies reflecting this awareness, including corporate newsletters, employee surveys and performance evaluation systems, or so-called "management by objectives," which enhances communication between supervisors and subordinates. It is assumed that these communication activities are responsible for the relatively high scores for the subscales of "teamwork," "sufficient flow of information about the organization and the job" and "supervisors' management."

We would now like to present other such measures taken by companies as discussed in academic literature and other useful articles.

Glaser (1994) described the intervention method for team-building. The intervention consisted of interpersonal communication and group process. The purpose of interpersonal communication was to develop honest and frank communication among team members, including how to respond to criticism, raise issues and praise one another. The idea behind group process was to create consensus among team members and collaborate to solve issues through discussing real managerial issues and conflicts. In all, four follow-up sessions were held to bring the intervention to a successful conclusion.

Another example of building teamwork is the cross-functional team method. In a cross-functional team, a group of people with different functional expertise work together towards a common goal. Pinto and Pinto (1990) examined relationships among cross-functional cooperation, communication within project teams, tasks and psychological outcomes. In their study, highly-cooperative project teams exhibited better psychological outcomes, such as enjoying teamwork with project members, which suggests that introducing cross-functional teams may enhance team-building. They also reported that highly cooperative project teams engaged in more informal communication compared to poorly cooperative project teams. Furthermore, the reason for communication in highly cooperative project teams was to accomplish a task rather than resolve interpersonal conflicts, suggesting that well-working cross-functional teams may contribute to improving communication among team members across divisions.

Management needs to consider the active use of various communication media to improve the flow of information within the organization. For example, holding weekly meetings at the workplace and utilizing internal newsletters are basic conventional methods for ensuring the flow of information in many organizations. Today, we are able to exchange information quickly and conveniently via such IT tools as email and intranet. Moreover, the introduction of social networking tools such as SNS allow people within the organization to communicate interactively. An article by Ibarra and Hansen (2011) discusses virtual off-site meetings through the use of IT communication tools. The organization featured in the article holds the company's annual management meeting, which only top management attends, off-site meeting. Since management realized that employees wanted to know what issues were discussed in the meeting, they set up a vertical off-site meeting in which the entire organization can participate by computer. This event initiated active information exchange and discussion within the organization. While holding off-site meetings using IT may pose difficulties for some organizations, uploading minutes of managerial meeting on the intranet and gathering employees' feedback is easier to accomplish.

Meanwhile, face-to-face communication is also important in addition to IT-based communication. Previous studies have shown that perceptions of mutual understanding and group decisions are better with face-to-face communication group than with computer-mediated communication groups (Adrianson and Hjelmquist 1991, Straus and McGrath 1994). Organizations need to think about setting up opportunities for employees to be together, such as town-hall meetings with top management, regular departmental meetings, social events and brownbag luncheons at the workplace, to allow employees to share information and exchange opinions. Besides opportunities for formal communication, measures which promote informal communication among employees are also important. One example is Google, which brings games and training machines into the office so that employees naturally come together and talk, the purpose of which is to encourage interaction between employees to stimulate new ideas (Takahashi et al. 2008). These are only a few examples for enhancing organizational communication.

Finally, we would like to address the strengths and limitations of this study. In terms of strengths, first, this is one of the few studies of which we are aware that examines the relationship of participative climate to employee health and work motivation. As we discussed earlier, participatory work approaches to improving employee well-being must be integrated into daily work for their effects to be long-lasting (Mikkelsen et al. 2000). However, this has not been demonstrated. Participative climate is the environment in which employee participation is rooted in daily work. Thus, the results of the present study demonstrated the suggestion Mikkelsen et al. pointed out in the above-mentioned study. In work–family literature, it has been reported that, rather than specific family-friendly policies, an organizational climate in which supervisors and co-workers understand and support family responsibilities is the strongest predictor of satisfaction with the work–family balance and job satisfaction among parents (Saltzstein et al. 2001). In a similar way, this study suggests that enhancing participative climate is as important as establishing a formal policy or structure for employee participation in the organization.

Second, traditionally, both stress and motivation research have been conducted separately (Demerouti and Bakker 2011), and there are few studies which treat both health and work motivation concurrently (Van Yperen and Hagedoorn 2003). In reality, the workplace requires effective guidance for both. Our results suggest that creating a participative climate can be a solution for this.

This study has several limitations. First, since this is a cross-sectional study, causal relationships remain unclear. Longitudinal studies are needed to further validate the causality of these relationships. Second, we used snowball sampling through labour unions to invite participants in the study. High response rate suggests that participants were cooperative and that they have a favourable perception of their work environment. Also, participants in this study were non-supervisory white-collar employees of large pharmaceutical companies in the Tokyo metropolitan area. Thus, findings should be generalized with caution.

We expect that future research will be conducted with various participants by industry, size and location of the company, and employee position to further verify the generalizability of the findings.

Conclusion

Succeeding in today's competitive global management environment may require workers to bear increased workload and demands. Given such, creating a participative climate at workplaces can be a viable solution for improving employee health, work motivation and organizational effectiveness. Thus, this study confirmed that participative climate can be an important part of creating healthy workplaces.

Acknowledgements

This study was supported by a FY2009 Grant-in-Aid for Scientific Research from the Ministry of Education, Culture, Sports, Science and Technology (Project number 21243033).

Appendix 1

Items of Participative Climate

TEAMWORK

1. People I work with try to resolve difference of opinion together.
2. People I work with function as a team.
3. People I work with confront problems in a constructive manner.

ORGANIZATIONAL TRUST AND FAIRNESS

4. The company responds to everyone with fairness and consistency.
5. There is a trusting atmosphere in the company.
6. The company discusses issues constructively with employees.

SUFFICIENT FLOW OF INFORMATION ABOUT THE ORGANIZATION AND THE JOB

7. I have sufficient information to understand the overall picture of my work.
8. When change occurs in the company and job, the reasons are clearly explained.
9. I obtain the necessary information to work well.

VALUING EMPLOYEES' OPINIONS

10. The company makes efforts to gather opinions from employees.
11. The company welcomes suggestions from employees.
12. The company assigns equal value to opinions from all employees regardless of their positions.

SUPERVISOR'S MANAGEMENT

13. Directions from my supervisor are clear.
14. My supervisor is a good listener.
15. My supervisor provides me feedback.
16. My supervisor delegates authority.

EFFICIENT MEETINGS

17. In meetings, everyone participates in the discussion.
18. In meetings, agendas stay on track.
19. Creativity is encouraged from everyone in meetings.
20. Decisions made in meetings are implemented.

Appendix 2

Result of Exploratory Factor Analysis of Work Demand and Work Control Items

	Quantitive work demand	Qualititive work demand	Work control
My workload is heavy.	0.87	0.03	0.06
My work is busy.	0.82	0.08	0.02
It is impossible to finish work without working overtime.	0.70	-0.04	-0.04
I don't have time to breathe since I always face looming work deadlines.	0.57	0.09	-0.17
Clients as well as internal/external work partners are demanding.	0.50	-0.21	0.10
My work always requires creativity and innovations.	-0.12	0.86	0.01
My work always requires new skills and knowledge.	-0.06	0.64	-0.03
I am required to achieve aggressive targets.	0.14	0.53	0.08
I can decide on my own work schedule.	-0.04	0.02	0.86
I can decide how to proceed with my own work.	0.03	0.02	0.71
Variance explained (%)	29.5	17.2	6.4

References

Adrianson, L. and Hjelmquist, E. 1991. Group processes in face-to-face and computer-mediated communication. *Behaviour and Information Technology*, 10(4), 281–96.

APA 2012. *Psychology Healthy Workplace Program*. Available at: http://www.phwa.org/resources/creatingahealthyworkplace/ [accessed 26 April 2012].

Arnold, K.A., Turner, N., Barling, J., Kelloway, E.K. and McKee, M.C. 2007. Transformational leadership and psychological well-being: The mediating role of meaningful work. *Journal of Occupational Health Psychology*, 12(3), 193–203. Available at: https://www.umanitoba.ca/faculties/management/faculty_staff/media/arnold_et_al.pdf [accessed 19 June 2012].

Bakker, A.B. and Demerouti, E. 2007. The job demands-resources model: State of the art. *Journal of Managerial Psychology*, 22(3), 309–28.

Bültmann, U., Kant, I., Kasl, S.V., Beurskens, A.J.H.M. and van den Brandt, P.A. 2002. Fatigue and psychological distress in the working population: Psychometrics, prevalence, and correlates. *Journal of Psychosomatic Research*, 52(6), 445–52. Available at: http://www.sciencedirect.com/science/article/pii/S0022399901002288 [accessed 21 April 2012].

Burton, J. 2010. *WHO Healthy Workplace Framework and Model: Background and Supporting Literature and Practices*. Available at: http://www.who.int/occupational_health/healthy_workplace_framework.pdf [accessed 2 December 2011].

Cartwright, S., Cooper, C.L. and Whatmore, L. 2000. *Improving Communications and Health in a Government Department*. London: Taylor & Francis.

Cooper, C.L. and Cartwright, S. 1994. Healthy mind, healthy organization – a proactive approach to occupational stress. *Human Relations*, 47(4), 455–71.

Cooper, C.L., Liukkonen, P., Cartwright, S. and European Foundation for the Improvement of Living and Working Conditions 1996. *Stress Prevention in the Workplace: Assessing the Costs and Benefits to Organisations*. Dublin: The European Communities.

Demerouti, E. and Bakker, A.B. 2011. The job demands-resources model: Challenges for future research. *SA Journal of Industrial Psychology*, 37(2), 1–9. Available at: http://www.sajip.co.za/index.php/sajip/article/view/974/1037 [accessed 20 April 2012].

Demerouti, E., Bakker, A.B., Nachreiner, F. and Schaufeli, W.B. 2001. The job demands-resources model of burnout. *Journal of Applied Psychology*, 86(3), 499–512. Available at: http://psycnet.apa.org/index.cfm?fa=buy.optionToBuyandid=2001-06715-012 [accessed 15 March 2012].

Denison, D.R. 1996. What is the difference between organizational culture and organizational climate? A native's point of view on a decade of paradigm wars. *Academy of Management Review*, 21(3), 619–54.

Fukui, S., Haratani, T., Toshima, Y., Takahashi, M., Nakata, A., Fukazawa, K., Ohba, A., Sato, E. and Hirota, Y. 2004. Measuring workplace climate: Reliability and validity of the 12-item organizational climate scale (OCS-12). *San Ei Shi*, 46(6), 213–22. Available at: http://www.jstage.jst.go.jp/article/sangyoeisei/46/6/213/_pdf/-char/ja/ [accessed 24 April 2012].

Glaser, S.R. 1994. Teamwork and communication. *Management Communication Quarterly*, 7(3), 282–96.

Glaser, S.R., Zamanou, S. and Hacker, K. 1987. Measuring and interpreting organizational culture. *Management Communication Quarterly*, 1(2), 173–98.

Grawitch, M.J., Gottschalk, M. and Munz, D.C. 2006. The path to a healthy workplace: A critical review linking healthy workplace practices, employee well-being, and organizational improvements. *Consulting Psychology Journal: Practice and Research*, 58(3), 129–47. Available at: http://psycnet.apa.org/journals/cpb/58/3/129/ [accessed 16 October 2011].

Grawitch, M.J., Trares, S. and Kohler, J.M. 2007. Healthy workplace practices and employee outcomes. *International Journal of Stress Management*, 14(3), 275–93.

Heaney, C.A., Israel, B.A., Schurman, S.J., Baker, E.A., House, J.S. and Hugentobler, M. 1993. Industrial relations, worksite stress reduction, and employee well-being: A participatory action research investigation. *Journal of Organizational Behavior*, 14(5), 495–510.

House, J.S. and Cottington, E.M. 1986. Health and the workplace. In *Applications of Social Science to Clinical Medicine and Health Policy*, edited by L.H. Aiken and D. Mechanic. New Jersey, NJ: Rutgers University Press, 392–416.

Ibarra, H. and Hansen, T.M. 2011. Are you a collaborative leader? *Harvard Business Review*, 89 (July–August), 68–74.

JILPT 2008. *Working Style of Japanese People in the Diversification of Work Arrangements*. Tokyo: The Japan Institute for Labour Policy and Training [in Japanese].

JTB Motivation Research Development Team 1998. *Theory and Practice of Eliciting Motivation and MSQ Method*. Tokyo: Kawade Shobo Shinsha [in Japanese].

Kawai, N. 2011. Does downsizing really matter? Evidence from Japanese multinationals in the European manufacturing industry. *The International Journal of Human Resource Management*, 1–19.

Kawakami, N. and Fujigaki, Y. 1996. Reliability and validity of the Japanese version of job content questionnaire: Replication and extension in computer company employees. *Industrial Health*, 34, 295–306.

Kelloway, E.K. and Barling, J. 2010. Leadership development as an intervention in occupational health psychology. *Work & Stress*, 24(3), 260–79.

Kinoshita, E. 2001. Examining 16 items short version CES-D scale. In *Nationwide Survey of Family Life*, edited by Shinji Shimizu. Tokyo: National Family Research of Japan [in Japanese].

Kobayashi, Y., Kaneyoshi, A., Yokota, A. and Kawakami, N. 2008. Effects of a worker participatory program for improving work environments on job stressors and mental health among workers: A controlled trial. *Journal of Occupational Health*, 50(6), 455–70. Available at: http://www.jstage. jst.go.jp/article/joh/50/6/455/_pdf [accessed 25 April 2012].

Lepine, J.A., Podsakoff, N.P. and Lepine, M.A. 2005. A meta-analytic test of the challenge stressor-hindrance stressor framework: An explanation for inconsistent relationships among stressors and performance. *Academy of Management Journal*, 48(5), 764–75.

Likert, R. 1967. *The Human Organization: Its Management and Value*. New York, NY: McGraw-Hill.

Locke, E.A. and Latham, G.P. 1990. Work motivation and satisfaction: Light at the end of the tunnel. *Psychological Science*, 1(4), 240–46.

Lowe, G.S., Schellenberg, G. and Shannon, H.S. 2003. Correlates of employees' perceptions of a healthy work environment. *American Journal of Health Promotion*, 17(6), 390–99. Available at: http://www.rcrpp.org/documents/23701_fr.pdf [accessed 22 February 2012].

Messenger, J.C. 2004. *Working Time and Workers' Preferences in Industrialized Countries: Finding the Balance*. London: Routledge.

Mikkelsen, A., Saksvik, P.O. and Landsbergis, P. 2000. The impact of a participatory organizational intervention on job stress in community health care institutions. *Work & Stress*, 14(2), 156–70.

Ministry of Health, Labour and Welfare 2007. *Guidelines for Promoting Employee Health in the Workplace*. 2nd revision. Tokyo: Ministry of Health, Labour and Welfare [in Japanese].

Ministry of Health, Labour and Welfare 2008. *Survey of Health Conditions of Working People*. Available at: http://www.mhlw.go.jp/toukei/itiran/roudou/saigai/anzen/kenkou07/index.html [accessed 20 February 2012].

Nyberg, A., Alfredsson, L., Theorell, T., Westerlund, H., Vahtera, J. and Kivimäki, M. 2009. Managerial leadership and ischaemic heart disease among employees: The Swedish WOLF study. *Occupational and Environmental Medicine*, 66(1), 51–5. Available at: http://oem.bmj.com/content/66/1/51.full [accessed 22 April 2012].

Offermann, L.R. and Hellmann, P.S. 1996. Leadership behavior and subordinate stress: A 360' view. *Journal of Occupational Health Psychology*, 1(4), 382–90. Available at: http://psycnet.apa.org/ journals/ocp/1/4/382/ [accessed 19 June 2012].

Parker, C.P., Baltes, B.B., Young, S.A., Huff, J.W., Altmann, R.A., LaCost, H.A. and Roberts, J.E. 2003. Relationships between psychological climate perceptions and work outcomes: A meta-analytic review. *Journal of Organizational Behavior*, 24(4), 389–416. Available at: http://onlinelibrary.wiley. com/doi/10.1002/job.198/pdf [accessed 22 April 2012].

Pinder, C.C. 2008. *Work Motivation in Organizational Behavior*. 2nd edition. New York, NY: Psychology Press.

Pinto, M.B. and Pinto, J.K. 1990. Project team communication and cross-functional cooperation in new program development. *Journal of Product Innovation Management*, 7(3), 200–212. Available at: http://www.sciencedirect.com/science/article/pii/073767829090004X [accessed 19 April 2012].

Probst, T.M. 2005. Countering the negative effects of job insecurity through participative decision making: Lessons from the demand-control model. *Journal of Occupational Health Psychology*, 10(4), 320–29.

Ryan, J.C. and Hurley, J. 2007. An empirical examination of the relationship between scientists' work environment and research performance. *RandD Management*, 37(4), 345–54.

Saltzstein, A.L., Yuan, T. and Saltzstein, G.H. 2001. Work-family balance and job satisfaction: The impact of family-friendly policies on attitudes of federal government employees. *Public Administration Review*, 61(4), 451–67.

Sikorska-Simmons, E. 2005. Predictors of organizational commitment among staff in assisted living. *Gerontologist*, 45(2), 196–205. Available at: http://gerontologist.oxfordjournals.org/content/45/2/196.full.pdf [accessed 12 October 2011].

Sikorska-Simmons, E. 2006. Linking resident satisfaction to staff perceptions of the work environment in assisted living: A multilevel analysis. *The Gerontologist*, 46(5), 590–598. Available at: http://gerontologist.oxfordjournals.org/content/46/5/590.full.pdf [accessed 12 October 2011].

Sparks, K., Faragher, B. and Cooper, C.L. 2001. Well-being and occupational health in the 21st century workplace. *Journal of Occupational and Organizational Psychology*, 74(4), 489. Available at: http://onlinelibrary.wiley.com/doi/10.1348/096317901167497/abstract [accessed 19 June 2012].

Spreitzer, G.M. 1996. Social structural characteristics of phychological empowerment. *Academy of Management Journal*, 39(2), 483–504.

Straus, S.G. and McGrath, J.E. 1994. Does the medium matter? The interaction of task type and technology on group performance and member reactions. *Journal of Applied Psychology*, 79(1), 87–97.

Takahashi, K., Kawai T., Nagata, M. and Watanabe, M. 2008. *Why Atmosphere of Your Workplace is Unpleasant*. Tokyo: Kodansha [in Japanese].

Tesluk, P.E., Vance, R.J. and Mathieu, J.E. 1999. Examining employee involvement in the context of participative work environments. *Group and Organization Management*, 24(3), 271–99.

Tsutsumi, A., Nagami, M., Yoshikawa, T., Kogi, K. and Kawakami, N. 2009. Participatory intervention for workplace improvements on mental health and job performance among blue-collar workers: A cluster randomized controlled trial. *Journal of Occupational and Environmental Medicine*, 51(5), 554–63. Available at: http://journals.lww.com/joem/Abstract/2009/05000/Participatory_Intervention_for_Workplace.5.aspx [accessed 21 November 2011].

van der Klink, J.J., Blonk, R.W., Schene, A.H. and van Dijk, F.J. 2001. The benefits of interventions for work-related stress. *American Journal of Public Health*, 91(2), 270–76.

Van Yperen, N.W. and Hagedoorn, M. 2003. Do high job demands increase intrinsic motivation or fatigue or both? The role of job control and job social support. *Academy of Management Journal*, 46(3), 339–48.

Worrall, L. and Cooper, C. 1997. *The Quality of Working Life 1997 Survey of Managers' Changing Experiences*. London: Institute of Management.

Yamazaki, Y. and Asakura,T. 1999. *Health Science as a Way of Life*. Tokyo: Yushindo [in Japanese].

Yamazaki, Y., Kawai, K., Tsuno, Y., Itagaki, T., Mashiko, T. and Seki, K. 2011. *Study of the Work-related Stressors among Medical IT Employees*. Paper to the 70th Conference of Japanese Society of Pubic Health, Akita, Japan, 19–21 October 2011 [in Japanese].

Ylipaavalniemi, J., Kivimäki, M., Elovainio, M., Virtanen, M., Keltikangas-Järvinen, L. and Vahtera, J. 2005. Psychosocial work characteristics and incidence of newly diagnosed depression: A prospective cohort study of three different models. *Social Science and Medicine*, 61(1), 111–22.

Available at: http://www.sciencedirect.com/science/article/pii/S0277953604005969 [accessed 20 November 2011].

Zimmerman, M.A. and Rappaport, J. 1988. Citizen participation, perceived control, and psychological empowerment. *American Journal of Community Psychology*, 16(5), 725–50.

Leadership Interventions

12 *Transformational Leadership Training for Managers: Effects on Employee Well-being*

MEGHAN DONOHOE and E. KEVIN KELLOWAY

Abstract

The way that organizational leaders behave, and in particular the way that leaders interact and deal with their direct reports, has been consistently shown to be a robust predictor of employees' psychological and physical well-being (Kelloway and Barling 2010). In this chapter, we present case studies to demonstrate that transformational leadership can be trained across a multitude of industries and, more importantly, is likely to result in changes in followers' perceptions, attitudes, development and performance. In addition, there is not one best way to train transformational leadership. Organizations can make use of group training or individual counselling sessions to help leaders change their leadership style and embrace the behaviours of transformational leadership. Leadership training is a proven intervention that promises to be an effective means of improving employee well-being in organizations.

Introduction

FORMS OF LEADERSHIP

Table 12.1 summarizes the different forms of leadership and their characteristics. Organizational leaders have the capacity to enhance or hinder the well-being of their subordinates (Kelloway and Barling 2010). Not only are leaders in a position of power to issue rewards, promotions or punishment, but they also serve as role models for others in the organization. The daily decisions made by leaders can have a variety of consequences on followers, such as increased stress due to work overload (Kelloway et al. 2005), or decreased workplace injuries from the modelling of safe work behaviours (Cree and Kelloway 1997). As has been described elsewhere (Gilbreath 2004), these observations would not come as a surprise to any working person. However, research has now begun to both identify the ways in which organizational leadership affects well-being and to validate the effectiveness of leadership training as an intervention designed to increase both employee and leader well-being (Kelloway and Barling 2010).

The behaviour of leaders in organizations is key to understanding the effect of leaders on employee well-being. Engaging in abusive or aggressive behaviours, or demonstrating a lack of appropriate leadership skills, can have devastating consequences on employee health and well-being (Kelloway et al. 2005). On the other hand, leaders who display transformational leadership behaviours by stimulating and encouraging employees to think for themselves, who raise expectations, and who display consideration for individuals, enhance employee trust and promote individual well-being (Arnold et al. 2007). Although it is common to speak of "transformational" or "poor" leaders, increasingly we recognize that leaders can engage in a variety of behaviours (Kelloway et al. 2006, Mullen et al. 2011) and it is the relative frequency of good or bad leadership behaviours that seems to have an effect on well-being. We suggest that such behaviours can be classified into four distinct forms – abusive, passive, contingent reward and transformational leadership.

Abusive Leadership

Abusive leadership is a highly ineffective approach to leadership. Leaders who are abusive or aggressive may engage in behaviours that involve both physical acts of violence and nonphysical aggression. While the former is relatively uncommon in the workplace, aggressive or punitive behaviours at work do occur. These include behaviours such as ridiculing or yelling at employees, name calling, punishing employees by withholding promotions or special assignments, threatening job loss or pay cuts, blaming employees for mistakes they didn't make, or excluding employees by withholding information (Kelloway and Barling 2010, Kelloway et al. 2005). Abusive supervisors can have detrimental effects on the well-being of followers, with researchers linking abusive supervision to employee burnout (Grandey et al. 2007), decreased feelings of commitment to the organization (Tepper 2007), and increased levels of strain (Harvey et al. 2007). Organizational outcomes of abusive leaders are also a major concern from an organizational perspective, ranging from absenteeism due to stress and high levels of employee turnover (Schat and Kelloway 2003).

Passive Leadership

Passive leadership is commonly defined as a combination of laissez-faire and management-by-exception leadership, both of which are associated with an overall lack of leadership skills (Bass 1990). Leaders who manage by exception may do so either passively or actively. Passive management-by-exception leaders wait uninvolved for mistakes, errors or anything outside of the normal to occur, before taking corrective action. This passive approach to leadership has been linked to lower performance in business units (Howell and Avolio 1993). Managers may also manage actively by exception, whereby they actively monitor situations for mistakes and errors, or anything that is outside the normal work standard. When a mistake is made, the manager only takes steps to correct the mistake when it is absolutely necessary. Laissez-faire leadership is a style of leadership that is generally considered the least effective. These managers will avoid management role responsibilities altogether and make themselves scarce when problems occur (Bass and Avolio 1994, Kelloway et al. 2005). The absence of leadership guidance shown by laissez-faire leaders can have strong negative impacts on worker safety and the occurrence of workplace injuries among employees (Kelloway et al. 2006). Ultimately, leaders who lack positive leadership skills contribute to negative work characteristics that are related to increased stress and strain among employees, including overloading employees with work, propagating conflict and ambiguity in the workplace, and treating employees unjustly (Kelloway et al. 2005).

Contingent Reward

Contingent reward is viewed as a much less detrimental form of leadership compared to abusive and passive leadership styles. It subscribes to the notion of reward and punishment. Good performance is rewarded, through recognition, pay raises and promotions among other things, while lesser or poor performance is penalized through demotion, disciplinary action and job loss. While in theory this seems to be an effective leadership style, all too often leaders engaging in contingent reward behaviours subscribe to passive leadership traits (Bass 1990, Barling et al. 2011). The reward and punishment of follower behaviours often falls prey to inconsistencies when leaders apply these principles differently across employees, or when not all behaviours are met with a reward or punishment. Frequently, leaders emphasize negative consequences and spend their time focused on rectifying mistakes rather than also recognizing accomplishments.

When contingent rewards are appropriately provided, employees can be expected to alter their behaviours so as to meet expectations – in this sense contingent rewards may be thought of as good management. However, for behaviours that inspire and motivate leadership beyond expectations, we turn to Bass's (1990) conceptualization of transformational leadership.

What is Transformational Leadership?

Transformational leadership is the most researched style of positive leadership in the scientific literature (Barling et al. 2011). Four dimensions characterize transformational leadership, known as the 4Is: idealized influence, inspirational motivation, intellectual stimulation and individual consideration (Bass 1990, Avolio 1999).

Leaders demonstrate idealized influence when they act in a fair and ethical manner, engendering trust and respect from their employees. Inspirational motivation is demonstrated by leaders who motivate employees with compelling stories and symbols, and who encourage employees to go beyond their current performance and achieve their maximum potential. A leader who responds to the unique needs of each and every employee in a way that is meaningful and appropriate for each individual demonstrates individualized consideration. This leader will recognize the contributions made by each individual employee and celebrate their achievements, for example by sending them an

Table 12.1 Types of organizational leadership styles

Abusive Leader
Engages in hostile verbal and nonverbal behaviours, but are generally not physically violent.
Passive Leader
Active management by exception: actively monitors situations for mistakes and errors, or anything that is outside the normal work standard. When a mistake is made, the leader takes steps to correct the mistake when necessary.
Passive management by exception: waits passively for mistakes and errors, or anything outside of the normal to occur and then takes corrective action.
Laissez-faire: avoids management role responsibilities and is often absent when problems occur.
Contingent Leader
Rewards good performance and punishes poor performance.
Transformational Leader
Idealized influence: engenders trust and respect, does the right thing, and acts as a role model.
Inspirational motivation: encourages employees to achieve levels of performance beyond their own expectations, using stories and symbols to communicate their vision.
Individual consideration: treats employees as individuals, showing compassion, responding to needs, and recognizing and celebrating their achievements.
Intellectual stimulation: engages rationality, challenges assumptions and encourages new ways of thinking, and helps employees answer their own questions.

email of appreciation for a project well done, or recognizing the employee in front of his or her peers for accomplishing an assigned task. Finally, intellectual stimulation is demonstrated by leaders who continually challenge employees to think for themselves in new and innovative ways, and encourage them to challenge assumptions and work towards finding better answers to their own questions.

Does Transformational Leadership Make a Difference?

Many leaders may insist that they are not necessarily passive or abusive leaders, and they also don't feel the need to engage in transformational leadership behaviours. Overall, these leaders feel they are effective at achieving organizational goals, managing subordinates, and at the end of the day they are able to get the job done. If you generally think of yourself as an effective leader, does transformational leadership really matter?

The answer is Yes! Both practice and research have established that transformational leadership is related to increased organizational effectiveness (e.g. Judge and Piccolo 2004) and employee well-being (Kelloway and Barling 2010). While some leaders may believe they are effective at getting employees to complete the tasks at hand, transformational leaders are not only getting the job done, they are doing so in a way that promotes employees health and well-being, enhances employee attitudes, stimulates teamwork, and surpasses the status quo.

EMPLOYEE ATTITUDES

Employees with transformational leaders have reported feeling more satisfied with their jobs (Hater and Bass 1988, Koh et al. 1995, Purvanova et al. 2006), expressed increased feelings of fairness (Pillai et al. 1999), as well as increased trust (Pillai et al. 1999, Podsakoff et al. 1996) and satisfaction with their leaders (Hater and Bass 1988, Koh et al. 1995). A useful way to gauge employee attitudes towards their work is to ask employees to describe what they do in their daily job. For example, when researchers Purvanova et al. (2006) asked aerospace wire technicians about their job, the employees whose leaders were not transformational described their role as connecting the red wire with the yellow wire. However, the technicians with transformational leaders described their job as making wire harnesses for airplanes. One technician even reported that whenever a plane crash occurred, employees waited anxiously to find out if the plane involved in the crash was using one of their wire harnesses. Not only did these employees see a greater vision for their work, they felt a keen sense of responsibility associated with the task. Although all the technicians were doing the exact same job, the employees with the transformational leader saw their work as challenging, meaningful and engaging even though the tasks were routine and low in complexity.

Transformational leadership behaviours have also been linked to increased organizational citizenship behaviours (Koh et al. 1995, Piccolo and Colquitt 2006, Purvanova et al. 2006). Employees demonstrate organizational citizenship behaviours when they go above and beyond the requirements of their jobs to help another employee or contribute in some way to the organization's objectives. Employees are also less likely to leave the organization (Bycio et al. 1995) and feel a greater sense of personal commitment to their company when their leaders are transformational (Barling et al. 1996, Bycio et al. 1995, Koh et al. 1995).

WORK PERFORMANCE

There is general consensus among researchers that transformational leadership is related to organizational effectiveness and work performance (Judge and Piccolo 2004, Lowe et al. 1996). How do transformational leaders motivate employees to perform effectively? Transformational leaders use individual consideration and intellectual stimulation to inspire creativity and innovation (Bass 1985, Burns 1978, Howell and Avolio 1993, Jung 2000, Sosik et al. 1998). When employees feel as though their leader values their ideas and views, they are more likely to continue to produce and share these ideas for the betterment of the organization. On the same note, leaders who intellectually stimulate their followers will challenge employees to think outside the box, for example, to help find more efficient and cost-effective solutions to problems, or innovative ways of aligning organizational processes. Ultimately, these leadership behaviours are positively related to the achievement of targeted business unit performance goals (Howell and Avolio 1993), sales performance (Barling et al. 1996), project team performance (Keller 2006, Lim and Ployhart 2004), and even leader performance (Judge and Piccolo 2004). Of particular importance are findings that transformational leaders generally play central roles in advice and influence networks within the organization (Bono and Anderson 2005). As transformational leaders become integral and influential parts of the organizational web, they can have a significant impact on the organization, a scenario which Dvir and colleagues (2002) liken to a "falling domino effect": transformational leaders positively influence the performance of direct reports, who then have an influence on their own direct reports. These behaviours can trickle down and expand across the entire organization.

SAFETY

The involvement of leaders in organizational safety initiatives is imperative for the safety and well-being of employees (Mullen et al. 2011). Transformational leadership in particular is related to both safety compliance and safety participation among employees, including behaviours such as supporting workplace safety, attending safety meetings and helping co-workers with safety related tasks (Inness et al. 2010, Mullen et al. 2011). Alternatively, abusive and passive leadership can have detrimental and even reverse effects on safety. Abusive leaders who coerce or mock employees displaying positive safety behaviours may, in fact, intimidate employees into performing their tasks unsafely for fear of being further humiliated or alienated (Mullen 2004). It is also not enough to only encourage safe work practices some of the time. Passive leaders who inconsistently supported employee safety accounted for fluctuations in safety consciousness, safety climate, safety related events and injuries (Kelloway et al. 2006). Leading a workplace that is safe for all employees requires the consistent promotion of safe work practices (Teed et al. 2008). Transformational leaders are adept at instilling safe working behaviours in employees, especially when they emphasize safety compliance and participation in their leadership behaviours. Specifically, when transformational leaders intellectually stimulate employees to think about and champion ways to enhance their own safety, employees are more likely to be conscious of their safe behaviour and engage in safe work practices (Barling et al. 2002, Kelloway et al. 2006, Mullen et al. 2011).

HEALTH AND WELL-BEING

It comes as no surprise that leadership is one of the key determinants of employee well-being (Kelloway and Day 2005). A vast number of studies have examined the relationship between transformational leadership behaviours and employee health. The employees of transformational leaders reported lower levels of job related stress (Offermann and Hellmann 1996, Podsakoff et al. 1996, Sosik and Godshalk 2000), job strain, burnout and depression (e.g. Kelloway and Barling 2010, Kuoppala et al. 2008, Lee and Ashforth 1996, Moyle 1998, Rooney and Gottlieb 2007, Van Dierendonck et al. 2004). Furthermore, transformational leadership is linked to enhanced employee well-being. Specifically, psychological well-being is associated with feeling motivated, cheerful, joyous, energetic, enthusiastic and lively (Arnold et al. 2007, Bono et al. 2007, McKee and Kelloway 2009, Nielsen et al. 2008). In addition, transformational leadership is associated with spiritual well-being, that is, feelings that life is a positive experience and having a meaningful relationship with a higher power (McKee et al. 2011). Furthermore, transformational leaders support a sense of workplace spirituality by helping employees experience meaning in their work, and enhancing employee understanding of the connection between their own values and that of the organization and community (McKee et al. 2011).

Given the positive impact that transformational leaders can have on employee attitudes, performance, safety behaviours, health and well-being, managers must make an effort to enact positive and effective leadership behaviours. Yet, if managers do not naturally exhibit transformational leadership behaviours, is it possible to teach them how to become transformational leaders? The following section explores research in practice on training transformational leadership behaviours.

Can Transformational Leadership be Trained?

Researchers were keen to uncover whether training managers on the behaviours of transformational leadership could in fact change their existing leadership style. Several researchers teamed up with actual organizations to test this question through empirical study.

CASE 1: INVESTING IN TRANSFORMATIONAL LEADERSHIP

Researchers: Barling et al. (1996)

Organization: Large Canadian Bank

One of Canada's largest banks participated in an experiment examining the effectiveness of transformational leadership training for branch managers. If the training was indeed effective at enhancing transformational leadership behaviours, the researchers expected three specific outcomes:

1. that subordinates would notice the changes in their manager's leadership behaviours;
2. that subordinates would feel increased commitment to their organization as a result of the improved leadership behaviours; and
3. the financial performance of the work unit would be enhanced.

Twenty male and female branch managers were grouped into either a training intervention group, or a control group that received no training. Employees who reported directly to each manager completed questionnaires rating their commitment to the organization, and their manager's leadership behaviours (on the Multifactor Leadership Questionnaire). Branch level financial performance was measured based on sales of personal loans and credit cards. After the initial questionnaires and financial performance measures were gathered, the branch managers in the training group participated in leadership training that involved both classroom training and individual booster sessions. Following the training, the researchers tested the leadership styles of managers again by having subordinates complete the same scales as before. They found that managers who received transformational leadership training were judged by their subordinates to be higher on intellectual stimulation, charisma and individual consideration compared to those managers who did not receive training. Furthermore, subordinates of managers with the training showed increased organizational commitment. Training also had an impact on increased personal loan sales, and a small impact on increased credit card sales.

This study provides evidence that transformational leadership can result in changes in employees' perceptions of manager leadership behaviours, employees' own commitment to the organization, and some aspects of financial performance.

CASE 2: A TREATMENT FOR LEADERSHIP

Researchers: Kelloway et al. (2000)

Organization: Health Care Corporation

Time constraints plague many managers in busy organizations. Therefore it was important for researchers to understand what type of leadership training is the most effective at teaching transformational leadership behaviours. Department managers from a health care corporation participated in an experiment comparing the effectiveness of classroom leadership training versus counselling leadership training. The researchers assigned 40 managers from nursing and administrative support services to one of four groups: classroom leadership training, leadership counselling, classroom training and counselling combined, and a control group. One hundred and eighty employees were asked to rate their supervisor's leadership style prior to the training sessions. The managers assigned to the classroom leadership training attended a one-day workshop, similar to that received by the bank employees in the previous case (Barling et al. 1996). The managers assigned to the counselling training met individually with a researcher for a one-hour session to review employee ratings of their leadership style. Both the training workshop and the counselling session tasked managers with the development of specific and attainable goals focused on transformational leadership behaviours that managers could undertake on a daily and weekly basis. Six months following the training interventions, the employees once again completed an assessment of their supervisor's leadership styles.

Both leaders who participated in the training and those that participated in the feedback sessions were rated higher on transformational leadership than those who did not participate in either session. Of particular interest was that the researchers found that managers who had both workshop training and counselling sessions combined did not have higher ratings of transformational leadership than either intervention on their own. Thus, the training workshop and the counselling feedback are interchangeable when it comes to effectively training transformational leadership behaviours. In short, what mattered is that leaders changed their behaviours – not whether they changed as a result of training or coaching. For leaders that have limited time or budgets, it may be enough to train managers on transformational leadership behaviours in classroom workshops, as opposed to the more expensive and time-consuming individual counselling sessions.

CASE 3: ENHANCING DEVELOPMENT AND PERFORMANCE THROUGH LEADERSHIP

Researchers: Dvir et al. (2002)

Organization: Military

The Israel Defense Forces participated in an experimental leadership intervention to examine the impact of transformational leadership behaviours of military leaders on the development and performance of their platoon followers (both officers and recruits). Fifty-four military leaders received either transformational leadership training or eclectic leadership training over three days. The transformational leadership training emphasized the difference between transformational leadership behaviours and other leadership styles, and the importance of continuous development to higher levels of motivation, morality and empowerment. The eclectic leadership training discussed basic team leadership behaviours including goal setting, trust building, group cohesion and leading by personal example. Following the training, the 54 military leaders were assigned to lead basic training platoons through a four-month infantry basic training course. The researchers assessed the impact of the different leadership trainings by collecting leadership ratings and developmental data from 90 direct followers of the leaders (officers), and 724 indirect followers (new recruits) before and after the training.

The researchers found that the platoons with transformational leaders outperformed the platoons with eclectic leaders in every performance area in the basic training course, including written tests, practical tests, fitness, obstacle course completion and marksmanship. Furthermore, followers of the transformational leaders reported increased feelings of empowerment (as indicated by increased self-efficacy) and demonstrated extra effort compared to the followers in the control group. Thus, the impact of transformational leadership on the enhanced development and performance of followers was significantly greater than the impact of eclectic or general leadership.

The above cases demonstrate that transformational leadership can be trained across a multitude of industries, and more importantly, is likely to result in changes in followers' perceptions, attitudes, development and performance. In addition, there is not one best way to train transformational leadership. Organizations can make use of group training or individual counselling sessions to help leaders change their leadership style and embrace the behaviours of transformational leadership.

Developing Transformational Leaders

Organizations frequently encounter the choice to either train current leaders, or recruit and hire leaders who already have the leadership competencies needed (Bamberger and Meshoulam 2000). Recruitment and selection of leaders outside of the organization is both expensive and time-consuming, and can have negative impacts on the attitudes and commitment of leaders within the organization. Alternatively, training and developing existing leaders in the organization on transformational leadership behaviours is not only cost-effective for the organization, but also empowering for current leaders. Thus, leadership development and training activities are highly recommended as effective leadership interventions by practitioners and researchers alike (Avolio et al. 2009, Kelloway and Barling 2000).

There exist a number of different training formats for effectively enhancing transformational leadership behaviours, including group classroom training, and one-on-one counselling sessions (Kelloway et al. 2000). The assortment of available training opportunities offers flexibility for organizations to train leaders in a way that fits with organizational objectives and limitations. That being said, research has specified several important characteristics that should be included in any type of transformational leadership training. To begin, a training intervention should review the concepts of transformational leadership theory to highlight the differences between the various styles of leadership. For example, Kelloway et al. (2000) asked participants to identify the characteristics of the best and worst leaders they have encountered. These characteristics were then matched to transformational leadership behaviours and poor leadership behaviours. Second, participants should be provided with feedback on their current leadership style. Leaders and subordinates can complete assessments, such as the Multifactor Leadership Questionnaire (MLQ), to gain a well-rounded understanding of current leadership behaviours. Enhancing self-awareness of current leadership styles will contribute to greater self-reflection on how to improve leadership behaviours. Third, participants should be given the opportunity to discuss and role-play the newly learned transformational leadership behaviours. Role-playing activities allow participants a chance to practice changed leadership behaviours and identify the specific behaviours that are consistent with the mission of their organization. When transformational leadership behaviours are linked to the objectives of the organization, the transfer and integration of these behaviours within the context of the workplace is much greater. Lastly, goal setting is a key behaviour to help instill transformational leadership behaviours. Participants must have the opportunity to set specific, challenging, yet attainable goals for themselves to enhance the implementation of transformational leadership behaviours in their daily role. For example, Kelloway et al. (2000) challenged participants to set between three and five specific goals that would allow them to demonstrate transformational leadership behaviours daily. Participants were then encouraged to revisit and adjust their goals over time.

In addition to the inclusion of the above characteristics of effective transformational leadership training, organizations must also pay particular attention to three design elements when considering the implementation of leadership training. Specifically, the intensity of the training intervention, the specific intervening variables used, and the logistics associated with the evaluation of training (Kelloway et al. 2000) all require some consideration.

INTERVENTION INTENSITY

Leadership training interventions do not need to be long, drawn-out courses to enact effective change (Dvir et al. 2002, Kelloway et al. 2000, Mullen and Kelloway 2009). Published intervention studies have had success with training interventions of varying time lengths, ranging from a half day (three-hour) intervention (Mullen and Kelloway 2009) to upwards of five days of training (Dvir et al. 2002). Given that the current literature on transformational leadership training has yet to establish an optimal time length for training interventions, this affords organizations the flexibility to undertake leadership training in a variety of different formats.

SPECIFY INTERVENING VARIABLES

Although leadership training involves the development of positive leadership behaviours for participating managers, it is important not to forget the real purpose of leadership training. That is, we want leaders to be more effective so that they can elicit positive outcomes from their followers (such as productivity, job satisfaction, positive attitudes, etc.). Thus, although the direct effect of leadership development is to enhance the skills of leaders, the ultimate targets of leadership training are in fact the employees, who don't actually participate in the training at all (Avolio 1999). Since leadership training has an indirect impact on employee outcomes, it is important for organizations to understand the elements that might impact the effectiveness of the training between the time the training is delivered, to the time it trickles down to subordinates.

For example, for training to be effective, subordinates must first perceive the changes in their manager's leadership behaviours, and then be motivated to enact changes in their own behaviour as a result (such as improving their job performance). Thus, it is useful for organizations to consider the mechanisms through which effective leadership training, and subsequently behaviours, can improve employee outcomes. This will allow improved measurement of the effectiveness of the training. In addition, identifying intervening variables can help set realistic expectations as to the amount of time it will take for the true effects of the training to be perceived, by both followers and the organization. Research interventions have commonly measured the effects of leadership training three months post intervention, an amount of time considered to be reasonable enough to observe the indirect outcomes of training (Barling et al. 1996, Mullen and Kelloway 2009).

It is interesting to note that while the main purpose of leadership interventions is to improve outcomes among employees, the behavioural outcomes of leaders may also be enhanced. For example, Mullen and Kelloway (2009) found that leaders who were trained on safety specific transformational leadership behaviours reported that their own attitudes and confidence related to safety was improved as a result of the training. Thus, leadership training can also directly influence leader outcomes, whether it may be increased safety attitudes, or greater feelings of personal well-being (Kelloway and Barling 2010).

LOGISTICS OF EVALUATION

Finally, the evaluation of leadership training interventions can present a challenge for organizations. Since employee outcomes should be measured both before and after the implementation of training, many times it can be difficult to gather responses from all the followers of participating leaders. Situations such as employee attrition and lack of survey response may contribute to a reduced ability to effectively evaluate the intervention. Similarly, organizations may go through other changes (e.g. reorganizations, downturns in business, re-assignment of staff) that can affect their ability to conduct a rigorous evaluation of the training.

The critical component of evaluating leadership training is that the leaders' behaviour is evaluated by the same group of employees both prior to, and after, the training. When re-assignments or reorganizations make this impossible, it is difficult to determine whether differences between the pre-test and post-test scores are attributable to:

1. a real change in leadership behaviour (i.e. an effect of training);
2. different expectations in the raters; or
3. both.

Such difficulties may lead organizations to abandon any attempt on evaluation and, in our experience, it is frequently difficult for organizations to see the point of doing the post-training evaluation. However, we believe that ignoring this component of training has detrimental consequences in that the presence of a formal evaluation creates a powerful expectation for change among leaders. Armed with the results of the initial assessment (pre-test) and knowing that they will be assessed again creates a culture in which leaders actively work at changing their transformational leadership behaviours in the desired direction. Moreover, organizations that implement such post-tests also send a message about the desire for and need for change. Ultimately, for transformational leadership training to be effective, mechanisms need to be in place to support first the learning of new transformational behaviours, and second the opportunity for leaders to implement these behaviours in their natural work environment.

The timing of evaluation must also be considered. It is common for organizational researchers to suggest that the identification of an appropriate time lag is critical and needs to be based on theoretical concerns. Yet, there are few organizational theories, including theories of leadership, that are helpful in selecting such a time lag (Kelloway and Francis 2012). Published accounts have suggested a three-month (e.g. Barling et al. 1996) to six-month (Kelloway et al. 2000) time lag between the training and the evaluation. We believe that three months may be the minimum time frame required for changes in leadership to be

1. implemented consistently;
2. recognized by employees as a change; and
3. to "trickle down" to affect employee attitudes and behaviours.

At the same time, we have no rationale or information regarding the upper limit for a time frame. In the absence of such, organizational constraints may dictate the upper limit of the evaluation period and, in our experience, even a three-month time frame results in considerable loss of data through inability to ensure that the same individuals are rating leaders (see, for example, Mullen and Kelloway 2009).

Can Leaders Actually Change Their Behaviours?

Encouraging managers to first learn and then implement new leadership behaviours can be a difficult task. The common approach for making changes to leadership behaviours is to task managers with making a few infrequent, yet major, changes to their behaviours. However, this approach has a weak impact on behaviour change and subordinate outcomes over the long term (Kelloway and Barling 2000). Not only is it difficult to implement a large change to a behaviour or leadership style, but if that particular behaviour does not fit with their own personal style or role within the workplace, then the behaviour will not be integrated and soon be dropped completely. As a result, subordinates may begin to question the ability of their leader to commit to change and become cynical of every effort the leader makes to improve their leadership style.

A more effective approach to successful leadership behaviour change is making small changes and maintaining these changes over time. Small changes in leadership behaviours create greater impacts over time, both on the ability of the leader to truly change their style, and on the perceptions of subordinates (Kelloway and Barling 2000). What do these small changes look like? The changes should be something that the leaders can incorporate into their daily leadership role, and should exemplify the 4Is of transformational leadership.

To enhance idealized influence, leaders should focus on making decisions that exemplify the ethical or right thing to do, as opposed to the most cost-effective or quick solution. This could involve taking the necessary steps to ensure the work environment is safe by replacing old or broken equipment on a regular basis. Leaders can also demonstrate idealized influence by making decisions that are transparent and consistent across all employees. For example, following established policies and procedures and communicating how each decision is made can help employees learn what they can expect from their leader, and further builds respect and trust. "Leaders who are seen by their employees as people who can be counted on to 'do the right thing' epitomize idealized influence, and in return are justly rewarded with their employees' trust" (Kelloway and Barling 2000: 358).

Leaders can display inspirational motivation in small ways by communicating their trust in employees' ability to accomplish their goals and contribute to the organization as a whole. Leaders who show employees that they believe in them, through messages such as "I know you can do it," can have a powerful impact on the self-efficacy of employees and inspire them to try harder and go further.

Hand-in-hand with inspiring motivation, leaders must also stimulate employees to think in novel ways about their work. Instead of always providing the answers to employee queries, transformational leaders intellectually stimulate their followers by engaging employees to come up with their own solutions to their problems. This can be done by simply asking employees, "what do you think we should do?" or "what would you advise in this situation?"

Finally, leaders can demonstrate individualized consideration by making time to pay attention to the individual concerns of each employee on a regular basis. The act of responding to the individual needs of employees can be as simple as taking a few moments each day to walk around and talk with employees. Appreciating the individual efforts of each employee is also a necessary behaviour of individual consideration, and can take the form of personally thanking employees for a job well done, sending a short email of consideration, or even writing a thank you card. It's no wonder that many subordinates and executives alike have admitted to keeping thank you cards they've received in the past for many, many years. The behaviours outlined above are small gestures that take very little time to carry out. Leaders who have been successful in implementing these behaviours have carefully and thoughtfully identified the small behaviours that they plan to enact and designate time to carry out these behaviours frequently. In addition, the small changes should be consistent with the personality and character of the leader. If the behaviour to be implemented does not fit with how the leader views themselves or their workplace, then the new behaviour will fail. For example, if a leader is not one to display marked enthusiasm about the successes of subordinates, it would not be wise to set the goal to recognize employee efforts daily. Instead, a small change in behaviour could involve sharing appreciation and praise with employees at the completion of larger projects.

Conclusion: The Focus of Intervention

We began this chapter with a consideration of the leadership behaviours that affect employee well-being. In particular we suggested a continuum of leadership behaviours ranging from abusive leadership, through passive leadership and contingent reward to transformational leadership. Given that all of the forms of leadership have the potential to affect well-being one might appropriately ask what should be the focus of leadership development? Should we focus on eliminating or reducing negative leadership behaviours (such as passive and abusive leadership behaviours) or should we focus on enhancing more positive leadership behaviours.

We believe that the answer is clearly the latter – the appropriate focus of leadership training should be to enhance transformational leadership. This recommendation is based on both pragmatic and empirical considerations.

First, in terms of behavioural change, we suggest that it is easier and more effective to train people in what to do as opposed to what not to do. Interventions designed to reduce workplace incivility, for example, can quickly identify and ban a laundry list of inappropriate behaviours. However, even if one is successful in eliminating negative behaviours, they are often quickly replaced by new forms of the same behaviour. In contrast, interventions designed to promote "respectful workplaces" (Leiter et al. 2011) have been shown to be effective. In the same way, leadership training focused on the promotion of positive leadership behaviours is, we suggest, the more effective approach.

Second, as reviewed above, there are now several rigorous studies demonstrating the effectiveness of transformational leadership training as a means to alter leadership behaviours. Simply put, this is an approach that has been shown to work, is well grounded in the research literature and presents a practical and cost-effective approach to leadership development in organizations.

Concluding Thoughts

The way that organizational leaders behave, and in particular the way that leaders interact and deal with their direct reports, has been consistently shown to be a robust predictor of employees' psychological and physical well-being (Kelloway and Barling 2010). Leadership training is a proven intervention that promises to be an effective means of improving employee well-being in organizations.

References

Arnold, K.A., Turner, N., Barling, J., Kelloway, E.K. and McKee, M.C. 2007. Transformational leadership and psychological well-being: The mediating role of meaningful work. *Journal of Occupational Health Psychology*, 12(3), 193–203.

Avolio, B.J. 1999. *Full Range Leadership Development: Building the Vital Forces in Organizations*. Thousand Oaks, CA: Sage Publications.

Avolio, B.J., Reichard, R.J., Hannah, S.T., Walumbrwa, F.O. and Chan, A. 2009. A meta-analytic review of leadership impact research: Experimental and quasi-experimental studies. *Leadership Quarterly*, 20, 764–84.

Bamberger, P. and Meshoulam, I. 2000. *Human Resource Management Strategy*. Thousand Oaks, CA: Sage Publications.

Barling, J., Christie, A. and Hoption, C. 2011. Leadership. In *APA Handbook of Industrial and Organizational Psychology*, edited by S. Zedeck. Washington, DC: American Psychological Association, 183–240.

Barling, J., Loughlin, C. and Kelloway, E.K. 2002. Development and test of a model linking safety-specific transformational leadership and occupational safety. *Journal of Applied Psychology*, 87, 488–96.

Barling, J., Weber, T. and Kelloway, E.K. 1996. Effects of transformational leadership training on attitudinal and financial outcomes: A field experiment. *Journal of Applied Psychology*, 81(6), 827–32.

Bass, B.M. 1985. *Leadership and Performance beyond Expectations*. New York, NY: Free Press.

Bass, B.M. 1990. From transactional to transformational leadership: Learning to share the vision. *Organizational Dynamics*, 18(3), 19–36.

Bass, B.M. and Avolio, B.J. 1994. *Improving Organizational Effectiveness through Transformational Leadership*. Thousand Oaks, CA: Sage Publications.

Bono, J.E. and Anderson, M.H. 2005. The advice and influence networks of transformational leaders. *Journal of Applied Psychology*, 90(6), 1306–14.

Bono, J.E., Foldes, H., Vinson, G. and Muros, J.P. 2007. Workplace emotions: The role of supervision and leadership. *Journal of Applied Psychology*, 92(5), 1357–67.

Burns, J.M. 1978. *Leadership*. New York, NY: Harper & Row.

Bycio, P., Hackett, R.D. and Allen, J.S. 1995. Further assessment of Bass' (1985) conceptualization of transactional and transformational leadership. *Journal of Applied Psychology*, 80, 469–78.

Cree, T. and Kelloway, E.K. 1997. Responses to occupational hazards: Exit and participation. *Journal of Occupational Health Psychology*, 2, 304–11.

Dvir, T., Eden, D., Avolio, B.J. and Shamir, B. 2002. Impact of transformational leadership on follower development and performance: A field experiment. *Academy of Management Journal*, 45, 735–44.

Gilbreath, B. 2004. Creating healthy workplaces: The supervisor's role. In *International Review of Industrial and Organizational Psychology, Volume 19*, edited by C. Cooper and I. Robertson. Chichester: John Wiley & Sons, 93–118.

Grandey, A.A., Kern, J. and Frone, M. 2007. Verbal abuse from outsiders versus insiders: Comparing frequency, impact on emotional exhaustion, and the role of emotional labour. *Journal of Occupational Health Psychology*, 12, 63–79.

Harvey, P., Stoner, J., Hochwarter, W. and Kacmar, C. 2007. Coping with abusive supervision: The neutralizing effects of ingratiation and positive affect on negative employee outcomes. *The Leadership Quarterly*, 18(3), 264–80.

Hater, J.J. and Bass, B.M. 1988. Superiors' evaluations and subordinates' perceptions of transformational and transactional leadership. *Journal of Applied Psychology*, 73, 695–702.

Howell, J.M. and Avolio, B.J. 1993. Transformational leadership, transactional leadership, locus of control, and support for innovation: Key predictors of consolidated-business-unit performance. *Journal of Applied Psychology*, 78(6), 891–902.

Inness, M., Turner, N., Barling, J. and Stride, C.B. 2010. Transformational leadership, and employee safety performance: A within-person, between-jobs design. *Journal of Occupational Health Psychology*, 15(3), 279–90.

Judge, T.A. and Piccolo, R.F. 2004. Transformational and transactional leadership: A meta-analytic test of their relative validity. *Journal of Applied Psychology*, 89(5), 755–68.

Jung, D.I. 2000. Transformational and transactional leadership and their effects on creativity in groups. *Creativity Research Journal*, 13(2), 185–95.

Keller, R.T. 2006. Transformational leadership, initiating structure, and substitutes for leadership: A longitudinal study of research and development project team performance. *Journal of Applied Psychology*, 91(1), 202–10.

Kelloway, E.K. and Barling, J. 2000. Knowledge work as organizational behavior. *International Journal of Management Reviews*, 2(3), 287–304.

Kelloway, E.K. and Barling, J. 2010. Leadership development as an intervention in occupational health psychology. *Work & Stress*, 24(3), 260–279.

Kelloway, E.K. and Day, A.L. 2005. Building healthy workplaces: What we know so far. *Canadian Journal of Behavioural Sciences*, 37(4), 223–35.

Kelloway, E.K. and Francis, L. 2012. Longitudinal research methods. In *Research Methods in Occupational Health Psychology*, edited by M. Wang, R. Sinclair and L. Tetrick. New York, NY: Routledge, 374–94.

Kelloway, E.K., Barling, J. and Helleur, J. 2000. Enhancing transformational leadership: The roles of training and feedback. *The Leadership and Organizational Development Journal*, 21, 145–9.

Kelloway, E.K., Mullen, J.E. and Francis, L. 2006. Divergent effects of passive and transformational leadership on safety outcomes. *Journal of Occupational Health Psychology*, 11, 76–86.

Kelloway, E.K., Sivanathan, N., Francis, L. and Barling, J. 2005. Poor leadership. In *Handbook of Workplace Stress*, edited by J. Barling., E.K. Kelloway and M. Frone. Thousand Oaks, CA: Sage Publications, 89–112.

Koh, W.L., Steers, R.M. and Terborg, J.R. 1995. The effects of transformational leadership on teacher attitudes and student performance in Singapore. *Journal of Organizational Behavior*, 16, 319–33.

Kuoppala, J., Lamminpaa, A., Liira, J. and Vainio, H. 2008. Leadership, job well-being, and health effects: A systematic review and meta-analysis. *Journal of Occupational and Environmental Medicine*, 60(8), 904–15.

Lee, R.T. and Ashforth, B.E. 1996. A meta-analytic examination of the correlates of the three dimensions of job burn out. *Journal of Applied Psychology*, 81, 123–33.

Leiter, M.P., Spence-Laschinger, H.K., Day, A. and Gilin-Oore, D. 2011. The impact of civility interventions on employee social behavior, distress, and attitudes. *Journal of Applied Psychology*, 96(6), 1258–74.

Lim, B.C. and Ployhart, R.E. 2004. Transformational leadership: Relations to the five-factor model and team performance in typical and maximum contexts. *Journal of Applied Psychology*, 89, 610–21.

Lowe, K.B., Kroeck, K.G. and Sivasubramaniam, N. 1996. Effectiveness correlates of transformation and transactional leadership: A meta-analytic review of the MLQ literature. *Leadership Quarterly*, 7, 385–425.

McKee, M. and Kelloway, E.K. 2009. *Leading to Wellbeing*. European Academy of Work and Organizational Psychology, Santiago de Compostella, Spain, 13–16 May.

McKee, M., Kelloway, E.K., Driscoll, C. and Kelley, E. 2011. Exploring linkages among transformational leadership, workplace spirituality and well-being in health care workers. *Journal of Management, Spirituality & Religion*, 8(3), 233–55.

Moyle, P. 1998. Longitudinal influences of managerial support on employee well-being. *Work and Stress*, 12(1), 29–49.

Mullen, J.E. 2004. Factors influencing safety behavior at work. *Journal of Safety Research*, 35, 275–85.

Mullen, J.E. and Kelloway, E.K. 2009. Safety leadership: A longitudinal study of the effects of transformational leadership on safety outcomes. *Journal of Occupational and Organizational Psychology*, 82, 253–72.

Mullen, J., Kelloway, E.K. and Teed, M. 2011. Inconsistent style of leadership as a predictor of safety behaviour. *Work & Stress*, 25(1), 41–54.

Nielsen, K., Randal, R., Yarker, S. and Brenner, S. 2008. The effects of transformational leadership on followers' perceived work characteristics and psychological well-being: A longitudinal study. *Work & Stress*, 22, 16–32.

Offermann, L.R. and Hellmann, P.S. 1996. Leadership behavior and subordinate stress: A 3608 view. *Journal of Occupational Health Psychology*, 1, 382–90.

Piccolo, R.F. and Colquitt, J.A. 2006. Transformational leadership and job behaviors: The mediating role of job characteristics. *Academy of Management Journal*, 49, 327–40.

Pillai, R., Schreisheim, C.A. and Williams, E.S. 1999. Fairness perceptions and trust as mediators for transformational and transactional leadership: A two-sample study. *Journal of Management*, 25, 897–933.

Podsakoff, P.M., MacKenzie, S.B. and Bommer, W.H. 1996. Transformational leader behaviors and substitutes for leadership as determinants of employee satisfaction, continent, trust and organizational citizenship behaviors. *Journal of Management*, 22, 259–98.

Purvanova, R.K., Bono, J.E. and Dzieweczynski, J. 2006. Transformational leadership, job characteristics, and organizational citizenship performance. *Human Performance*, 19(1), 1–22.

Rooney, J. and Gottlieb, B. 2007. Development and initial validation of a measure of supportive and unsupportive managerial behaviors. *Journal of Vocational Behavior*, 71(2), 186–203.

Schat, A. and Kelloway, E.K. 2003. Reducing the adverse consequences of workplace aggression and violence: The buffering effects of organizational support. *Journal of Occupational Health Psychology*, 8, 110–22.

Sosik, J. and Godshalk, V. 2000. Leadership styles, mentoring functions received, and job- related stress: A conceptual model and preliminary study. *Journal of Organizational Behavior*, 21, 365–90.

Sosik, J.J., Kahai, S.S. and Avolio, B.J. 1998. Transformational leadership and dimensions of creativity: Motivating idea generation in computer-mediated groups. *Creativity Research Journal*, 11, 111–21.

Teed, M., Kelloway, E.K. and Mullen, J.E. 2008. *Young Workers' Safety: The Impact of Inconsistent Leadership*. Work, Stress, and Health Conference, Washington, DC, 6–8 March.

Tepper, B.J. 2007. Abusive supervision, workplace hostility, leadership behavior, employee deviance. *Journal of Management*, 33(3), 261–89.

Van Dierendonck, D., Haynes, C., Borrill, C. and Stride, C. 2004. Leadership behavior and subordinate well-being. *Journal of Occupational Health Psychology*, 9(2), 165–75.

13 How Positive Psychology and Appreciative Inquiry Can Help Leaders Create Healthy Workplaces

SARAH LEWIS

Abstract

In this chapter, the author explains how key areas of positive psychology, such as flourishing organizations, positivity, strengths and psychosocial resources contribute to our understanding of healthy workplaces. Appreciative Inquiry is then introduced as a positive methodology for organizational change and development, and recent work on the perspective and behaviour of appreciative leaders is outlined. The case for the mediation between positive organizations and healthy people being the effect of the social and emotional environment on our basic physiological processes will also be explained.

Introduction: Context of Positive Psychology

The majority of people spend at least some of their life working in organizations, many of them in large corporations or government agencies. Such large conglomerates are so prevalent in our lives that it is easy to overlook the fact that they are relatively recent forms of societal organization which emerged during the Industrial Revolution, aided by the timely invention of the skyscraper safe lift (Bellis 2012), and by the establishment of the corporation as a legal entity (Micklethwait and Woolridge 2003). Ever since we have been puzzling over exactly what they are and how best to make them work.

Fredrick Taylor, one of the first to attempt to understand the nature of large organizations, took a very mechanistic view of them. Given that he was by training an engineer, this may be no surprise. His attitude towards the people who worked in them, sadly necessary to the success of the organization, was to regard them as essentially components of the machine. One of his principles states, "There are scientific laws of administration, from which human values and emotions can be excluded" (Taylor 1912). He also clarifies that, in the context of work, "people are rational, economic actors," by which he means to suggest that the nature of the work is unimportant, only the relationship between effort and pay is of interest to the worker (ibid.). It was not thought, in these early days of organizational science, that the nature or context of the work, or the company kept while working, had any effect on the human body. Repetitive strain injury was not yet identified as a medical phenomenon, any adverse effects of prolonged periods of boredom induced by repeating the same action again and again were not considered a concern pertinent to the running of a business; and the importance of the group dynamic to individual performance remained unrecognized.

One hundred years of psychological discovery, sociological investigation into organizations, and advances in medical diagnosis later and we understand much better the systemic influence of mind and matter, brain and body. We understand much better, for example, the adverse effects of physical trauma on the human psyche and functioning, manifest at its worst in post-traumatic stress disorder. We understand something of the nature of mental ill-health and its effect on the physical phenomena of sleeping and eating as well as on patterns of thought. We also understand more keenly how our preconceptions affect how we make sense of what we see (Damasio 2005): in many respects we see what we are expecting to see or are looking for.

Yet the way we think about organizations hasn't fully kept up with these advances, and many working in organizations still retain a model of organizations as places (and spaces) where an understanding of people as essentially no more than rationally and logically functioning parts of a well-oiled machine, is both correct and sufficient. Many of us are still asked to be only half our full psychological selves when we enter our workplace: "Please leave emotional states and reactions at the door"; "Please act as if you had no considerations for the next eight hours other than those pertaining to the work of the organization"; "Please act as if the only function of communication is to convey information and please disregard as irrelevant any awareness conveyed by attitude, look or tone of voice about how the person conveying the message feels about you, themselves, or indeed the message." And so on.

But this approach, adhered to perhaps more in hope than belief (sometimes it feels as if it would be so much easier if people were emotionless automatons), we now know to be mistaken. We now know that the human body (not to mention soul) responds to the nature of particular jobs as well as to the emotional tenor of interactions and workplace climate. For example, we know that a lack of autonomy, role overload or lack of role clarity, unclear lines of accountability and other elements of job or organization design all contribute to more or less stressful places to work (Hackman and Oldman 1976, French and Caplan 1972).

Positive psychology as a branch of psychological research has long roots, but was only named as a distinct field of endeavour by Martin Seligman (1999) in his APA inaugural address where he first coined the term to cover all research directed towards the discovery of, or learning about, human excellence or flourishing. Human excellence or flourishing stands in contrast to human inadequacy or languishing (Seligman 1999). He issued a rallying call that resources be devoted to this under-investigated field, and an enthusiastic response from academics and researchers ensued.

Positive Organizational Scholarship

Kim Cameron established a line of research known as positive organizational scholarship (Cameron et al. 2003) focused on discovering what distinguishes the best from the rest amongst organizations. Initially interested in the effects of downsizing on organizations, he began to notice that some organizations suffered much less negative fallout from this difficult organizational change than others (Cameron 1998). In an echo of Seligman's language about people (Seligman 2011), he uses the term "flourishing" to describe organizations that are both good for people and good at business. These organizations that do well, both commercially and by their people, are regarded as good places to work. We can describe these as healthy workplaces where people can flourish.

He has identified three distinguishing features of such organizations: positive deviance, an affirmative bias and virtuous practices (Cameron 2003). Positive deviance refers to a greater than usual interest in identifying and learning from exceptionally good practice. Most organizations concentrate on attempting to eliminate bad practice, errors, failures, substandard performance and other negative deviations from the required minimum standard. However Cameron noticed that an unintended consequence of effective control processes can be to also inadvertently eliminate positive deviations from the required minimum standard. His exceptional organizations were more able than others to avoid this unhelpful consequence of deviation correction; instead actively seeking out and learning from their positive deviations.

Flourishing organizations recognize that the factors that contribute to success are not necessarily the same ones that contribute towards failure. They have learnt that it is a mistake to assume that if the factors that contribute to failure are eliminated, flourishing will automatically occur. Rather, they have discovered that the factors that move performance from standard to excellent are a different set from those that move an organization from standard to substandard. To learn how success, excellence, extraordinary performance and error-free behaviour occur, we need to study them as discrete phenomena.

Second, flourishing organizations demonstrate a bias towards growth through affirmation. In other words they endeavour to notice and positively connote the good in people, teams and their organization. Affirmation is an interesting concept. Being affirmed in our essential goodness is very life enhancing. Our concept of ourselves, our possibility, our potential grows in the reflection of the positive affirmation of others. As others see us, so we begin to see ourselves. Of course mistakes, errors and unhelpful behaviour still have to be addressed, but the overall experience of people in a flourishing organization is one of being affirmed; we might say valued and appreciated for who they are and what they bring to the endeavour.

One organizational practice that utilizes this recognition of the developmental power of affirmation is the creation of best-self feedback. Using this approach, rather than collect 360-degree feedback against a set of competences or some such measure, an individual asks those around him to give him or her feedback, through recounting specific experiences of him, or her, at their best. The person concerned takes all these stories about how and when they are seen at their best by others and analyses them for themes, consistent strengths, effectiveness, etc. While this process provides useful concrete data about how they are seen to add value, be successful, etc., it also creates a very affirming view of themselves as seen by others towards which they might aspire to grow.

Finally he noticed that in the best organizations people behave differently towards each other. He noticed more what he calls "virtuous practices," that is people being helpful, patient, enthusiastic, encouraging, humble, appreciative, grateful, forgiving and so on towards each other. Again these behaviours are very life-affirming. In the company of others who accept us as imperfect people, doing our best and sometimes getting it wrong, we are more able to be our true selves, take risks and be brave in our endeavours. In this way we learn and become more able. Only the other day I was with a group in a slightly besieged organization examining the concept of "winning relationships," very hesitantly one group named a theme they had identified from the stories of winning relationships of "not feeling that someone is looking over your shoulder, waiting for you to make a mistake." Sadly the defensive emotional tone referenced here colours a lot of organizations where people's energy then goes into covering their backs or defending their actions rather than into creating great solutions or generating great ideas. This is very relevant to our sense of well-being.

The work of Kahnman (2011) referenced by Achor (2011) in his eminently readable and authoritative book *The Happiness Advantage* makes it clear that the brain effort commonly known as "will power," that is, the brain energy needed to override the zillions of default connections in our brains, is a limited resource, easily depleted. It becomes obvious then that energy diverted into having to police what we say for fear of being interpreted negatively, or having it used as a stick to beat us with, is not available for the important organizational tasks. It is also very tiring and lessens our resources for looking after ourselves. As the energy goes to this form of physic defence, so less is available to help us resist the temptation to eat sugary food, drink alcohol or otherwise self-medicate. A link can be made, through the medium of energy depletion of the brain, from oppressive organizational cultures to features of self-neglect, self-harm and self-sabotage. Oppressive work cultures are likely to breed more unhealthy behaviour and flourishing cultures the reverse. Research in other areas of positive psychology enable us to understand something of how this all works.

It was Barbara Fredrickson who devised the research question, "what good are positive emotions?" (1998). It is a brilliant question that has led to much fruitful discovery. She reasoned that we know how negative emotions work: anger, frustration, depression, fear, etc. essentially act to ready the body for a small choice of actions and relieve the brain of having to choose between them. Colloquially known as the flight/fight/freeze response, this response is technically the production of a reduced thought-action repertoire. Negative emotions work to reduce the need to think in situations where we are in some way under threat. But why do we have positive emotions?

Building on the work of earlier researchers in this field, such as Alice Isen (1990), Fredrickson and colleagues have identified a number of effects of positive emotions, for example when we are feeling good we are, in broad terms: more sociable, more creative, more able to deal with complexity, able to think faster, more flexible in our thinking, more motivated and tenacious, have greater verbal fluency and are generally able to think better (Johnson et al. 2010). We also know that this is due to the effect of the neurotransmitters dopamine and serotonin, released when we experience positive emotions, enhancing the neural connectivity that facilitates better information retention, retrieval and manipulation: when we are happier we become brighter (Ashby et al. 1999). Sadly the effects don't last as positive emotions are temporary states. However there are also long-term benefits that accrue over time.

Fredrickson and colleagues have developed a theory of positive emotions known as the "broaden and build" theory. The theory suggests that, in the moment, the flood of chemicals in the brain released when we experience positive emotions encourages us to widen, or to broaden, our field of view and interest. Free from worry and danger, we are at liberty to explore our environment, be adventurous, be curious and discover things. In this way we increase our resource base. For instance we might make new friends, discover new information, overcome a particular challenge that has been blocking us, make new thinking connections, invest in learning new skills, "have a go" at something we don't usually do. In this way we broaden our experience, knowledge, skills and networks and so build up our resource base. This is the resource base that we call on in difficult times. There is an upward virtuous circle here: the more we feel positive, the more resourceful we become, the more resilient that makes us, the more quickly we are likely to recover from difficult situations, the more we experience feeling positive. Positivity is very connected to resilience, which is of course connected to stress.

It is usually somewhere around here in the exposition of positivity that people start to challenge, talking about the dangers of a Pollyanna view of the world. For those who are unfamiliar with children's fiction from 1913, Pollyanna was an orphan who took her optimistic attitude everywhere in her disaster-strewn life through her invention of "the glad game" which involved finding silver linings in the most unlikely of scenarios. Pollyannaism, a term coined by Matlin and Stang in 1978, is a concept used to describe optimism that tips over into naïveté or refusing to accept the facts.

The call for alertness against the danger of Pollyannaism as the outcome of a search for positivity is well made. Positivity is not about pretending all is well when it isn't; it isn't about ignoring bad news, it isn't about managers touring the office spraying insincere praise and compliments on all they pass. It is about the ratio of genuine experiences of positive to negative emotional states in someone's life, and the recognition that we can affect this ratio for ourselves and others by making conscious choices about upon what, in our multifaceted world, we choose to focus (Achor 2011).

This increasing understanding of the importance of the balance of positive to negative emotional states to human well-being and happiness comes from a number of research areas. Two of the best known and most important are team functioning and marriage. Losada and Heaphy (2004) discovered that the highest performing teams in their study could be distinguished from the others by, most importantly, the ratio of positive to negative comments made during group meetings. Their non-linear dynamics method of analysis also allows them to say with confidence that it is the behaviour in the group that predicts the performance in the wider world. In other words, that the relationship is positively causal not just correlational. The distinguishing line appears to be 3:1 positive to negative experiences, while the upper limit above which a group might move into a dysfunctional positive pattern appears to be around 12:1. Similarly, Gottman in his study of successful and other marriages found that the positive to negative ratio is the simplest and quickest predictor of the likely success of a marriage relationship (Gottman 1994).

Interestingly some research shows that insincere positivity has severe negative effects for the person being insincere (Rosenberg et al. 2001). And, we posit that it is the quickest way to cultivate cynicism in an organization. This means that for someone to raise their general levels of positivity they have to genuinely retrain their brain to focus on spotting the good, the right, the beautiful, the praiseworthy, the working, the awe-inspiring, and so on. Many of us, brought up to spot flaws, solve problems and offer critiques, have to learn a whole new way of searching through what the world offers us to start spotting things that will raise our feelings of positivity on a moment to moment basis. For example, only yesterday I was walking home from the shops, deep in thought about something, not feeling any way particularly, when I happened to notice in a garden on the opposite side of the road where I was walking a beautiful damask red rose bush against some other green foliage. I briefly focused on this beautiful colour amongst the various shades of grey and brown that tend to make up my part of town, and allowed myself to savour the moment, and, importantly, the tiny yet perceptible blip of pleasure it gave me. The better our brain gets at singling out these kinds of experiences amongst the zillion we process every day, and bringing them to our conscious attention, the more blips of positivity we will experience, and the more able we will be to appreciate and genuinely offer a positive view on things.

Once we are past the Pollyanna argument, people can rush to point out that negative emotions can't be ignored. Indeed. Fredrickson is very clear that negative emotions have their place (Fredrickson 2009). In essence negative emotions are important as they tell us that something is wrong in our world that needs attention. However, the quicker we can move to a more positive emotional state, the more able we are to make changes that will relieve the situation. Interestingly, some of her research into Americans' responses to the 9/11 shock suggest that resilient people experience the same negative emotions as less resilient people, shock, fear, outrage, etc. but also experience more positive emotions, love, gratitude, joy (Fredrickson et al. 2003). This ability to experience positive moments in times of greatest crisis is at the core of resilience and recovery from trauma.

It is important to note that none of this is to be taken to read that "positive thinking can destroy cancer cells" or even worse that "if you get cancer its your own fault for not being positive enough." Cultivating a positive outlook may help you cope with the many negative aspects of having cancer, i.e. to be more emotionally resilient, but to my knowledge there is no documented evidence that it has any direct effect on cancerous cells (Ehrenreich 2009). We believe it is very important to emphasize this as people do get confused about positive thinking and positive psychology. Positive psychology states that, all things being equal, happier people tend to live longer, healthier lives, than unhappy people. So by training ourselves to notice and experience the good things in our lives we can increase our experience of happiness and seemingly positively affect our general health. But it is important to remember that as bad things can happen to good people, so negative events arise in the lives of positive people, through no fault of their own.

Part of the explanation for why feeling good seems to boost physical well-being is related to the effects of positive emotions on our physiology. Happier people tend to have stronger immune systems and better cardiovascular health (Diener and Biswas-Diener 2008). Positive emotions "undo" the physiological effects of negative emotions, that is, they bring heart-rate and blood pressure back to the baseline up to 20 seconds faster (Fredrickson et al. 2000). So while the instruction to "calm down" may only cause your blood pressure to rise even higher, someone managing to make you laugh in amongst all the calamity may restore you to your resting state much quicker. As an aside it is clear that the humour of stressful situations, often known as "gallows" humour, serves a useful purpose in maintaining human functioning in distressing situations: "If I didn't laugh I'd cry" says it all. And in terms of performance most of us are more functional when we're laughing than when we're crying.

Another key concept emerging from positive psychology, related to positivity, is that of character strengths (Peterson and Seligman 2004). Peterson and Seligman wanted to create a well-being (human flourishing) diagnostic tool to act as a counterweight to the diagnostic tool of mental ill-health (human languishing) known as the DSM (a weighty tome by all accounts, currently in its fifth iteration) which is used to diagnose, classify and suggest appropriate treatment for a range of mental health or lifestyle problems. They undertook some grounded research to establish what the best of being human has looked like across various cultures, religions and times. They identified 24 character strengths that they then further classified into six virtues. Peterson recently defined these as "the subset of personality traits on which we place moral value," indicating that they are seen as aspects of personality, developed from our genetic inheritance and our life experiences (Peterson and Park 2011).

Other people have also developed taxonomies of strengths, notably Buckingham and others from the Gallup research (Buckingham and Clifton 2002), Brewerton from a works strengths perspective (Brewerton and Brook 2010), and Linley, also from an organizational perspective (Linley et al. 2010). While these differ in their surface taxonomies, they agree on a few key principles. Strengths are well-developed, habitual ways of behaving that when expressed through positive endeavours are seen as good, beneficial or admirable. So "cowardly" doesn't appear as a strength, but "courage" does. Note though that courage can of course be expressed through heroism on the battlefield or heroism as a Mafioso or drugs baron. Despite this, courage is generally regarded as a positive attribute while cowardliness is not. So strengths have a general moral value of being "good things."

All also agree that using our strengths is experienced as life-enhancing or affirming. Or, to put it another way, when we are feeling at our best we are probably using our strengths. Using our strengths is relatively effortless. Using our strengths is an energizing, self-reinforcing and positive experience. Great performance is thought to be predicated on the use of well-honed strengths. It is recognized that strengths can be misused, used to excess, or used inappropriately; therefore we need to be clear that *innate strength plus skill in its use* equals excellent performance or talent.

The physiological explanation for this observed relationship between strengths and performance is that the expression of strengths involves activation of well developed, dense, neural pathways, memorably described by Buckingham and Coffman as "the four lane highways" of our brain (Buckingham and Coffman 2001), our weaknesses, by contrast, being "stony pathways." It is a good analogy. We can whizz information around our motorway network: information flows around our brain, comes to mind when called, and combines easily together to create new thoughts or insights. We are thought to be at our most authentic when using our strengths: there is less translation between thought and action: we aren't "trying." So while we are at our most authentic when calling on our strengths, we may appear inauthentic when working hard to perform in a way that plays against our strengths. This is important information for leaders. We know that authenticity is highly valued in leaders by followers (Avolio et al. 2010). It follows that leaders need to find a way of leading that is based clearly on their personal strengths perhaps more than on a generalized template of "leadership behaviours or personality."

Strengths, Skills, Traits and States

Note that strengths are different to skills. We can be very skilful at something that we have had to learn, but it may not be a strength. We know because we don't find exercising this skill energizing: we may experience the well known "losing the will to live" feeling instead. So while I may have mastered the skill of basic book-keeping, it plays to none of my strengths. However the strength of a love of learning I take everywhere with me, and can call upon to make even the most tedious thing interesting in the early, learning, stages: even book-keeping.

There is an interesting puritanical streak in the Western psyche, or perhaps only particularly the UK and United States, that suggests that only things that are a struggle are worth having: no pain no gain. That anything that is easy is worthless. This leads to two obvious consequences. First, we ignore, overlook or downplay our strengths (because they come easily to us). Second, we focus our efforts on developing our weaknesses. Taken together, these can mean that people don't find the way to maximizing their potential, striving instead to become adequate where they are poor, rather than excellent where they are good. Continually struggling, and failing, to improve in an area in which one has no natural ability can be demoralizing and frustrating. Neither state in excess is good for our well-being.

By contrast, using our strengths is good for our well-being and health. For example, it has been found that various character strengths such as hope, kindness, social intelligence, self-control or perspective can buffer against the negative effects of stress and trauma, preventing or mitigating disorders in their wake (Peterson and Park 2011), while the strengths of love, hope, gratitude, curiosity and zest are particularly correlated with psychological well-being (Park et al. 2004).

Strengths are established by habit. This means there is there is always the possibility of developing new strengths as we grow, although establishing new habits is hard, effortful work that takes time (Achor 2011). Better to first capitalize on the strengths we are blessed with. That said, I think long-term strategies to enhance our ability to see the positive in adversity or disappointment, or to bring more awareness of gratitude into our lives, are probably a good investment of effort. More so for me, for example, than struggling to attend better to detail all the time!

If strengths are essentially psychological *traits* and so hard to change, then we might look to psychological *states* that are known to help enhance resilience and performance at work as offering potential for active development of well-being at work. Luthans et al. (2007) have done just this, identifying four key psychological states that enhance resilience and performance at work: hope, optimism, self-efficacy or confidence, and resilience. Taken together these four states are known as an individual's psychological capital (PsyCap), and PsyCap is proving to be a predictor of performance, satisfaction and absenteeism rates (ibid.). It is also positively related to organizational commitment and the intention to remain with the organization. All of these states are also associated with good life experiences and outcomes for individuals. There is, for instance, an established relationship between hopefulness and mental and physical health (Taylor 2011).

Taylor (2011) takes the four PsyCap states and adds social support to produce her list of psychosocial resources that enhance health and well-being. The evidence of the benefit of these psychosocial resources to health and well-being is overwhelming. Optimism is related to greater psychological well-being (Kubzansky et al. 2002), lower vulnerability to infection (Cohen et al. 2003), faster recovery from illness (Schier et al. 1989), and slower course of advancing disease (Reed et al. 1999). Self-efficacy, personal control and mastery, have all been shown to be related to better psychological health (Rodin et al. 1985), lower incidence of coronary heart disease (Karasek et al. 1982), better self-rated health (Seeman and Lewis 1995), better functional status and lower mortality as well as reduced mental and physical health risks (Lachman and Weaver 1998). A positive sense of self is related to lower autonomic and cortisol responses to stress (Seeman and Lewis 1995).

Taylor argues that social support is the most significant and reliable psychosocial predictor of health outcomes with effects on health on a par with smoking behaviour and lipid levels. Indeed healthy relationships with others have been shown time and again to be the best, most consistent and strongest predictor of human happiness. The message that comes through loud and clear from all the research on happiness and psychological well-being is the positive difference that social support makes (Bormans 2012). The key distinguishing factor of the happiest 10 per cent of people is the quality of their social support network (Diener and Seligman 2002). And, to follow the circle around, happier people cope with stressful events with less strain on their physiology and enjoy better health.

Taylor explains how these psychosocial factors work together to reduce the experience of stress in three ways. First, they help people appraise potentially stressful situations in more benign ways, often by reframing them; for example by re-casting a problem as an opportunity. Second, they are helpful in dealing with the taxing events that follow the perception or experience of some threat or negative event. We can understand from this perspective how hope helps create the tenacity to keep going, how social support helps with moment to moment regrouping and re-energizing in the fact of ongoing threat, and how resilience or coping skills feature in the ability to solve problems and to self-manage emotional states. And, third, people with good psychosocial resources are less likely to turn to maladaptive coping strategies such as drinking or drug-taking.

Once again the inoculating effects of psychosocial resources against stress and poor health, as with positivity and strengths, comes from its effects on our physiology. People with strong psychosocial resources have lower biological responses to stress (McEwen 1998). The two main systems involved in stress, the sympathetic nervous system and the HPA (hypothalamic-pituitary-adrenal) system, respond less and recover quicker. These are both activated during stress to facilitate the fight/flight response. Protective in the short term, overused they can become permanently "on" and so lay the groundwork for more chronic heart or immune system illnesses. In essence the research is clear that "individual differences in psychosocial resources influence the brain's response to threat, which in turn reduces neuro-endocrine responses to stress" (Taylor 2011). At the conclusion of her article Taylor asks how we can use this knowledge to engineer a positive social environment at work. Appreciative Inquiry offers a partial answer to this question (Lewis et al. 2007).

Appreciative Inquiry

Appreciative Inquiry, an organizational development methodology created by Cooperrider and Srivastva (1987), draws on or creates all the elements we have identified here as being relevant to healthy workplaces. It is a relational process, actively building relationships across the organization, so enhancing the social capital of the organization and the social support for the individuals. It also generates hope and optimism, so enhancing the psychosocial resources and psychological capital of the organization. It is predicated on the recognition of positive emotions as a powerful source of motivation and energy for change. And it works on a basis of creating growth from expanding existing strengths as much, if not more, than eliminating weaknesses.

In this way, while Appreciative Inquiry (AI) emerged from different roots to positive psychology, it can be seen as an approach that offers a pragmatic methodology for putting some of the findings from positive psychology to practical effect in an organization. To be effective at an organizational level, Appreciative Inquiry requires appreciatively oriented leadership. Diana Whitney, Amanda Trosten-Bloom and Kae Radar (2010) are consultants and academics at the Taos Institute who have worked extensively with organizations and their leadership to implement appreciative practices and ways of being. Recently they distilled their learning about what makes for appreciative leadership. When they pooled their observations they noticed that those who choose AI as their vehicle for positive change had four things in common: they were willing to engage with other members of their organization or community to create a better way of doing business or living; they were willing to learn and change; they truly believed in the power of the positive; and they cared about people, often describing the work of the organization or business in terms of helping people to learn, grow and develop.

Following this they conducted 100 appreciative interviews with participants in the appreciative leadership programme and held 10 focus groups, which led to the identification of five core strategies. This is consultant-led research and so the five key behaviours are packaged in a memorable way as: inquiry, illumination, inclusion, inspiration and integrity. Inquiry refers to an ability to ask appreciative and generative questions of people, groups and organizations, recognizing the potential of a good question to create shift, change and creativity. Illumination refers to the ability to highlight what is desired and to light the way forward by constant concrete reference to the strengths being exhibited and the good being achieved. Inclusion is key to Appreciative Inquiry generally as a process and refers to the ability to involve others to co-create plans, ideas and activities for the future. Inspiration refers to the ability to awaken desire, ambition, hope, tenacity, etc. in others. And integrity they summarize as "making choices for the good of the whole."

A special issue of the *AI Practitioner* recently focused on appreciative and positive leadership (2011). The editors, Lewis and Moore, identified a number of themes common to these accounts of appreciative leadership, which we can use to add to our understanding of what appreciative leadership looks like. They found that, from an appreciative perspective, leadership is clearly understood to be a relational activity. Rather than being seen as the *task* of directing in a disconnected way from the front of the organization, it is clearly seen as the *process* of being in a reciprocal relationship with others. They noticed also that appreciative leadership encompasses a certain humility: it's not about having all the answers, rather it's about creating conditions where answers can emerge.

It is also understood as being a balance between control and direction, with a general shift away from a more traditional emphasis on control to one of clearly indicating the general direction of desired travel. In this way such leadership invites others to step up to offer leadership in different ways, to different people at different times. Leadership becomes a relational property of the system that enhances its growth as much as a formal position of control.

Lewis and Moore also identify inclusion and integrity as key components of appreciative leadership. Appreciative leaders can be seen to clearly value everyone's voice and experience: they choose to take an inclusive and valuing attitude towards others.

Knowledge is seen as an asset to be shared, which grows in power as it is multiplied rather than diminishes. Shared information is seen as an essential component of great decisions and collective ways forward.

Integrity is emerging from various directions as a fundamental element of authentic or positive leadership, including in an examination of authentic leadership (Avolio et al. 2010) where the possession of a moral compass was identified as a key necessary component. In addition, appreciative leaders are seen to: search for what works; celebrate everyday miracles; find and celebrate everyday heroes; be bold and take risks, and to get out the way sufficiently to allow others to act.

Conclusion

So where does that leave us in terms of practical guidance for those looking to create healthy workplaces? It seems to me some of the lessons from positive psychology, as outlined above, are: cultivate a positive and flourishing organizational culture; help people understand, use and develop their strengths; ensure the positivity ratio of interactions within your organization are 3:1 positive to negative or above; adopt positive psychology influenced management processes (for more on this see Lewis 2011 or Cameron 2008); adopt an appreciative inquiry approach to organizational development and become an appreciative leader.

References

Achor, S. 2011. *The Happiness Advantage: The Seven Principles that Fuel Success at Work*. London: Virgin Books.

Ashby, G.F., Isen, A.M. and Turken, U. 1999. A neuropsychological theory of positive affect and its influence on cognition. *Pyschological Review*, 106(3), 529–50.

Avolio, B., Griffith, J., Wernsing, T.S. and Walumbwa, F.O. 2010. What is authentic leadership development? In *Oxford Handbook of Positive Psychology and Work*, edited by P. Linley, A.S. Harrington and N. Garcea. Oxford: Oxford University Press, 39–52.

Bellis, M. 2012. History of the elevator. *About.com, Inventors*. Retrieved on 6 July 2012 from http://inventors.about.com/od/estartinventions/a/Elevator.htm.

Bormans, L. (ed.). 2012. *The World Book of Happiness*. London: Marshall Cavendish Editions.

Brewerton, P. and Brook, J. 2010. *Strengths for Success: Your Pathway to Peak Performance*. London: Strengths Partnership Press.

Buckingham, M. and Clifton, D. 2002. *Now, Discover Your Strengths: How to Develop Your Talents and Those of the People You Manage*. London: Freepress Business.

Buckingham, M. and Coffman, C. 2001. *First Break All the Rules*. London: Simon & Schuster.

Cameron, K. 1998. Strategic organizational downsizing: An extreme case. *Research in Organizational Behavior*, 20, 185–229.

Cameron, K. 2003. Organizational virtuousness and performance. In *Positive Organizational Scholarship: Foundations of a New Discipline*, edited by K. Cameron, J. Dutton and R. Quinn. San Franciso: Berrett Koehler, 48–65.

Cameron, K. 2008. *Positive Leadership: Strategies for Extraordinary Performance*. San Francisco: Berrett-Koehler.

Cameron, K.S., Dutton, J.E. and Quinn, R.E. (eds). 2003. *Positive Organizational Scholarship: Foundations of a New Discipline*. San Francisco: Berrett Koehler.

Cohen, S.W.J., Turner, R.B., Alper, C.M. and Skoner, D.P. 2003. Emotional style and suspectibility to the common cold. *Psychosomatic Medicine*, 65, 652–7.

Cooperrider, D. and Srivastva, S. 1987. Appreciative inquiry in organizational life. In *Research in Organizational Change and Development: Volume 1*, edited by R. Woodman and W. Pasmore. Greenwich, CT: JAI Press, 129–69.

Damasio, A. 2005. *Descartes' Error: Emotion, Reason and the Human Brain*. London: Penguin

Diener, E. and Biswas-Diener, R. 2008. *Happiness: Unlocking the Mysteries of Psychological Wealth*. Malden, MA: Blackwell Publishing.

Diener, E. and Seligman, M. 2002. Very happy people. *Psychological Science*, 13, 81–4.

Ehrenreich, B. 2009. *Brightsided: How the Relentless Promotion of Positive Thinking has Undermined America*. New York: Metropolitan Books.

Fredrickson, B. 1998. What good are positive emotions? *Review of General Psychology*, 2(3), 300–319.

Fredrickson, B. 2009. *Positivity*. New York: Crown.

Fredrickson, B., Mancuso, R.A., Branignan, C. and Tugade, M.M. 2000. The undoing effect of positive emotions. *Motivation and Emotion*, 24, 237–58.

Fredrickson, B.L., Tugade, M.M., Waugh, C.E. and Larkin, G. 2003. What good are positive emotions in crises? A prospective study of resilience and emotions following the terrorist attacks on the United States of America September 11th 2001. *Journal of Personality and Social Psychology*, 84, 365–76.

French, J.R.P. and Caplan, R.D. 1972. Organizational stress and individual strain. In *The Failure of Success*, edited by A. Marrow. New York: AMACOM.

Gottman, J.M. 1994. *What Predicts Divorce? The Relationship between Marital Processes and Marital Outcomes*. Hillsdale: Erlbaum.

Hackman, J.R. and Oldman, G.R. 1976. Motivation through the design of work: Test of a theory. *Organizational Behaviour and Human Performance*, 16(2), 250–279.

Isen, A.M. 1990. The Influence of positive and negative affect on cognitive organization: Some implications for development. In *Psychological and Biological Approachs to Emotions*, edited by N. Stein, B. Leventhal and T. Trabasso. Hillsdale: Erlbaum, 75–94.

Johnson, K.J., Waugh, C.E. and Fredrickson, B. 2010. Smile to see the forest: Expressed positive emotion broadens attentional scope and increase attentional flexibility. *Cognition and Emotion*, 24, 299–321.

Kahman, D. 2011. *Thinking Fast and Slow*. London: Allen Lane.

Karasek, R.A., Theorell, T., Schwartz, J., Pieper, C. and Alfredsson, L. 1982. Job psychological factors and coronary heart disease: Swedish prospective findings and U.S. prevalence findings using a new occupational inference method. *Advances in Cardiology*, 29, 62–7.

Kubzansky, L.D., Wright, R.J., Cohen, S., Weiss, S., Rosner, B. and Sparrow, D. 2002. Breathing easy: A perspective study of optimism and pulmonary function in the normative aging study. *Annals of Behavioral Medicine*, 24, 345–53.

Lachman, M.E. and Weaver, S.L. 1998. The sense of control as a moderator of social class differences in health and well-being. *Journal of Personality and Social Psychology*, 74, 763–73.

Lewis, S. 2011. *Positive Psychology at Work: How Positive Leadership and Appreciative Inquiry Create Inspiring Organizations*. Chichester: Wiley-Blackwell.

Lewis, S. and Moore, L. (eds). 2011. Positive and appreciative leadership. *Appreciative Inquiry Practitioner*, 13(1), 4–6.

Lewis, S., Passmore, J. and Cantore, S. 2007. *Appreciative Inquiry for Change Management: Using AI to Facilitate Organizational Development*. London: Kogan Page.

Linley, A., Willars, J. and Biswas-Diener, R. 2010. *The Strengths Book: Be Confident, be Successful, and Enjoy Better Relationships by Realising the Best of You*. Coventry: Capp Press

Losada, M. and Heaphy, E. 2004. The role of positivity and connectivity in the performance of business teams: A nonlinear model. *American Behavioral Scientist*, 47(6), 740–765.

Luthans, F., Youssef, C. and Avolio, B. 2007. *Psychological Capital: Developing the Human Capital Edge*. Oxford: Oxford University Press.

Matlin, M.W. and Stang, D.J. 1978. *The Pollyanna Principle: Selectivity in Language, Memory, and Thought*. Cambridge, MA: Schenkman Pub. Co

McEwen, B.S. 1998. Protective and damaging effects of stress mediators. *New England Journal of Medicine*, 338, 171–9.

Micklethwait, J. and Wooldridge, A. 2003. *The Company: A Short History of a Revolutionary Idea*. London: Phoenix.

Park, N., Peterson, C. and Seligman, M.E.P. 2004. Strenghts of character and well-being. *Journal of Social and Clinical Psychology*, 1, 118–29.

Peterson, C. and Park, N. 2011. Character strengths and virtues: Their role in well-being. In *Applied Positive Psychology: Improving Everyday Life, Health, Schools, Work and Society*, edited by S.I. Donaldson, M. Csikszentmihalyi and J. Nakamura. New York: Routledge, 51.

Peterson, C. and Seligman, M.E.P. 2004. *Character Strengths and Virtues: A Handbook and Classification*. New York: Oxford University Press.

Reed, G.M., Kemeny, M.E., Taylor, S.E. and Visscher, B.R. 1999. Negative HIV-specific expectancies and AIDS related breavement as as predictors of symptom onset in asymtopmatic HIV-positive gay men. *Health Psychology*, 18, 354–63.

Rodin, J., Timko, C. and Harris, S. 1985. The construct of control: Biological and psychosocial correlates. *Annual Review of Gerontology and Geriatrics*, 5, 3–55.

Rosenberg, E.L., Ekman, P., Jiang, W., Babyak, M., Coleman, R.E., Hanson, M., O'Connor, C., Waugh, R. and Blumenthal, J.A. 2001. Linkages between facial expressions of anger and transient myocardial ischemia in men with coronary artery diesease. *Emotion*, 1, 107–15.

Schier, M.F., Matthews, K.A., Owens, J., Magovern, G.J., Lefebvre, R.C., Abbott, R., Craver, A. and Charles, S. 1989. Dispositional optimism and recovery from coronary artery bypass surgery: The beneficial effects on physical and psychological well-being. *Journal of Personality and Social Psychology*, 57, 1024–40.

Seeman, M. and Lewis, S. 1995. Powerlessness, health and mortality: A longitudinal study of older men and mature women. *Social Science and Medicine*, 41, 517–25.

Seligman, M. 1999. *Presidential Address*. American Psychological Association's 107th Annual Convention, Boston, 21 August.

Seligman, M.E.P. 2011. *Flourish: A New Understanding of Happiness and Wellbeing – and How to Achieve Them*. London: Nicholas Brealey.

Taylor, F. 1912. Scientific management. Reproduced in Pugh, D. 1997. *Organizational Theory, fourth edition, Selected Readings*. London: Penguin.

Taylor, S.E. 2011. How psychosocial resources enhance health and well-being. In *Applied Positive Psychology: Improving Everyday Life, Health, Schools, Work and Society*, edited by S.I. Donaldson, M. Csikszentmihalyi and J. Nakamura, New York: Psychology Press, 72.

Whitney, D., Trosten-Bloom, A. and Rader, K. 2010. *Appreciative Leadership: Focus on what Works to Drive Winning Performance and Build a Thriving Organization*. New York: McGraw-Hill.

PART 5

Implementing Interventions

14 Interventions to Prevent Mental Health Problems at Work: Facilitating and Hindering Factors

NATHALIE JAUVIN, RENÉE BOURBONNAIS, MICHEL VÉZINA, CHANTAL BRISSON and SANDRINE HEGG-DELOYE

Abstract

While the study of the implementation of preventive interventions for mental health problems at work is increasingly popular, it is as yet difficult to identify the factors that facilitate or hinder their success. This chapter aims to partly address this gap by reviewing the factors that facilitate or hinder the implementation of preventive interventions in the area of mental health in the workplace, based on three intervention research projects carried out in Quebec (Canada) over the past ten years. Among the factors revealed by the analyses, three were more specifically related to the context of the interventions, while eight were more specifically related to the intervention process itself. Depending on the specific ways in which these factors manifest themselves, they can either facilitate or hinder the successful implementation of an intervention process. The authors explain in detail the impact of these factors and offer overall considerations for the implementation of such interventions.

Introduction: Statement of the Problem

Mental health problems at work, as is now widely documented, entail high costs for organizations (Brun et al. 2003, Hemp 2004, Parsonage 2007). These costs include, in particular, a high rate of absenteeism and presenteeism, a high level of employee turnover, and a considerable loss of productivity (Sultan-Taïeb 2012). In 2003, a report on mental health in the workplace in Quebec concluded that the direct and indirect costs of absenteeism at work represented 17 per cent of total payroll expenditures (Brun et al. 2003). Presenteeism at work also generates considerable costs; in fact, Parsonage (2007), a British researcher, maintains that the cost of presenteeism is approximately one and a half times higher than that of absenteeism. Hemp (2004), an American researcher, has estimated that the productivity of an employee engaged in presenteeism can drop by up to 33 per cent. In 2007–2008, the ECQOTESST survey (Quebec survey on conditions of work and employment and on occupational health and safety) reported that 18 per cent of workers in Quebec showed a high level of psychological distress while, worldwide, 12 per cent of work-related illness and injury is attributable to mental health problems (Vézina et al. 2011).

Faced with these facts, an increasing number of research teams from various countries, including Canada and Denmark, have sought to identify the organizational factors behind mental health problems at work so as to be able to act at the level of primary prevention on the mental health of workers. These studies, based on workplace interventions, have increasingly documented the effect of psychosocial constraints in the workplace on the prevalence of mental health problems (Mikkelsen et al. 2003, Bourbonnais et al. 2006a, 2006b, 2011, Gilbert-Ouimet et al. 2011, Hasson et al. 2012) and have reported on interventions that may help reduce these constraints (Kobayashi et al. 2008, Tsutsumi et al. 2009, Czabala et al. 2011, Hasson et al. 2012).

The psychosocial constraints most often addressed in these intervention studies are those based on Karasek's and Siegrist's work stress models (Karasek and Theorell 1990, Siegrist 1996), which have gained international renown. Most of these studies have reported positive changes for workers, but not all the interventions have had the anticipated effects and some have not been successful at all (Robson et al. 2001, Mikkelsen and Gundersen 2003, Aust and Ducki 2004, Nielsen et al. 2006, Biron et al. 2010, Gilbert-Ouimet et al. 2011).

It is as yet difficult to identify the factors that facilitate or hinder the successful implementation of preventive interventions for mental health problems at work. Indeed, there appears to be no synthesis of knowledge on the processes that might lead to the success or failure of such interventions (Zwerling et al. 1997, Shannon et al. 1999). Research teams in Canada and England have begun the work of synthesizing this knowledge, but much work remains to be done (Baril-Gingras et al. 2007, Robson et al. 2007).

Nevertheless, a close reading of the literature reveals that most researchers who have published work on this question mention some factors that have facilitated or limited, or even prevented, the interventions from being successfully carried out.[1] The two facilitating factors most often mentioned are: initial support from decision-making bodies (management and unions) and the active participation of senior administrators, managers, and employees (Kompier et al. 1998, Trudel et al. 2000, Robson et al. 2001, Saksvik et al. 2002, Mikkelsen and Gundersen 2003, Baril-Gingras et al. 2006, Nielsen et al. 2006, Whysall et al. 2006, Robson et al. 2007, Kobayashi et al. 2008, Tsutsumi et al. 2009, Van and Pillay-Van 2010). Kompier et al. (2000) emphasizes the importance of choosing interventions that were initiated by the workplace. Based on his observations, when the request for an intervention comes from the outside, it is very difficult to start the process off in a rigorous manner with the support of the decision-making bodies, as communication tends to be difficult, and there tends to be very limited shared management throughout the entire intervention (Kompier et al. 2000). The need for a rigorous process – including, right from the start of the intervention, risk assessment, developing solid arguments, choosing valid targets, transparency in the actions taken,

---

1   The terms used by researchers to describe the factors that facilitate or hinder these interventions vary. For example, Kompier (2000) addresses "stimulating and obstructing factors"; Biron et al. (2010) mention the "contextual influence"; Whysall et al. (2006) lists all the barriers encountered without categorizing them; and Baril-Gingras et al. (2007) deals with the context and processes which influence the achievement of interventions.

time set aside specifically for the participation of all the actors involved – is considered by many research teams to be a major factor for the success of these interventions (Robson et al. 2001, Tsutsumi et al. 2009, Czabala et al. 2011).

Several authors also mention factors that can hinder the success of interventions. Among these are: resistance to change linked to a conservative organizational culture; mistrust developed through past experiences which ended in failure; and communication problems related to environmental changes taking place within an organization (political and/or economic reorganization) but also changes in personnel (shift work, rotating shifts) (Mikkelsen et al. 2000, Mikkelsen and Gundersen 2003, Nielsen et al. 2006, Whysall et al. 2006, Robson et al. 2007, Biron et al. 2010).

Thus, the state of knowledge on the factors that facilitate or hinder the implementation of interventions aimed at preventing mental health problems remains fragmented. Although a close reading of the literature makes it possible to identify some factors that have facilitated or hindered the implementation of preventive interventions in the area of mental health problems at work, to our knowledge, no author has, to date, conducted a comprehensive review of these factors. Knowledge on this subject is thus not yet highly organized. This chapter aims to partly address this gap by reviewing the factors that facilitate or hinder the implementation of preventive interventions in the area of mental health in the workplace, based on three intervention research projects carried out in Quebec (Canada) over the past ten years.

Context and Methods

Three participatory intervention projects aimed at the primary prevention of mental health problems in the workplace have been developed by RIPOST[2] and GIROST[3] researchers and their partners over the last 10 years. Observations have emerged from these projects relating to the factors that have facilitated or hindered the implementation of these interventions. All of these studies were subject to rigorous scientific evaluation. Evidence of these factors was catalogued using various qualitative research strategies and merited a more in-depth analysis by integrating the knowledge developed more specifically in three "participatory intervention" research projects which shared a similar method. This chapter thus presents all the facilitating and hindering factors brought out in these three projects.

The three projects are briefly described in Table 14.1 and Table 14.2. All three projects were conducted in Quebec (Canada), between 2000 and 2009. The first partnership-based intervention research project was conducted among the employees of a short-term hospital centre (n=674 employees) in Quebec City (Project A). The second study was conducted among more than 1,300 employees in a public organization in the insurance sector in Quebec (Project B). The third project was carried out in three detention centres in Quebec (small, medium and large), and involved 445 correctional officers (Project C).

All three of these intervention projects shared the same goal, that of acting at the level of primary prevention on mental health problems in the workplace by carrying

---

2   Recherches sur les interrelations personnelles, organisationnelles et sociales du travail (Research Team on Personal, Organizational and Social Interrelations at Work).

3   Groupe interdisciplinaire de recherche sur l'organisation et la santé au travail (Interdisciplinary Research Group on Workplace Organization and Health).

Table 14.1   Brief description of the projects: Common methods

Research design (common methodological aspects)	Theoretical framework of the intervention research projects
Quasi-experimental design including a control group and pre- and post-intervention measures *A priori* **risk assessment** Demand-control model (Karasek) and effort-reward imbalance model (Siegrist) **Effects of the intervention measured approximately six and 36 months after implementation** **Mixed methods** Quantitative (questionnaires) and qualitative methods **Common variables used in the three projects** *Psychosocial constraints*: psychological demands/decision latitude/social support at work/effort–reward imbalance	**Three-phase intervention research process developed by Goldenhar et al. (2001)** 1) Identifying risks *a priori* and targeting priorities for intervention. 2) Implementing an intervention which is consistent with these priorities. 3) Studying the effectiveness of the intervention. **Participatory approach** involving intervention groups (IGs) composed of both management and employee representatives and researchers

out an intervention[4] that addressed the workplace environment rather than focusing on individuals and their capacity to adapt. They also used a very similar methodological approach, as shown in Table 14.1: a similar quasi-experimental design, including a control group, risk assessment using validated models, a mixed methods approach (quantitative and qualitative), and several common study variables including psychosocial constraints at work.

Karasek's demand-control model and Siegrist's effort–reward imbalance model, the theoretical models used in all three projects, allowed for the identification of the psychosocial constraints at work whose effects on mental and physical health had been the most highly documented (Karasek and Theorell 1990, Siegrist 2002). The demand-control model consists of two components, psychological demands and decision latitude. According to this model, job strain occurs when high psychological demands are combined with low decision latitude. A third dimension was later added to this model: social support at work from colleagues and supervisors (Johnson and Hall 1988). The effort–reward imbalance model focuses on the imbalance between the effort spent and the rewards received (Siegrist 2002). The adverse effects of this imbalance have also been observed at the emotional and physiological levels (van der Doef and Maes 1998, van der Doef and Maes 1999, Stansfeld and Candy 2006).

The theoretical framework for these three intervention research projects was also based on the same three-phase intervention research process developed by Goldenhar and colleagues (2001), that is, the development phase, the implementation phase and the effectiveness phase. The three projects thus aimed, more specifically, to:

---

4   In all three projects, the intervention was defined as being any organizational change carried out by the management or the organization, introduced with the goal (or having the explicit consequence) of improving the situation regarding one of the four psychosocial constraints targeted.

1. produce knowledge that could be applied to the development of preventive interventions so as to reduce the psychosocial constraints in the workplace environment;
2. systematically document the intervention implementation process; and
3. measure the effects of the intervention on the prevalence of the four targeted variables related to the psychosocial environment in the workplace and on various health indicators.

The information collected allowed for the a posteriori analysis of the factors that had facilitated or hindered the implementation of the interventions in each organization.

All three projects were centred on actor participation, which entailed, for each of the intervention processes, the creation of intervention groups (IG) composed of researchers, employee and union representatives, and management representatives. The goals of the intervention groups were as follows:

1. to identify the constraints in the psychosocial environment in the workplace;
2. to seek solutions (interventions) to act on these constraints in order to lessen or even eliminate them;
3. to establish the order of priority to be given to the solutions found and determine their feasibility;
4. to disseminate this information to all individuals represented by the group; and
5. to receive comments and reactions on the information shared about the project's progress (consultation, feedback).

The main goal of having members participate was to allow the organization to take charge of the process of identifying and analysing the constraints, solving problems and identifying solutions aimed at preventing mental health problems. The researchers who participated in these groups facilitated the organization's appropriation of the process.

The similarities between the three projects thus constituted a solid basis for the a posteriori integrated analysis of the factors that had facilitated or hindered the implementation of the interventions.

Given that the three studies were spread over several years, lessons were learned which could be applied to improve the intervention process from one project to another. Thus, for example, specific communication measures, such as the creation of a monthly news bulletin, were added to the third project (detention centres) so as to better inform all workers of the activities being carried out in the context of this intervention project. In addition, based on previous experiences, the intervention groups also included in the third project the signature of formal commitment agreements by senior administrators, so as to anchor the intervention process more firmly.

Methods

In order to identify all the factors that had facilitated or hindered the successful implementation of the interventions in the three research projects, we conducted an in-depth examination of various sources, from which relevant information was collected and compared: research reports, doctoral theses, minutes of meetings held during the three projects, interviews with some researchers, etc. Table 14.2 presents the references

Table 14.2 Pertinent sources of information and main results of the intervention effects

Research projects considered for analysis	Pertinent sources of information used to identify factors which facilitated or hindered the process	Main results	Publications related to the research
PROJECT A: HOSPITAL CENTRE A hospital centre in Quebec (Canada) *674 caregivers at a short-term HC (experimental hospital centre)* One control hospital centre (2000–2004)	Observation in the care units. Interviews with key informants. Follow-up of changes recommended by the IGs. Focus groups (with personnel involved in the participatory intervention project; three years after the start of the project).	One year after the intervention: a significant difference in the means of all psychosocial factors except one (decision latitude) and a significant reduction in sleeping problems and work-related burnout in the experimental hospital. Three years after the intervention: the means of all adverse factors except one (psychological demands) and all health indicators were significantly more favourable in the experimental hospital than in the control hospital.	St-Arnaud and Gignac 2005. Bourbonnais et al. 2006a. Bourbonnais et al. 2011.
PROJECT B: PUBLIC ORGANIZATION A large public organization in Quebec City (Canada) *Over 1,300 employees; 12 different departments involved* Two control organizations (2003–2007)	Logbooks (kept by the senior manager of each department). Employees' reports of the changes (second follow-up questionnaire). Focus groups (8–14 employees).	Improvement in three of the four organizational risk factors (demands, social support and reward). Improvement in all three health indicators (psychological, musculoskeletal and cardiovascular).	Brisson et al. 2006. Vézina et al. 2009. Simard 2009. Gilbert-Ouimet et al. 2011. Hasson et al. 2012.
PROJECT C: DETENTION CENTRES Correctional facilities under Quebec jurisdiction (Canada) *445 correctional officers from three experimental centres: small, medium, large* 15 control facilities (2004–2009)	Observation (detention centres). IGs' reports of changes in the three experimental centres and the control centre. Analysis of the implementation process and its appropriation by the centres. Records on the progress of each project. Key informants' reports of changes, and observation (individual interviews).	Positive effects on reward and social support Significant positive impact on relationships between workers and supervisors (social support from supervisors, decrease in psychological harassment and intimidation). Positive effect on social support from colleagues. No significant change in health indicators.	Bourbonnais et al. 2005. Bourbonnais et al. 2007. Bourbonnais et al. 2012. Dussault et al. 2010 and 2012.

consulted for each of the three projects. This table also presents, for each of the projects, the tools used to collect data on the facilitating and hindering factors: participant observation, focus groups, interviews with key informants, logbooks, records describing the progress of the intervention projects, etc.

Inductive analysis was used to identify, for all three projects, the factors that had facilitated or hindered the development and implementation of the interventions or their appropriation by the organization. An initial list of factors was generated by the analysis. A second level of analysis then made it possible to combine and group the factors into categories and sub-categories, which are presented in the next section.

Results and Analyses

An initial content analysis which grouped qualitative data from the three participatory intervention research projects led to the identification of a set of factors which influenced both the progression of the process and it impacts. These factors are presented globally for the three studies as conditions that either facilitated or hindered the development and implementation of participatory interventions aimed at improving the psychosocial environment and mental health in the workplace. In some cases, the factors identified are found in both groups ("facilitating factors" and "hindering factors") depending on whether they had a positive or an adverse effect on the intervention process.

A second analysis subsequently allowed us to further classify these factors into two distinct categories which emerged from the analysis: contextual variables and process variables. The factors in the "context" category were specific to each workplace and were related, not to the intervention itself, but rather to pre-existing dimensions (existing before the intervention research project began or present when it started, e.g. unfavourable economic context, support for the project from senior management or union officials) or new contextual dimensions (outside the project, e.g. organizational restructuring) which interfered with the intervention process. The factors in the "process" category were directly related to the development or implementation of the interventions (e.g. instability in committee membership), and thus to the intervention method itself.

Table 14.3 presents the 11 facilitating or hindering factors brought out by the analyses. Among these factors, three were more specifically related to the context of the interventions, while eight were more specifically related to the intervention process itself. Depending on the specific ways in which these factors manifest themselves, they can either facilitate or hinder the successful implementation of an intervention process. The specific ways in which each of these factors manifested themselves in the intervention research projects – for a presentation purpose, under their positive form – are presented in the third column of this table. In the section following Table 14.3, each of the factors will be explained more fully, accompanied, when pertinent, by references to the literature.

FACILITATING AND HINDERING FACTORS RELATED TO CONTEXT

The analysis revealed the importance of taking into account the contextual factors present in the organization both at the time an intervention was being considered and, if it was chosen, throughout its development and implementation. These contextual factors, as mentioned above, were not related to the intervention itself but rather to pre-existing

Table 14.3   Factors which facilitated or hindered the intervention process

Category of factors	Factors	Manifestations
Context	1. Request came from the organization	• Need for intervention was identified by organization. • Recognition of problems by partners and real willingness to intervene.
	2. Initial support from decision-making bodies	• Initial support from senior management and involvement of all levels of management. • Initial support from union officials (or employee representatives). • Formal commitment by decision-making bodies. • Leadership on the part of decision-makers and actors concerned by the intervention to be implemented. • Lack of pre-existing mistrust within the organization, in particular, disenchantment caused by previous experiences and degree of scepticism regarding the process.
	3. Openness to shared management	• Organizational culture that is open to developing interventions with partners, from an egalitarian perspective. • Changes in usual mode of operation, resulting in openness to develop interventions, based on a participatory approach. • Lack of tensions in labour relations or, when they arose, proper management of them by the partners.
Process	4. Rigorous theoretical and methodological framework	• Rigorous approach allowing for intervention priorities and solutions to be identified, and providing valid measures of psychosocial and mental health factors. • Detailed assessment of risks existing in the organization, using recognized tools. • Involving all parties in identification of intervention objectives or desired changes. • Deciding which interventions should be given priority based on results of this assessment.
	5. Active participation of employees in the entire process	• Crucial involvement of employees in process of identifying problems and solutions: they are the experts in their organization.

Table 14.3 Continued

Category of factors	Factors	Manifestations
	6. Rigorous process to ensure effective operation of intervention groups (IGs)	• Using rigorous and transparent process to select members of intervention groups. • Presence within IGs of members who were recognized for their leadership qualities, credibility, and communication and listening skills. • Clear definition of roles, responsibilities and duties, shared among all members of IG. • Seeking the greatest stability possible in IG membership. • Aspiring to meet the expectations of all IG members. • Consensus within the IG or the organization regarding the priority of interventions. • Points of view of all or most subgroups concerned by the intervention represented by members of IG.
	7. External mediation	• Presence of "neutral" individuals who could lead discussions and act as mediators when needed.

Process	8. Support from management for the implementation of the interventions	• Financial support when needed to implement proposed changes.
		• Freeing up employees for the time needed to plan, develop and implement interventions.
		• Measures allowing employees to take the time needed to participate in IG activities.
		• Appropriate or sufficient management of changes arising within the organization:
		  – Mergers, major organizational transformations taking effect during the intervention process.
		  – Adoption, along the way, of new policies or ways of doing things which had an impact on the intervention.
		  – Deterioration in working conditions (e.g. increased workload).
		  – High turnover among employees and managers.

	9. Rigorous implementation and follow-up of changes	• Rigorous implementation of proposed changes among targeted groups of workers; respecting commitments. • Implementing a process by which participants could follow up and provide feedback on proposed changes; reaching consensus on priorities and feasibility.
	10. Communication and information-sharing mechanisms	• Introduction of communication tools used to inform all employees and managers on progress of the process. • Introduction by IGs of modes of consultation involving all workers.
	11. Appropriation of the process by the organization	• Appropriation of process and proposed changes by the organization, thus ensuring long-term viability of process.

dimensions (at the start of the intervention process) or new contextual dimensions (outside the process).

More specifically, the analysis of the three intervention projects brought out three contextual factors which, overall, proved to facilitate the development and implementation of the interventions within the organizations. However, some obstacles related to these contextual factors were also identified.

One: Request came from the organization

When a project is initiated by the organization itself, the organization is more inclined to develop and support the interventions chosen. An intervention project should thus be based on a request from the organization in which it will be implemented rather than on an initiative imposed from outside. This request should come from employee representatives or management, or, better yet, from both groups. This usually begins, as was the case in the projects analysed, by a mutual recognition by management and union officials of the problems that exist and, therefore, the need to intervene. Kompier et al. (2000) emphasized the importance of this first factor, pointing out, based on a study on interventions carried out among bus drivers, that when the initiative for an intervention came from outside the company, it was more difficult for the process to take shape than when the initiative came from inside.

Two: Initial support from decision-making bodies

The intervention must be a priority for both the management and the union (or, when there is no union, employee representatives) and this priority must be recognized by all levels of management concerned. In fact, a close analysis of the three research projects showed that initial support from senior management was a crucial factor for the success of the interventions. Indeed, management's commitment to support the implementation of the action plans put forward by the intervention groups assured the employees that things would move forward and that their points of view would be heard. Moreover, commitment on the part of union officials was also shown to be essential in order for workers to support the intervention process. This mutual commitment was expressed, in concrete terms, by the signing of agreements, which ensured the long-term viability of the process even when there was some turnover among management personnel within the various decision-making structures involved.

This second factor, moreover, has often been identified in the literature on preventive workplace interventions, in particular by Kompier et al. (2000) who emphasized the importance of commitment on the part of management but also on the part of key stakeholders and all participants. Mikkelsen and Gundersen (2003), in a study conducted at a sorting terminal in the Norwegian Postal Service, also identified the importance of initial support from management and union representatives. Whysall et al. (2006), having analysed 24 interventions to tackle occupational ill-health, also identified, among "facilitators and barriers involved in implementing such interventions," the importance of gaining managerial support and commitment. The importance of union support for the implementation of interventions has also been identified by other authors (Aust and

Ducki 2004, Baril-Gingras et al. 2006). As pointed out by Whysall et al. (2006), on the other hand, a lack of involvement on the part of managers can prove to be a serious obstacle to the implementation of the desired changes.

Leadership on the part of decision-makers is an important component of this support. There needs to be sufficient leadership right from the start within the organization, making it possible to introduce the desired changes down the road. Such leadership ensures that the intervention project will be given priority and that the organization and its representatives will truly engage in the search for solutions to problems and the identification of possible avenues for action. Responsibility for the effective implementation of interventions must therefore first be *entrusted* to managers and then delegated to *persons with recognized leadership skills in the workplace under study*.

The presence of pre-existing mistrust within the organization can also constitute a significant obstacle to the implementation of interventions. It is thus important to identify and address this mistrust before engaging in the process. In fact, for various reasons that have nothing to do with the intervention process itself, the organization can already be affected by a high level of mistrust and may therefore not welcome the intervention project enthusiastically. This mistrust can be caused by past experiences related to intervention projects or changes that turned out poorly, or by pre-existing tensions between the management and the union, or can, as was the case in the correctional setting, for example, reflect an organization that has been contaminated by a culture of mistrust. As pointed out by Kristensen (2000), workers sometimes oppose an intervention because they believe it might interfere with the bargaining system already in place within the organization, while management representatives can oppose it out of fear that it will threaten their managerial right and, consequently, cause them to lose their monopoly over information.

Three: Openness to shared management

The organization in which an intervention project will be carried out, must, from the outset, be open to shared management and therefore to collaboration between management and union officials (or employee representatives). Indeed, in the three organizations in which the research projects were conducted, it was observed that the actors agreed to open the door to a participatory approach which was not necessarily consistent with established traditions, in particular in the correctional setting where the traditional management was very hierarchical, almost militaristic. Thus, as pointed out by Baril-Gingras et al. (2006), the organization must be open to new ways of doing things if it is considering this type of intervention. The literature, moreover, demonstrates that, to date, the interventions that are most likely be effective are those that adopt a participatory approach to the analysis and resolution of problems (Semmer 2006, DeJoy et al. 2010). Some authors, on the other hand, maintain that it is not advisable to involve actors in different hierarchical positions in the identification of problems (Chambaud and Richard 1984, Lescarbeau 1994). However, in the three projects studied, the involvement of management representatives proved, overall, to guarantee the success of the intervention process.

Nevertheless, shared management is not always easy, especially in a more antagonistic setting. Tensions in labour relations, moreover, were one of the obstacles identified by the researchers who conducted the three intervention projects. In fact, labour relations can, in a global way and for reasons unrelated to the project, become so acrimonious that they can impede the work of the IGs. When such tensions arise, they must be managed in an appropriate way by the partners so as to avoid having them hinder the intervention process. A critical period in labour relations in which the union and management are publicly confronting one another can have a serious impact on the project or at least poison relations between the members of the joint intervention group. These labour relations issues were also reported by Whysall et al. (2006) who revealed that, in some cases, priority bargaining issues had be addressed before it was possible to proceed with the interventions planned.

FACILITATING AND HINDERING FACTORS RELATED TO PROCESS

Eight factors brought out by the integrated analysis of the three intervention projects were classified in the "process" category, that is, were directly related to the development or implementation of the interventions and thus to the intervention method itself.

One: Rigorous theoretical and methodological framework

In light of the three projects studied, it seems clear that the success of an intervention process relies to a great extent on the use of a rigorous theoretical and methodological framework based on tested and recognized instruments for measuring psychosocial constraints as well as reliable mental health indicators (Bourbonnais et al. 2006b). These tools make it possible to accurately evaluate the situation in the workplace as it stands at the start of the project and to monitor its development over time, as the interventions are implemented. Before interventions are developed, it is important to devote the necessary time and energy to risk assessment. This assessment, based on tested theoretical models and recognized tools, ensures that the intervention targets and solutions identified will be pertinent and based on psychosocial work factors known to have an impact on workers' health. Highly reliable data will be less easily contested than data collected through a less rigorous process. As early as 1998, having reviewed 10 intervention projects carried out in industries in the Netherlands, Kompier et al. (1998) identified, among the five factors for the success of the approaches studied, what they called "an adequate diagnosis or risk analysis."

The length and complexity of a rigorous research project can, however, slow down the implementation of interventions. The three projects studied were conducted over several years and it was sometimes difficult to maintain the momentum and the interest of participants over such a long period, especially in a context of high turnover among the employees and managers involved. It is thus essential to clearly explain the pertinence of this kind of rigour.

Two: Active participation of employees in the entire process

The active participation of employees in all phases of the intervention process is highly essential and was a key factor for the success of the interventions developed in the context of the three projects analysed. Workers are effectively the experts in their workplace and it is advisable to include them in the entire process. Other authors have also pointed out the important nature of this factor (Mikkelsen et al. 2000, Goldenhar et al. 2001), including Bond and Bunce (2001) who reported that the participation of employees can contribute to the improvement of mental health and productivity and reduces absenteeism from work. In fact, while the involvement of employees has been shown to have positive effects on the workplace setting, the literature has shown, on the other hand, that where interventions have been carried out without any real involvement by employees in the process of implementing the changes, there has been a drop in employee support for these changes (Lourijsen et al. 1999, Dahl-Jørgensen and Saksvik 2005).

The representation of all employees' groups concerned is, moreover, an important issue with respect to the participation of workers. Ideally, insofar as is possible, the IG should include members who represent the different realities of workers (working hours, sectors, age groups, gender, etc.) The exclusion of one or more employee groups, as reported by Mikkelsen and Gundersen (2003), can lead to conflict and hinder the intervention process.

Three: Rigorous process to ensure effective operation of intervention groups (IGs)

The participatory approach developed by the researchers was based on the creation of intervention groups (IGs) made up of employee and management representatives and researchers. Since these IGs were the central mechanism of the intervention process, it was essential to use great rigour in forming them and to ensure that the members of these groups had the support of those they represented. The selection process, therefore, had to be particularly rigorous and transparent. As our analysis brought out, relying on a consultation process to form the IGs can prove to be an effective way to ensure that these groups will have the support of the workplace. As pointed out by Barbier (1996), in a participatory intervention process, the members of the group should be willing, interested and motivated by their determination to solve the problems identified. These representatives should have credibility and recognized leadership skills as well as strong listening and communication skills.

Clearly defined roles and assigned responsibilities within the IGs is an important component for the effective operation of these groups, as was observed a posteriori in the three intervention projects studied. Each member of the IG must have clearly defined roles and responsibilities so as to ensure rigour in the process but also to allow for true participation and commitment within the committee. The importance of this factor has been highlighted by a number of authors, including Nielsen et al. (2006) in a study conducted among employees in hospital canteens, and Kompier et al. (2000) in the context of a retrospective analysis of interventions carried out among bus drivers.

Because of the length of the intervention projects and natural turnover in personnel within the organizations, there was a degree of instability among the membership of the IGs. This created significant challenges in terms of maintaining the level of motivation within these groups as well as with respect to the transfer of information. Changes that took place within the organizations (e.g. changes in orientation and major restructuring of management, turnover within the union executive) explained much of this instability. Biron et al. (2010) also referred to this significant obstacle in a study they conducted in Great Britain.

It was not always easy to reach a consensus on the priority to be given to the interventions considered and to meet all the expectations of all members of the IGs. The IGs assigned priority to interventions, first, in accordance with the constraints identified by the process, and then, according to their feasibility and the internal resources available. Some interventions could therefore not be chosen due to the lack of conditions needed to carry them out. This led to a degree of dissatisfaction among some workers or within some subgroups who felt that they had not been sufficiently heard and that the choice of interventions did not meet their expectations. This brings up the challenge referred to by Gélinas and Gagnon (1983) of properly addressing the various expectations of the participants so as to construct a common understanding of the reality at hand and to identify coherent possible solutions which will have unanimous support when they are implemented.

It is also essential for it to be understood, from the outset, that the intervention cannot, in itself, solve all problems. It can be dangerous to create too many expectations which will inevitably lead to disappointment. In the same vein, Baril-Gingras et al. (2006) emphasized the importance of choosing the desired changes based on what is possible for the organization as well as the internal resources available.

Four: External mediation

The participation of researchers or other individuals from outside or who are neutral is crucial to the success of an intervention process. The analysis of the three research projects studied showed that the involvement of individuals from outside with experience in interventions lent a degree of seriousness to the process. These individuals often also acted as "facilitators," "translators" or "mediators" for the parties. In addition to leading meetings, these researchers helped to explain, in terms of recognized theoretical concepts, the irritants identified during the risk assessment phase, thus helping the organizations to better grasp the significance of the impact of the prevailing working conditions on mental health. In the three projects studied, this neutral individual from outside also sometimes played a mediating role, which proved to be essential to the effective operation of the IGs. When, at the end of the projects, these individuals withdrew from the process, it proved to be difficult to continue to move forward in the absence of someone who could motivate people and organize and lead the meetings. We will return to the importance of support and sustainability in factor 8 below.

Five: Support from management for the implementation of interventions

Managerial support is necessary in order for an intervention project to be successfully implemented. This support can be in the form of financial support when necessary or, as was more often the case in the three projects studied, can entail freeing up employees and managers – and replacing them when necessary – so that they can participate in IG activities. In fact, most of the interventions implemented did not require a significant financial investment but rather openness on the part of management to new ways of doing things or to "pilot" projects that did not correspond to the usual ways. To this end, support from management was essential. The importance of managerial support has, moreover, often been referred to in the literature as a critical factor for the successful implementation of organizational interventions (Kristensen 2005, Semmer 2006). As noted by Kristensen (2005), when a project lacks the support of corporate management, this has a negative impact on the scope of the organizational changes implemented.

Moreover, support from management depends on a set of conditions and can be irremediably threatened by changes that take place within an organization. Indeed, the workplace is far from being an isolated laboratory and it is impossible to carry out interventions in the workplace without potentially being affected by the changes that arise, especially when these changes are major. This is what occurred in the hospital network, which underwent several changes, including very extensive budget cuts, just when the intervention process was underway. As pointed out by Heany et al. (1993), it is, however, important to document these developments so as to be able to later distinguish between effects related to the project itself from those which can be attributed to administrative decisions from outside the project which may have affected the dimensions of work organization targeted by it. In some cases, as noted by Biron et al. (2010), the intervention can be perceived as having negatively affected working conditions which were already made more difficult by major changes that had taken place within the organization. Since organizations are regularly exposed to change, it is important for those in charge to manage these changes properly so as to lessen their impact and ensure that the interventions that have been planned or developed are not negatively affected by them. Administrative decisions from outside the intervention project can also sometimes hinder or prevent the implementation of the solutions identified by the intervention groups, or affect the degree to which they are appropriated by the organization over the long term. Furthermore, it can be difficult to distinguish between the effects of adopting new policies or ways of doing things from the effects of the interventions themselves. In their project involving canteen workers, for example, Nielsen et al. (2006) described the negative impact of the adoption of a non-smoking policy in the workplace just when the intervention projects were being implemented.

Tight budgets and a lack of human resources were one of the major obstacles cited in the intervention projects studied. A lack of resources acted as a brake on the implementation of some projects. Several solutions that had been put forward by the IGs could not be carried out due to budget cuts and the ensuing consequences. In a deficit situation, as was the case in the project conducted in a hospital setting, the managers appeared to be more preoccupied by financial pressures than by the preventive interventions that were to be developed and implemented. In a period of tight budgets, it sometimes proved to be difficult to ensure the participation of workers and managers in the IGs without creating additional pressures due to the already severe lack of personnel. In the hospital centre, for example, the excessive workload did not allow caregivers enough time to exchange information and consult their peers and thus to properly identify the constraints that needed to be addressed. The necessary debates did not, therefore, take place. In a study of 24 interventions carried out in workplaces, Whysall et al. (2006) also identified as an obstacle a lack of budgetary resources, entailing, in particular, a lack of human resources.

Another important obstacle to the implementation of the interventions studied was unquestionably the deterioration in working conditions during the period covered by the research project: increased workload, high turnover among employees and managers, heavier caseloads, etc. In the hospital centre and the prison setting, during the period in which the intervention was being carried out, working conditions were marked by heavier caseloads, budget cuts and restructuring, all of which added to the constraints already present. The challenge of intervening was that much greater for the researchers who had invested in an organization in which the working conditions, unrelated to the research project itself, continued to decline. The study conducted at a private British company by Biron et al. (2010) clearly illustrates the extent to which deteriorating working conditions can lead to the failure of an intervention.

High turnover among managerial personnel can also hinder the intervention process, making it necessary to constantly motivate new individuals who are called on to participate in it. This involves communicating information to the new managers and renewing the initial commitment to the project. The commitment agreements signed at the start of the project (see factor 2) can help to make up for the negative effects of this turnover. In their study, Biron et al. (2010) clearly documented the importance of this constraint and its negative impacts on the implementation of the interventions developed. Nielsen et al. (2006) also referred to the negative impact that replacing the project manager can have on the intervention process, mentioning nonetheless that the arrival of a new manager can, in some cases, have a positive effect on this process.

Six: Rigorous implementation and follow-up of changes

Rigorous implementation of the changes planned by the IG is essential, otherwise the situation may deteriorate due to the hopes raised by the project and the involvement and effort invested in it by the workplace. In fact, as reported by Aust and Ducki (2004), it is important to consider that organizational interventions lead to expectations on the part of workers and that when these expectations are not addressed to a sufficient degree, this has a deleterious effect on the quality of interpersonal relations between staff and management. Prudence and rigour are therefore mandatory.

In the three projects studied, the activities of the intervention groups led to the identification of solutions for several of the constraints identified by the members of the committee and by the workers consulted during the process. In some cases, during an initial phase, it appeared to be beneficial to rigorously implement measures that could be applied within a short timeframe so as to demonstrate the effectiveness of the intervention process. However, our analysis also showed that the process should not be rushed, since the search for effective solutions is contingent, first of all, on the capacity to properly define the problems. This phase requires time and the solutions chosen will not necessarily be easy to implement. Thus, as brought out by a close analysis of the projects, in some cases, quick solutions were tried for complex problems, while, in other cases, seemingly appropriate solutions were identified but the workers did not support them because they had not participated in the process leading to their selection.

In order to ensure rigour in the process, it is necessary to monitor the development and implementation of the changes involved. Thus, in order to avoid pitfalls related to implementation, an evaluation of the intervention process needs to be included in each step of the action research. Continuous evaluation makes it possible to carry out the needed adjustments along the way so as to better achieve the objectives set out. As mentioned by Kobayashi et al. (2008) and Hasson et al. (2012) the degree of implementation of planned interventions also greatly influences the effect of the intervention.

Seven: Communication and information-sharing mechanisms

Information-sharing mechanisms are vital to the success of organizational interventions. The IGs must disseminate information to the workers in a fairly systematic way, through appropriate means which will facilitate the transfer of information and help all members of the organization appropriate the knowledge. Furthermore, this approach favours the emergence of avenues for action and facilitates the implementation of the chosen measures in the workplace. It is, however, important to ensure that all groups are reached through these information-sharing mechanisms regardless of the organizational structure (work shifts, work sites, etc.). In the three intervention projects studied, the dissemination of information to workers was facilitated by the use of both formal and informal communication mechanisms within the organization, by the research team and the members of the IGs, throughout the entire research process.

Formal presentations were held in the form of meetings with the groups targeted by the interventions and managers, short texts were published in internal print media (organization, union) and, in one of the projects, a written communication tool, usually published monthly, was created and disseminated to all members of the organization through the intranet. According to Goldenhar et al. (2001), the dissemination of information in a form that is readily understood constitutes a crucial step in the implementation of interventions. Kompier et al. (2000) concur, considering good communication to be an important stimulating factor during the implementation phase of preventive interventions. Whysall et al. (2006) also identified communication as a facilitating factor for the implementation of interventions.

Eight: Appropriation of the process by the organization

It is important for the organization to take charge of the intervention process and to appropriate it in order to ensure its long-term viability. In the three projects studied, this proved to be a difficult challenge when the researchers eventually withdrew from the organizations in which the interventions were being carried out. It is not possible for an organization to take charge of the process unless it is provided with the tools and support needed to do so, and unless the process has been shown to have positive effects.

Even so, some organizations find it very difficult to take charge of an intervention process without outside support. In two of the three projects studied, after the researchers withdrew, management personnel (unit heads, HR) were given the mandate to carry the process through, but without much success because the co-management structure was quickly eliminated from the intervention process and the workers stopped supporting it. The IGs thus became inoperable, although some measures that had been implemented continued to be applied because they had been integrated into the work routines and new ways of doing things (e.g. daily statutory team meetings in some work units or more transparent and systematic consultation processes). Ensuring that the organization will appropriate the intervention process after the research project is finished thus remains a significant challenge in terms of ensuring that interventions will have the most positive impact possible. In fact, as pointed out by Mikkelsen et al. (2000), organizational interventions will be successful in the long run insofar as they have provided an opportunity to acquire participatory skills and are carried on after the implementation phase, having been integrated into the organization's formal structures and daily work routines.

Conclusion

The success of preventive interventions in the area of mental health at work unquestionably depends on a set of factors that relate either to the context inherent in the workplace or to the intervention process itself. Our analysis of the factors leading to the success of interventions that were common to the three intervention projects studied brought out three contextual factors and eight factors related to process. Depending on the specific ways in which these factors manifest themselves, they can either facilitate or hinder the successful implementation of an intervention process.

First, it is imperative for an organization that engages in a preventive intervention process to clearly identify the three contextual factors that can interfere with this process and, if necessary, take prior action to properly prepare the ground before the process begins. One example might be to seek clear initial support from the organization's decision-making bodies. It is also advisable to keep these factors in mind throughout the entire process, since the contextual conditions are likely to change over the course of the process.

Second, it is essential to pay particular attention to the factors related to process, which more directly concern the intervention method itself. Among the eight factors identified, a certain number appeared to be particularly crucial. First, it is important for employees to actively participate in the intervention process. Although this may seem obvious, many organizations implement interventions which have been developed in isolation and imposed from outside, without involving or even consulting the workers. Yet, workers are, in fact, the real experts in their workplace environment. Openness to shared management is also essential, as is support for the process from both senior management and, when applicable, union officials. Without real support, which can be expressed in terms of both human and financial resources when needed, an intervention process cannot really get off the ground, let alone meet the objectives set for it. This support can also be expressed by a clear message of commitment on the part of the organizational leaders who are investing in the process. A rigorous process is also very important, from the very start of the project until the outside experts have withdrawn and the organization itself has taken charge. This rigour entails, in particular, the use of a solid theoretical and methodological framework based on reliable tools, and close monitoring of the development and implementation of the interventions, as well as their continuous evaluation. Lastly, a follow-up by the experts, when they withdrew from the organization, could also be envisaged to ensure better results.

Implementing interventions to prevent mental health problems at work constitutes a significant challenge which, as we have seen, can lead to successful outcomes. How can this be achieved? Overall, the success of an intervention process hinges on three main principles: properly preparing the ground, encouraging the participation of all individuals concerned by the process, and maintaining a high degree of rigour throughout the entire process. By respecting these principles and ensuring that it has the means to do so, an organization will have created the optimal conditions needed to carry out effective interventions to prevent mental health problems in the workplace.

References

Aust, B. and Ducki, A. 2004. Comprehensive health promotion interventions at the workplace: Experiences with health circles in Germany. *Journal of Occupational Health Psychology*, 9(3), 258–70.

Barbier, R. 1996. *La recherche action*. Paris: Anthropos.

Baril-Gingras, G., Bellemare, M. and Brun, J.-P. 2006. The contribution of qualitative analyses of occupational health and safety interventions: An example through a study of external advisory interventions. *Safety Science*, 44(10), 851–74.

Baril-Gingras, G., Bellemare, M. and Brun, J.-P. 2007. Conditions et processus menant à des changements à la suite d'interventions en santé et en sécurité du travail: l'exemple d'activités de formation. *PISTES*, 9(1).

Biron, C., Gatrell, C. and Cooper, C.L. 2010. Autopsy of a failure: Evaluating process and contextual issues in an organizational-level work stress intervention. *International Journal of Stress Management*, 17(2), 135–58.

Bond, F.W. and Bunce, D. 2001. Job control mediates change in a work reorganization intervention for stress reduction. *Journal of Occupational Health Psychology*, 6(4), 290–302.

Bourbonnais, R., Brisson, C. and Vézina, M. 2011. Long-term effects of an intervention on psychosocial work factors among healthcare professionals in a hospital setting. *Occupationnal Environmental Medecine*, 68(7), 479–86.

Bourbonnais, R., Jauvin, N., Dussault, J. and Vézina, M. 2007. Psychosocial work environment, interpersonal violence at work and mental health among correctional officers. *International Journal of Law and Psychiatry*, 30(4–5), 355–68.

Bourbonnais, R., Jauvin, N., Dussault, J. and Vézina, M. 2012. Evaluation of an intervention to prevent mental health problems among correctional officers. In *Improving Organizational Interventions For Stress and Well-Being*, edited by C. Biron, M. Karanika-Murray and C. Cooper. New York: Routledge, 187–214.

Bourbonnais, R., Malenfant, R., Vézina, M., Jauvin, N. and Brisson, I. 2005. Work characteristics and health of correctional officers. *Revue d'Epidemiologie et de Santé Publique*, 53(2), 127–42.

Bourbonnais, R., Brisson, C., Vinet, A., Vézina, M., Abdous, B. and Gaudet, M. 2006a. Effectiveness of a participative intervention on psychosocial work factors to prevent mental health problems in a hospital setting. *Occupational Environmental Medicine*, 63(5), 335–42.

Bourbonnais, R., Brisson, C., Vinet, A., Vézina, M. and Lower, A. 2006b. Development and implementation of a participative intervention to improve the psychosocial work environment and mental health in an acute care hospital. *Occupational Environmental Medicine*, 63(5), 326–34.

Brisson, C., Cantin, V., Larocque, B., Vézina, M., Vinet, A. and Trudel, L. 2006. Intervention research on work organization factors and Health: Research design and preliminary results on mental health. *Canadian Journal of Community Mental Health*, 25(2), 241–59.

Brun, J.-P., Biron, C., Martel, J. and Ivers, H. 2003. Évaluation de la santé mentale au travail: une analyse des pratiques de gestion des ressources humaines. *Rapport de recherche R-342, Montréal, IRSST*, Études et Recherches, 1–99.

Chambaud, L. and Richard, L. 1984. Le projet santé agricole: éléments d'analyse et de réflexion ou les difficultés de l'approche recherche-action dans le réseau de la santé communautaire. *Saint-Jean sur Richelieu, Hôpital du Haut-Richelieu*, 1–49.

Czabala, C., Charzynska, K. and Mroziak, B. 2011. Psychosocial interventions in workplace mental health promotion: An overview. *Health Promotion International*, 26 Suppl 1, i70–84.

Dahl-Jørgensen, C. and Saksvik, P.O. 2005. The impact of two organizational interventions on the health of service sector workers. *International Journal of Health Services*, 35(3), 529–49.

DeJoy, D.J., Wilson, M.G., Vandenberg, R.J., McGrath-Higgins, A.L., Griffin-Blake, C.S. 2010. Assessing the impact of healthy work organization intervention. *Journal of Occupational and Organizational Psychology*, 83, 139–65.

Dussault, J., Jauvin, N., Vézina, M. and Bourbonnais, R. 2010. Prévention de la violence entre membres d'une même organisation de travail – Évaluation d'une intervention participative. *Études et recherches*, Rapport R-661, Montréal, IRSST, 1–98.

Dussault, J., Jauvin, N., Vézina, M. and Bourbonnais, R. 2012. Preventing violence among employees of the same work organization-evaluation of a participatory intervention. *Studies and Research Projects*, report R_739, IRSST(1–96).

Gélinas, A. and Gagnon, C. 1983. Systémique, recherche-action et méthodologie souple. *Chicoutimi, GRIR/UQAC*, 11(1), 1–16.

Gilbert-Ouimet, M., Brisson, C., Vézina, M., Trudel, L., Bourbonnais, R., Masse, B., Baril-Gingras, G. and Dionne, C.E. 2011. Intervention study on psychosocial work factors and mental health and musculoskeletal outcomes. *Healthcare Papers*, 11(Sp), 47–66.

Goldenhar, L.M., LaMontagne, A.D., Katz, T., Heaney, C. and Landsbergis, P. 2001. The intervention research process in occupational safety and health: An overview from the National Occupational Research Agenda Intervention Effectiveness Research team. *Journal of Occupational and Environmental Medicine*, 43(7), 616–22.

Hasson, H., Gilbert-Ouimet, M., Baril-Gingras, G., Brisson, C., Vézina, M., Bourbonnais, R. and Montreuil, S. 2012. Implementation of an organizational-level intervention on the psychosocial environment of work: Comparison of managers' and employees' views. *Journal of Occupational and Environmental Medicine*, 54(1), 85–91.

Heany, C., Israel, B., Schurman, S.J., Baker, E.A., Elizabeth, A., House, J.S. and Hugentobler, M. 1993. Industrial relations, worksite stress reduction, and employee well-being: A participatory action research investigation. *Journal of Organizational Behaviour*, 14(5), 495–510.

Hemp, P. 2004. Presenteeism: At work but out of it. *Harvard Business Review*, 82(10), 49–58.

Johnson, J.V. and Hall, E.M. 1988. Job strain, workplace social support, and cardiovascular disease: A cross-sectional study of a random sample of the Swedish working population. *American Journal of Public Health*, 78(10), 1336–42.

Karasek, R. and Theorell, T. 1990. *Healthy Work: Stress, Productivity and the Reconstruction of Working Life*. New York: Basic Books.

Kobayashi, Y., Kaneyoshi, A., Yokota, A. and Kawakami, N. 2008. Effects of a worker participatory program for improving work environments on job stressors and mental health among workers: A controlled trial. *Journal of Occupational Health*, 50(6), 455–70.

Kompier, M. and Kristensen, T. 2001. Organizational work stress interventions in a theoretical, methodological and practical context. In *Stress in the Workplace: Past, Present and Future*. London: Whurr, 164–90.

Kompier, M.A.J., Aust, B., van den Berg, A.M. and Siegrist, J. 2000. Stress prevention in bus drivers: Evaluation of 13 natural experiments. *Journal of Occupational Health Psychology*, 5(1), 11–31.

Kompier, M.A.J., Geurts, S.A.E., Gründemann, R.W.M., Vink, P. and Smulders, P.G.W. 1998. Cases in stress prevention: The success of a participative and stepwise approach. *Stress Medicine*, 14(3), 155–68.

Kristensen, T.S. 2000. Workplace intervention studies. *Occupational Medicine: State of the Art Reviews*, 15(1), 293–305.

Kristensen, T.S. 2005. Intervention studies in occupational epidemiology. *Occupational and Environmental Medicine*, 62(3), 205–10.

Lescarbeau, R. 1994. *L'enquête feed-back*. Montréal: Les Presses de l'Université de Montréal.

Lourijsen, E., Houtman, I., Kompier, M. and Gründemann, R. 1999. The Netherlands: A hospital, "Healthy Working for Health." In *Preventing Stress, Improving Productivity: European Case Studies in the Workplace*, edited by M. Kompier and C. Cooper. New York: Routledge, 86–120.

Mikkelsen, A. and Gundersen, M. 2003. The effect of a participatory organizational intervention on work environment, job stress, and subjective health complaints. *International Journal of Stress Management*, 10(2), 91–110.

Mikkelsen, A., Saksvik, P. and Landsbergis, P. 2000. The impact of a participatory organizational intervention on job stress in community health care institutions. *Work and Stress*, 14(2), 156–70.

Nielsen, K., Fredslund, H., Christensen, K.B. and Albertsen, K. 2006. Success or failure? Interpreting and understanding the impact of interventions in four similar worksites. *Work and Stress*, 20(3), 272–87.

Parsonage, M. 2007. *The Impact of Mental Health on Business and Industry: An Economic Analysis*. Presentation by the Sainsbury Centre for Mental Health.

Robson, L.S., Shannon, H.S., Goldenhar, L.M. and Hale, A.R. 2001. Guide to evaluating the effectiveness of strategies for preventing work injuries. *NIOSH*, 119, 1–139.

Robson, L.S., Clarke, J.A., Cullen, K., Bielecky, A., Severin, C., Bigelow, P., Irvin, E., Culyer, A.J. and Mahood, Q. 2007. The effectiveness of occupational health and safety management system interventions: A systematic review. *Safety Science*, 45(3), 329–53.

Saksvik, P.Ø., Nytrø, K., Dahl-Jørgensen, C. and Mikkelsen, A. 2002. A process evaluation of individual and organizational occupational stress and health interventions. *Work and Stress*, 16(1), 37–57.

Semmer, N. 2006. Job stress interventions and the organization of work. *Scandinavian Journal of Work, Environment and Health*, 32(6), 515–27.

Shannon, H.S., Robson, L.S. and Guastello, S.J. 1999. Methodological criteria for evaluating occupational safety intervention research. *Safety Science*, 31(2), 161–79.

Siegrist, J. 1996. Adverse health effects of high effort low-reward conditions. *Journal of Occupational Health Psychology*, 1(1), 27–41.

Siegrist, J. 2002. Effort-reward imbalance at work and health. In *Research in Occupational Stress and Well Being, Vol. 2: Historical and Current Perspectives on Stress and Health*, edited by P. Perrewe and D. Ganster. New York: JAI Elsevier, 261–91.

Simard, C. 2009. La participation des travailleurs à une recherche en santé mentale au travail: Une histoire de confiance. *Thèse de doctorat, Université Laval, Québec*, 1–217.

Stansfeld, S. and Candy, B. 2006. Psychosocial work environment and mental health – a meta-analytic review. *Scandinavian Journal of Work and Environmental Health*, 32, 443–62.

St-Arnaud, L. and Gignac, S. 2005. Évaluation de la prise en charge d'une démarche participative visant la réduction des contraintes psychosociales de travail. *Équipe de recherche RIPOST, CSSS de la Vieille-Capitale, Québec*, 1–38.

Sultan-Taïeb, H. (2012). *Comment calculer le coût des problèmes de santé mentale au travail pour l'employeur?* RIPOST, Des interventions pour agir sur la santé mentale au travail.

Trudel, L., Montreuil, S. and Brisson, C. 2000. Mieux comprendre l'impact par l'analyse de l'implantation: Le cas d'un programme de prévention chez des travailleurs avec ordinateur. *The Canadian Journal of Program Evaluation*, 15(1), 57–82.

Tsutsumi, A.M., Nagami, M., Yoshikawa, T., Kogi, K. and Kawakami, N. 2009. Participatory intervention for workplace improvements on mental health and job performance among blue-collar workers: A cluster randomized controlled trial. *Journal of Occupational and Environmental Medicine*, 51(5), 554–63.

van der Doef, M. and Maes, S. 1998. The job demand-control(-support) model and physical health outcomes: A review of the strain and buffer hypotheses. *Psychology & Health*, 13, 909–36.

van der Doef, M. and Maes, S. 1999. The job demand-control(-support) model and psychological well-being: A review of 20 years of empirical research. *Work and Stress*, 13, 87–114.

Van der Hek, H. and Plomp, H.N. 1997. Occupational stress management programmes: A practical overview of published effect studies. *Occupational Medicine*, 47(3), 133–41.

Van, W.B. and Pillay-Van, W.V. 2010. Preventive staff-support interventions for health workers. *Cochrane Database of Systematic Reviews*, 17(3), 1–35.

Vézina, M., Brisson, C., Gilbert-Ouimet, M., Vinet, A., Trudel, L., Bourbonnais, R., Masse, B. and Baril-Gingras, G. 2009. Intervention study on psychosocial work factors and health. 4e congrès annuel pour la recherche sur la santé mentale et la toxicomanie en milieu de travail. Toronto.

Vézina, M., Cloutier, E., Stock, S., Lippel, K., Fortin, É., Delisle, A., St-Vincent, M., Funes, A., Duguay, P., Vézina, S. and Prud'homme, P. 2011. Enquête québécoise sur les conditions de travail, d'emploi et de SST (EQCOTESST). *Rapport de recherche RR-691, Montréal, IRSST*, Études et recherches, 1–50.

Whysall, Z., Haslam, C. and Haslam, R. 2006. Implementing health and safety interventions in the workplace: An exploratory study. *International Journal of Industrial Ergonomics*, 36(9), 809–18.

Zwerling, C., Daltroy, L.H., Fine, L.J., Johnston, J.J., Melius, J. and Silverstein, B.A. 1997. Design and conduct of occupational injury intervention studies: A review of evaluation strategies. *American Journal of Industrial Medicine*, 32(2), 164–79.

15 Implementation of an Organizational Intervention on Quality of Life at Work: Key Elements and Reflections

CAROLINE BIRON, FRANCE ST-HILAIRE and JEAN-PIERRE BRUN

Abstract

In this chapter, we draw on models of occupational stress prevention and on process evaluation to document certain elements that have emerged as important during an intervention project in a large public organization in Quebec. Whereas several authors have made calls for more attention to be paid to process and contextual issues (Cox et al. 2007, Nielsen et al. 2010a, Biron et al. 2012b), it is still rare to find studies describing the "black box" of the intervention. Indeed, researchers generally report results regarding the effectiveness of the intervention, but often leave out how the intervention got to be implemented. In this chapter, we aim to describe the early stages of an organizational stress intervention.

Introduction

At the conference of the European Academy of Occupational Health Psychology in Rome in 2010, Cary Cooper argued that: "We have enough scientific data on what causes people to fall ill in the workplace ... We know the problems, what we need to do now is to find solutions" (Cooper 2010, in Nielsen et al. 2010a). Costs associated with decreased performance due to sickness absenteeism and presenteeism are substantial. This alone should encourage more employers to develop approaches to improve quality of life, prevent stress and improve well-being at work (Cooper and Dewe 2008, Sainsbury Centre for Mental Health 2007). While we recognize that the problems associated with stress at work are costly, interventions to prevent them are still rare and mostly focused on the effects of stress (e.g. training on stress, time or conflict management) rather than elimination of risk factors at the source (Giga et al. 2003c). Moreover, although theoretical models suggest to researchers different elements to document and evaluate in an intervention, little practical guidance exists as to how to concretely operationalize these elements in the field. For example, several approaches recommend having a strong support from top management and a clear and transparent communication plan about the interventions that were already implemented or that will be undertaken (Giga et al. 2003b, Health and Safety Executive 2003, Jordan et al. 2003). However, few studies clearly show what these imply for the conduct of the intervention. How can we characterize "clear support from management"? What indicates whether one has this support or not? Does this only imply the approval of a budget or a policy on stress at work? What should be contained in a communication plan?

In this chapter, we draw on models of occupational stress prevention and on process evaluation to document certain elements that have emerged as important during an intervention in a public organization in Quebec. Specifically, the chapter aims:

1. to highlight some key elements of the intervention process in relation to each step of the risk management approach (Cox et al. 2000, Brun et al. 2009); and
2. illustrate, using a case study on quality of life at work (QLW), how these key elements about the process were considered.

The effectiveness of the intervention has not yet been evaluated, but the intervention involved a strong commitment from management, a diagnosis of psychosocial risks, and it generated a significant mobilization of staff and managers. These are the early stages of a preventive intervention and are rarely described in the literature (Brun et al. 2009, Cox et al. 2000, Giga et al. 2003a). Organizational interventions are particularly difficult to implement and to evaluate (Biron et al. 2012b). As mentioned by Burke (Chapter 1, this volume), organizational attempts to reduce psychosocial risks can sometimes fail, as organizations often conduct a diagnosis of psychosocial risks without implementing any changes following the risk assessment (Biron et al. 2010). This lack of attention to the process of developing and implanting interventions could explain why many interventions obtain mixed results or do not get the desired effect in terms of well-being improvements or organizational effectiveness (Biron et al. 2012c). This chapter aims to illustrate, using a case study in a ministry of the Quebec government, elements that appear to be particularly important in the early stages of a preventive approach. Subsequent steps (implementation and evaluation of its benefits) are not discussed in this chapter because we specifically want to focus on describing the early stages of the process (preparation, the diagnosis of psychosocial risks and development of a plan of actions inside work teams).

Organizational Stress Interventions

Several theoretical models converge on some key elements regarding organizational stress interventions. For example, Giga et al. (2003a) present a model based on case studies identified by Kompier and Cooper (1999) and on a report of the Robertson Cooper Ltd company in the UK. Their model summarizes preventive studies on stress at work during the last decade in addition to considering the recommendations of a panel of experts in the field (Jordan et al. 2003). In this model, they suggest that stress prevention is facilitated by the presence of an organizational culture where employer and employees are all involved in the instigation of a preventive approach. This approach includes a diagnosis of psychosocial risks, which is the basis for managers and workers to establish a plan of action. Giga et al. (2003a) emphasize the importance of a culture of prevention where the parties communicate, analyse, revise their plans and learn from their mistakes when the intervention does not produce the desired results. This type of culture reflects the key elements of the intervention process identified by Nytrø et al. (2000) and by Saksvik et al. (2002). Moreover, among the key factors suggested by Giga et al. (2003) as critical to the success of interventions are:

1. staff participation;
2. management commitment;
3. a diagnosis of psychosocial risks;
4. an integrative approach with preventive interventions at both individual- and organizational-levels;
5. a strategy behind the implementation of actions; and, finally
6. a clear communication plan.

Very much in line with this model, Biron et al. (2012a) recently published a book on the intervention process to prevent occupational stress. The volume contains interventions in organizations of different sizes and from different sectors and countries (Australia, Norway, Denmark, the United Kingdom, Netherlands, Canada). Biron et al. (2012a) conclude by stressing, among other things, the importance of employee participation, the sustained commitment of management, a clear strategy of the implementation of interventions, a multi-level approach integrating the elimination of the causes and consequences of stress at work, a diagnosis of psychosocial risks based on evidence, and the use of terminology intervention that make consensus.

In the same vein, Nielsen and Randall (2012) propose a model to evaluate the intervention process. Their model includes three components to consider:

1. the context of the intervention;
2. the strategy behind the development and the implementation of interventions; and
3. the mental models of the participants.

First, the context refers to both the organizational context in general (e.g. organizational changes which may hinder/facilitate the intervention, how the intervention is or is not aligned with the culture of the organization, earlier attempts to intervene) and the specific context (e.g. events can influence the conduct of the intervention, such as workload management, conflicting priorities, introduction of competing projects). The second component of the model is the strategy behind the development and the implementation of interventions. More specifically, to document the intervention, Nielsen and Randall (2012) suggest documenting the motivations underlying the approach (who initiates the process? For what purpose?), the activities of the intervention (i.e. are the participants exposed to the intervention? Do the changes implemented correspond to those expected?), and the implementation strategy (who are the stakeholders in the project? What are the drivers of change? Is the approach participatory? What are the roles of management, managers, employees, external consultants? Which means of communication were used?). The third and final component of the model refers to the mental models of the participants, that is to say, their readiness to change (do they feel the need to change?) and their perceptions of the intervention (are they mobilized and involved in the intervention? Can they raise questions about the conduct of the project? Do they believe that the intervention can improve their working conditions?). For example, in an intervention aimed at changing certain responsibilities, Randall et al. (2005), reported that supervisors failed to disseminate information to staff as they perceived this change as damaging to their own work conditions. This mental model or this perception of change hindered the implementation of the intervention.

In sum, recent research and theoretical models of job stress have a consensus on some key elements of the intervention process. These elements include four recurring items which are described in more detail below and which are then illustrated in a case study.

Key Elements of the Intervention Process

ONE: STAFF PARTICIPATION

All models converge on the importance of psychosocial risk assessment in relation to the needs of the members of the organization (Cooper and Cartwright 1997, Cox et al. 2000, Israel et al. 1996, Jordan et al. 2003). Their participation is an essential component of the intervention process. Risk assessment allows the psychosocial development of a prevention programme adapted to the needs of those who will benefit. As indicated by Brun et al. (2009), there are many questionnaires available to measure psychosocial risks and other forms of assessment are possible (group interviews, individual interviews with representatives of workers and employers, analysis of collective data). In the case of small businesses, or when the use of a questionnaire is not possible, a screening tool can be used. For example, thanks to the work carried out by the National Institute of Public Health (INSPQ), there is now a tool that lies between the administrative data, group interviews and questionnaires. This workplace characterization tool allows assessing, in a simple way, the presence of psychosocial risks in the working environment (Vézina et al. 2008). This tool is designed for organizations wanting to identify risk factors and assess mental health but for which a thorough risk assessment of psychosocial constraints in not an option due to the size of the sample or due to lack of resources to administer a questionnaire to all employees (for an example of small-scale interventions, see Hesselink et al. 2012). The information is collected through interviews with two or three key informants inside the organization (e.g. director, president of the union, a representative of occupational health and safety (OHS)). The choice of informants should be based on their credibility and their degree of influence on the other members of the organization. Note that the validity of the tool depends largely on the identification of good key informants. The workplace characterization tool is used primarily to screen, not to assess a specific diagnosis.

Very often, stress intervention studies imply that employee participation is a necessary component of an intervention. This, however, does not just involve having people completing questionnaires or participating in focus groups. Indeed, as pointed out by Nielsen and Randall (2012) and their notion of mental models presented above, the participation of employees must go beyond simple consultation to identify the principal risks of the work. They must be mobilized to participate in subsequent steps following the diagnosis. Providing employees with information on the rationale for intervention activities can engage them in the intervention. However, employees and managers must be able to influence these activities (Cox et al. 2000, Cox et al. 2002, Kompier et al. 2000, Semmer 2006). For example, Nielsen et al. (2007) show that participants who report having had the opportunity to actively influence the interventions derived more benefits from them. In fact, the participants who contributed to the development and selection of intervention activities reported greater improvements in terms of their working conditions, their job satisfaction and their level of stress following the intervention.

TWO: THE SUSTAINED COMMITMENT OF SENIOR MANAGEMENT

The past few years have seen an emerging tendency to recognize the existence of psychosocial risks (e.g. workload, conflict, harassment) as risks in the business as well as the technical, mechanical or chemical risks, for instance. Thus, more and more countries have legislated on this issue and require companies to identify risks that may affect the psychological integrity of employees (e.g. MacKay et al. 2004, Saksvik et al. 2007). Brun et al. (2009) point out that if, for the moment, mental health problems are rarely recognized as occupational accidents or disabilities, the national prevention organizations such as NIOSH and the United States and HSE in the UK strongly encourage companies to implement prevention strategies. It is in this spirit that Quebec created the "Healthy Enterprise" standard which involves the acquisition of a healthy lifestyle and maintaining a work environment conducive to health (Bureau de normalisation du Québec 2008). Canada has also implemented a recent standard entitled "Psychological Health and Safety in the Workplace" which defines 13 major risk factors classified according to five criteria standards (CSA 2013). Being concerned about employee health and well-being should however not be just a legal or medical matter. A healthy company is not only the responsibility of the health and safety or the human resources department. The term "healthy business" does not limit itself to occupational health and safety services and activities. Companies must exceed this reductionism vision of being strictly focused on preventing ill-health or improving well-being. As Kortum (Chapter 2, this volume) mentions, health related issues and the management of psychosocial risks is rarely considered as a company priority. Yet, issues such attracting competent staff, promoting talents, motivating employees, retaining competencies and increasing performance by decreasing absenteeism are often considered in companies' strategic plans. These issues are not competing with health and well-being, in fact improving health and well-being is very much likely to result in obtaining these desirable outcomes. This implies that health and well-being should not just be considered as a "project," but should instead be a management function, that is to say, they should be integrated into everyday practice, in management decisions and in organizational policies.

As indicated by Brun et al. (2009), top management commitment is considered as the first step in any stress intervention project (Giga et al. 2003a, Jordan et al. 2003). Unfortunately, the literature describing what this entails is rather scarce, and this important step is often described briefly in the methodology section of research papers, or as a necessary condition to ensure in stress prevention guidelines. Although it is one of the most crucial to the success of an intervention project, the commitment is often limited to obtaining the approval of stakeholders (i.e. management, union) for the project (Biron et al. 2010). This approval from stakeholders is not an indicator of real commitment, but rather an agreement on the project. However, most studies remain elusive about what actually constitutes a "solid" commitment of management. In the case study presented below, we illustrate the observed indicators of engagement in the project (Table 15.1).

THREE: A STRATEGY TO IMPLEMENT ACTIONS THAT PROVIDE SUPPORT TO MANAGERS

Nielsen and Randall (2012) emphasize the importance of a strategy to implement interventions. This strategy should adequately describe the roles of managers in the intervention. Indeed, their participation is also essential for the interventions to be implemented properly. Line managers have a clear effect on workers' psychosocial constraints, for example by being in a position to modulate elements such as job demands and control (Karasek and Theorell 1990), efforts and rewards (Siegrist 1996), and social support (Johnson et al. 1989). In addition, line managers play a key role in the implementation of preventive organizational interventions (Biron et al. 2010). Indeed, studies show that the leadership style is associated with a reduction in depressive symptoms (Munir et al. 2010) and an improvement in the general health of workers (Lohela et al. 2009). As argued by Kelloway and Donohoe (Chapter 12, this volume), training leaders should be considered as an intervention per se, since they are likely to act as catalysts in implementing work-related changes which improve health and well-being. On the contrary, inadequate (or harmful) leadership could lead to a deterioration of psychological health (Borritz et al. 2005), cardiovascular health (Nyberg et al. 2009) and well-being in general (Lohela et al. 2009) for employees. Although their role is critical to the success of preventive interventions, few studies have focused on describing how they can be supported in the intervention. Yarker et al. (2008) have done extensive research on the behaviours and competencies of line managers to prevent stress. Although some of these competencies are likely to be similar for the conduct of organizational interventions, it is unknown whether the competencies required of line managers during complex organizational interventions are the same as for managing employee stress. For example, organizational interventions require specific project management competencies, which might not be the same as organizing the daily workload. The case study presented in this chapter illustrates how managers were accompanied during the intervention. Although models of stress prevention all insist on the crucial role of line managers, little has been published on how they should be supported during the conduct of complex interventions.

FOUR: A COMMUNICATION PLAN

As indicated by Brun et al. (2009), a strategic approach for the prevention of mental health problems at work needs to, in order to achieve its objectives, be accompanied by an internal communication plan that goes beyond the mere transmission of information on the project and results (Giga et al. 2003a, Jordan et al. 2003). Indeed, if the plan allows communication to disseminate information, it can also reinforce the commitment of workers to bring behavioural change and establish a climate of exchange between all business partners. Too often the communication plan is designed in the spirit of disseminating information on the prevention approach. This is a laudable goal, but it must be supplemented by the attempt to establish a dialogue (two-way communication) (Noblet and LaMontagne 2009). Employees and managers have expectations, doubts, requests, suggestions or comments on the proposed approach. The intervention context must also provide means to facilitate this dialogue and create a climate of mutual listening between all employees and management. The two chapters on the CREW project (Chapters 9 and 10, this volume) are in line with this idea of establishing a mutual respectful dialogue. These means can be quite simple: luncheon, hierarchical departments visits, invitation to management team meetings, meeting with labour unions, etc. It should be noted that this bi-directional communication plays an important role to rally and engage all members of an organization in a project (Jordan et al. 2003).

The communication plan should also include a strategic component in the terminology that will be used throughout the project intervention. For various reasons (strategy, culture, history), companies use a variety of terms to refer to preventive interventions on stress and health: work stress, mental health, psychological well-being, motivation, quality of work life, promotion of health, happiness at work, wellness strategies, attraction/retention, psychological health. Biron et al. (2012) discuss the advantages and limitations of these terms, which are derived from different paradigms. For example, the use of more positive terms such as "wellness" can help in terms of marketing intervention projects in organizations, but people struggling with psychological disorders may feel excluded, or even stigmatized by the process. In contrast, terms such as "psychosocial risks" or "stress at work" can be threatening to managers and management who may feel that it opens a Pandora's box. According to Brun et al. (2009), it is important to choose a term that allows consensus in the organization and, most importantly, that will ensure that all stakeholders are clear on the fact that the interventions are about preventing problems before they appear (i.e. primary prevention), instead of on managing their consequences (i.e. secondary and tertiary prevention). Whether the terminology chosen is about reducing stress, or whether it is a more positive approach such as improving the quality of life, commitment and well-being at work, most researchers and practitioners agree that a healthy and motivated workforce is a competitive advantage for organizations. In this chapter, the term "quality of work life" is used because a pilot project in the organization showed that stakeholders were rather demotivated and unenthusiastic regarding "stress" or "psychological distress." In fact, they did not feel responsible for employees' distress, whereas they felt more engaged when the project was labelled "quality of life at work," and involved actions to keep teams motivated.

These four elements of the intervention process appear to be crucial, and in the next section we describe an intervention highlighting how these factors were taken into consideration in an organizational intervention conducted in a large public organization in Quebec.

Case Study: Quality of Life at Work in a Ministry of Quebec Public Service

DESCRIPTION OF THE ORGANIZATION

The intervention began in 2008–2009, initially, in three branches – as a pilot project – and then in 2010, throughout the organization. Thus, over a period of just one year, more than 500 teams and 10,000 employees and managers have implemented actions to improve local quality of life at work in the spirit of shared responsibility: organization, managers and employees. The next section of this chapter has two objectives:

1. first, to describe the steps and the process of an intervention to improve quality of working life (QWL) in a large organization of the Quebec public service; and
2. second, to highlight key elements contributing to the success of the approach in this organization.

IMPLEMENTATION OF THE INTERVENTION

Subsequently to the pilot projects mentioned above, the QWL approach followed five steps recommended by Brun et al. (2009):

3. preparation;
4. identification of the magnitude of the problems and psychosocial risks;
5. identification of concrete problems;
6. identification of solutions;
7. implementation and evaluation.

Figure 15.1 illustrates these steps. The intervention took place over a period of one year.

Figure 15.1 Steps of an organizational stress intervention (Brun et al. 2009)

Step 1: Preparation

Nielsen and Randall (2012) mention the importance of understanding and documenting the reasons why an organization wants to initiate an intervention. Intervention could, for example, be initiated in order to reduce costs, manage a crisis, improve productivity or to meet legal obligations. These patterns have an influence on the structure of the intervention and how it will be perceived by stakeholders. For example, Biron et al. (2010) show that interventions initiated by senior management with a view to comply with legal obligations did not trigger a strong ownership from line managers, who were slow to implement changes. Biron describes a "tick the box approach," where stakeholders indicate publicly that they are committed to the intervention, when there are in fact very few concrete changes noticeable in terms of employees' work conditions or environment. Egan et al. (2007) suggest that the interventions undertaken in order to increase the performance can have a deleterious impact on the health of employees while interventions specifically initiated to improve health at work tend to have a more positive effect. This stage of initiation also requires an understanding of the role played by stakeholders in the initiation part of the process: are managers, union representatives, senior management and health professionals involved? What were their roles?

For the participating organization, there were two main motives for this QWL intervention:

1. concerns about absenteeism; and
2. the adoption of a policy to promote healthy personnel.

These two motives reflect a preoccupation for productivity, but also for improving jobs. In terms of rates of absenteeism, the organization had one of the highest in the Quebec public service. The project also took place within the framework of a policy adopted in 2004, i.e. the policy on the health of people working in the Quebec public sector.

The approach was first prepared mainly through meetings with management committees (middle and senior) who had the mandate to subsequently inform their own managers. In addition, press releases were sent to employees and managers in order to present the approach and announce to them that a questionnaire would be sent.

Step 2: Identification of risks

Several factors must be taken into account at this stage because they have a direct impact on the evaluation of the development of the intervention and its implementation. Properly diagnosed problems will ensure that the interventions subsequently retained will stand a higher probability of being effective (LaMontagne et al. 2007, Nielsen et al. 2010b, Noblet and LaMontagne 2009). Thus, in order to assess the extent of the problem and portray the risks involved, all employees of the participating organization responded to a questionnaire designed to measure exposure to psychosocial risks (decision latitude, psychological demands, stress recognition imbalance, support) as well as the consequences of exposure to these risks. Since theoretical models suggest that chronic exposure to psychosocial stress affects the psychological and physical health (Bambra et al. 2007, Egan et al. 2007, Marmot et al. 1997, Siegrist 1996) psychological distress is often measured to document the extent of the problem, especially in Quebec where the distress measure allows a comparison with national data (Vézina et al. 2011).

Participants and measurement tools

Completion of the electronic questionnaire took place over a period of three weeks. During this period, two reminder emails were sent to participants who did not participate. A total of 7,723 workers completed the questionnaire (409 managers and 7,314 employees) representing a participation rate of 78.5 per cent. The questionnaire was designed to assess the state of QWL for the entire organization as well as for each of the teams that composed it. Moreover, motivation was measured as an outcome since it is known to be associated with exposure to psychosocial risks (Fernet et al. 2004). A scale measuring work motivation has been created using an abridged version of the scale developed by Blais et al. (1993). This scale measured different forms of motivation. Self-determined motivation varied along a continuum ranging from "not at all motivated" to "very motivated." The highest levels of self-determined motivation lead to more positive consequences, such as satisfaction, perseverance, positive mental health, high performance, while the lowest levels lead to more negative consequences, such as withdrawal from the work activity, absenteeism and poorer mental health (Rinfret and Sénécal 1996). Finally, it seemed important for the organization to assess the role and impact of team leaders. For management practices, we have used the scales developed by Yarker et al. (2008), which had been translated by a translation committee composed of three researchers.

Approach to risk assessment

The process of risk assessment has usually an implicit purpose, which is to raise awareness among all stakeholders in the intervention. For example, Brun et al. (2007) reported that hospital leaders were strongly alarmed by a survey showing that 54 per cent of staff received a score of psychological distress compared to 20 per cent in a nationally representative sample of Quebec workers. However, in the present case in the public organization, the project team felt rather reluctant to use a psychological distress indicator as an outcome measure. There was a thorough reflection within the project team regarding the choice of the indicator of the extent of problems. Following pilot projects, the project team chose to use a measure of motivation to work as managers feel more comfortable with the idea that they could have an influence on employees' motivation, whereas they felt quite threatened by the notion that they could be responsible for psychological distress. Moreover, this motivation measure was better suited to the strategic objectives of the organization that wanted to become an "Employer of choice."

A second key element regarding risk assessment lies in how to define the scope and the unit of analysis for an organization of this size. When the number of participants allowed, each team was given its own diagnostic – late spring, early summer of 2010 – on their level of motivation (target: increase motivation) and priority risk factors (psychosocial risks on which teams must act to achieve the target). A diagnosis at the team level is in line with the literature on stress and line managers (see Yarker et al. 2008), and it appears appropriate to produce diagnosis as close as possible to the reality of work teams. This should in turn facilitate the identification of specific problems experienced by them. However, in practice, a team-level diagnosis is difficult, first, for statistical reasons (e.g. insufficient number of respondents or small number of employees in a unit) and in order to maintain the confidentiality of responses. Moreover, changes can occur very rapidly at the team level, which undermines the validity of the risk assessment and its durability. On the other hand, a risk assessment conducted at the company level or at the departmental level might seem too impersonal and fail to engage workers and managers in the process. In the present case, most teams had a team-based risk assessment, but when this was not possible, the risk assessment portrayed the unit. To ensure that the diagnosis reflected the current reality in the context of the intervention, each of the teams evaluated both the accuracy and timeliness of diagnosis, on the one hand, to ensure its relevance and, second, to promote an intervention that would create willingness and desire to engage in the subsequent steps.

Steps 3 and 4: Identification of concrete problems and solutions

Once the risk assessment is completed, there is often a feeling of accomplishment in terms of project advancement, as if the risk assessment would on its own create the momentum necessary for actions to take place. In the Foresight report (2008), it is reported that a risk assessment alone could produce a valuable return on investment. However, one must keep in mind that in the long term, if no changes took place following the risk assessment, employees are likely to feel disengaged and cynical (Biron et al. 2010). In order to go beyond the risk assessment, it is important to support the organization and the teams in the translation of risk factors specific to each team. Activities identified should be based on the risk assessment results, but should also be concrete enough to meet the needs of stakeholders (Jordan et al. 2003). Identifying concrete problematic work situations will help for the following step, namely the development of solutions and action plans. In the present case in the public organization, based on the results of the questionnaire, maps (for risk factors and management practices) were produced for each team. More precisely, each team received a risk assessment with priority rankings (the highest factors being the ones on which the team must intervene to influence the work motivation). The teams also received a motivation score that could compare to the rest of the organization. After receiving their diagnosis (e.g. motivation score and psychosocial risks that reduce motivation), managers had to meet their team once to, first, identify concrete problems (e.g. risk factors for the workload, as the real problem may be frequent interruptions experienced by employees) and identify solutions, and then jointly agree on a simple realistic action plan to address the specific problems identified in their present respective working environment. In summary, each team had to meet twice. The first meeting was used to present the results of the questionnaire and identify priorities and areas for improvement whereas the second meeting was to agree on solutions to implement. These meetings were scheduled over a maximum period of four months. An action plan had to be developed and subsequently implemented.

Step 5: Implementation and evaluation

The fifth and final step in a strategic intervention is to implement actions and assess their impact. For Nielsen and Randall (2012), the implementation of interventions requires the thorough documentation of the activities implemented. This represents a fundamental requirement since one must determine if the intended beneficiaries of the intervention were actually exposed to it. As an illustration, Biron et al. (2010) describe a study in which while the intervention has not been implemented in full, it would have been wrong to conclude that it had no impact, since the problem came from the fact that it had not been implemented. To be sure to document this process, Nielsen and Randall (2012) suggest the following questions: what aspects of intervention have led to changes in the participants? Who are the people who have been exposed and involved in these activities? Have employees been actively involved in the development and implementation of activities?

In this project, a mechanism had been established to monitor progress and document intervention activities. This implied several meetings between the project team, managers and senior managers, as well as logbooks completed by managers on changes implemented in their unit. In addition, status reports were prepared by the human resources department to aggregate data across business units. These aggregated reports were regularly transmitted to senior management and branches to inform them of the progress in their respective units.

In principle, the assessment should focus on both the intervention process (how the intervention took place) as well as the impact of these (what are the effects of the intervention) (Rossi et al. 2000). In this case, at the time of going to press, the benefits have not yet been assessed, but a year after the start of the project some elements of the process were measured. A short questionnaire measuring satisfaction was administered (319/725 respondents, response rate of 44 per cent). These elements include satisfaction with:

- the intervention process, in general (85 per cent satisfied);
- the information provided on the intervention project (87 per cent satisfied);
- how the manager conducted meetings (an example item: I am satisfied with how my manager has managed the different points of views regarding the QWL project) (83 per cent satisfied);
- interventions implemented (66 per cent satisfied);
- comments on the approach (see Table 15.2).

To document the progress, roadmaps have not only been used to report on the progress of the project as a whole, but also to document the changes made in each team. At the end of the intervention process, 94 per cent of teams were met and 86 per cent of action plans were developed. Roadmap and action plans were important tools providing detailed guidance on the changes made and the activities of interventions (solutions) and team members affected by these activities. More specifically, the deadline has been met for the majority of teams. Indeed, at the end of the process, for five of the eight branches, all teams had been met, 94 per cent of meetings with staff, and 86 per cent of action plans were developed.

Discussion: Key Elements of the Intervention Process

As indicated by Nielsen and Randall (2012), to document the design and implementation of the intervention, several factors can be considered, both in regard to the initiation of the intervention, the development and design of intervention activities, the implementation strategy, the role of stakeholders and consultants, and communication. We return to the four elements that have emerged as critical in this case study. Table 15.1 summarizes the concrete indicators obtained in the framework of the case study presented above in relation to these four key elements of the intervention process.

Table 15.1 Four critical factors for the success of interventions and concrete indicators characterizing these factors

Critical factors of success	Concrete indicators

Firm commitment of senior management	• Monthly monitoring of the project by senior management QWL.
	• Clear expectations of the CEO to complete the project in timely.
	• Provision of financial and human resources QWL.
	• Thus, the intervention process QWL is part of a coherent and strategic approach of the participating organization referred to as "employer of choice."
	• Anchoring the approach in the strategic plan.
	• Respect the schedule, do not use lack of time as an excuse to prevent or hinder intervention.
	• Maintenance of interventions through organizational change/changes in the structure/changes in the folder holders.
	• Reasons for engaging in the process:
	  1) there were organizational indicators concern about absenteeism; and
	  2) it was the adoption of a policy to promote human health.

Participation and commitment of stakeholders (unions, managers, employees, etc.)	• Involving stakeholders early in the process, right from the preparation stage. • Quarterly meetings with managers and with unions. Parties are involved in decisions regarding the conduct of the intervention.
Communication plan	• Choice of terminology and more positive measures (quality of life at work, motivation, rather than stress or psychological distress). • The more positive terms are viewed favourably by managers who tend to engage instead of feel threatened by the project. • Conduct a roadmap for leadership and several meetings between the project team and management, and throughout the intervention process. • Roadmaps of each teams regularly communicated to management. • Content roadmaps: documenting progress, description of interventions implemented, satisfaction with the process and support available to teams.

Table 15.1 Continued

Critical factors of success	Concrete indicators
Forms of support to managers during the intervention	• All managers were well aware of the role they should play in the context of the intervention and they had support. • Appointment and training of resource persons to relay information in each unit. • A consultant specializing in organizational development for all managers. • Mailbox with any questions regarding the process. • Tools have been developed and made available to managers, such as guides for conducting the intervention, presentation templates for meetings with teams, models of action plan or memory aids.

Table 15.2 Comments from participants 12 months after the start of the intervention

Positive	Negative
• Improved level of employees: more attention and less critical. • I feel that the team is more cohesive. • It is a good programme that improves the structure of work and the contribution of individual awareness to work in a better climate. • This has forced people to express themselves and our manager to hear various complaints from employees. • I felt involved in this project by attending meetings organized about it. • Managers have given their time and energy to the smooth functioning of this project. • This approach helps me to assume my role as manager of the first level.	• Little practical application: lack of time or budget. • The problems do not come from our level but above us. • The process was too long. • The intentions are good but not enough action. • I do not feel that I am a leader in the process and it is difficult to see where all this will lead. • Several of my colleagues participate in this activity with cynicism.

EMPLOYEE PARTICIPATION AND SUPPORT TO MANAGERS

Nielsen and Randall (2012) emphasize the importance of the roles and behaviours of various stakeholders in the intervention strategy. Among these factors, it is desirable to use a participatory approach in which employees are heavily involved, because it is a determining factor in the success of the intervention (Nielsen et al. 2007). In this project, stakeholders (e.g. managers, unions, employees) were involved in the process throughout the project. Stakeholders were not only consulted during the risk assessment phase, but were also involved in decisions regarding the conduct of the operation. More specifically, this involvement entailed regular meetings with the steering committee of the organization as well as the expanded committee comprised of middle managers (about 50 people). In addition, all unions were also invited at the time of initiating the project and also at key stages of the intervention.

Managers received significant support in order to facilitate both the stages of identifying specific problems related to the diagnosis and the development of solutions. This support came from several sources, such as the formation of "multipliers" who were employees attached to branches occupying the role of contact with managers and ensuring the sustainability of the process. In addition, each manager could use the services of a consultant specializing in organizational development. In addition, the project coordinator administered a mailbox that was specifically used for the QWL approach in order to provide answers to any questions or support requests. Finally, tools have been developed and made available to managers, such as guides for conducting the intervention, presentation templates for meetings with teams, models of action plan or memory aids. Managers could use the material as is, adapt to their own needs, or even create their own tools to conduct the intervention process. Managers therefore had access to coaching, on topics ranging from simple advice to full support, throughout the process. All this helped to ensure consistency between the identified risks (diagnosis), the practical problems and developing solutions to ensure an effective response.

Line managers play a key role since they are usually the ones who are responsible for communicating and implementing changes following the assessment of psychosocial risks (Biron et al. 2010, Nielsen and Randall 2009, Randall et al. 2007). Moreover, Kompier et al. (2000) argue that line managers are responsible for implementing all the case studies in their review of the literature on intervention, and in several European countries. Line managers play a key role since they can either obstruct change – for example, by reducing the time allocated to the intervention – or facilitate it – for example, by actively supporting staff in the implementation of changes (Dahl-Jørgensen and Saksvik 2005, Nielsen and Randall 2009).

These principles have been respected in the course of the intervention in the organization participating in this project. Thus, all managers were well aware of their role in the intervention and had an important support to carry out the changes (see Table 15.1). Nevertheless, it appears that depending on the level of involvement and commitment of middle managers, intervention progress varied. Indeed, in some branches, teams were required to produce additional reports – in addition to the one required by senior management. Teams who had to produce advancement reports to a higher hierarchical level in addition to senior managers were more likely to have completed their action plans. Thus, it appears that – especially in a large organization composed of several levels of management – the challenge lies in the support and commitment of all levels, not just by line and senior management. Beyond securing the commitment of senior management, it is up to the intermediate levels (of management) to demonstrate their commitment towards the process and to transmit clear expectations to their managers. It has been observed that the branches who have supported the intervention and have developed their own monitoring tools and their own expectations have experienced more success and engagement from teams.

Management commitment

The support of senior management is a recurring key requirement in the stress literature. Usually, guidelines and research reports indicate that they have secured commitment, which was generally done by presenting a quantitative business case (i.e. demonstrating the high costs of ill-health). However, very little is being said about securing senior management commitment in the long-term (as a process, instead of as a unique step). As a result, it is unclear how to operationalize what a solid commitment implies. Given that senior management is generally remote from the intervention, it is difficult for them to follow closely the intervention developments (Nytrø et al. 2000). In the present case, different indicators illustrated a strong commitment from senior management. First, there was a clear desire of the CEO in place not only on the need for an intervention to improve QWL, but also on the necessity to complete it within a short period. Although this posed a major challenge for the project team as well as managers who had to integrate and manage the process quickly, the rhythm imposed demonstrated how senior management made this a priority. Senior management commitment also resulted in the granting of financial and human resources as well as monitoring the project almost monthly. Overall, time, money, human resources, regular monitoring and anchoring the project within the strategic operational plan of wanting to become an employer of choice facilitated the process and implied that line managers were supported in the implementation phase.

Communication plan

Nielsen and Randall (2012), Brun et al. (2009) and Jordan et al. (2003) all mention the importance of a communication plan. When employees are informed and clearly understand the intentions behind actions, they tend to be more committed to face it and want to take part in it (Nytrø et al. 2000). Nielsen and Randall (2012) suggest four questions to assess related information and communication:

1. Were the participants informed about the project?
2. Were the results of the psychosocial risk assessment reported?
3. Were participants informed of the progress, however small they may be?
4. Were small successes celebrated?

Brun et al. (2009) also emphasize the importance of having a communication plan throughout all the steps of the risk management cycle (Figure 15.1), including during the preparation phase. The communication plan involves organizational mass communications, but also more interpersonal communications. In the present case, information regarding the project goals, expectations and advancements were distributed in news capsules and by leaders on the intranet. Questionnaire results show that 94 per cent of teams have been met with a view to discuss their risk assessment. During the early stages, the first pilot projects have not only improved the response rates to the questionnaire and confirmed the feasibility of the approach, but also created a movement of enthusiasm in publicizing successes, allowing staff to observe and experience daily the positive change implemented in work teams. Finally, the QWL approach was part of a larger organizational programme to develop an employer brand integrating career management skills and physical and mental health. Thus, the intervention process QWL is part of a coherent and strategic approach to make the participating organization an employer of choice.

Conclusion

Despite the fact that this intervention was successful in terms of bringing stakeholders past the risk assessment phase, and in mobilizing a large number of teams in developing concrete, evidence-based action plans, two factors have hampered the intervention process. First, as it is always the case in stress intervention projects, competing priorities caused delays in some of the teams. Some areas of the organization experienced workload peaks during the year which has hindered the progress of the intervention and respect for the work schedule. Although stress intervention projects have to be implemented with flexibility with regards to regular operations, it remains difficult to integrate production and well-being. Although well-being should remain a priority even during peak time, intervention activities are often left aside when the workload becomes too heavy. This is a major concern for all intervention projects, as it remains challenging to integrate the logics of production and workers' well-being. Second, at the end of the intervention, while the teams were to implement the solutions, the participating organization has undergone a substantial change in its structure, its management team, and especially its CEO. These changes have created such upheaval in the composition of teams in the intervention process as well as the role of the participant organization in the machinery of government that we can not measure the changes that have taken place in teams and to evaluate the effects post-intervention. To date, only some elements of the process have been measured. The issue of sustainability of occupational health interventions has rarely been addressed in the literature, although the majority of authors report that concurring organizational change hindered the intervention process. Integrating occupational health with daily business remains a serious issue and one of the most challenging parts of occupational health interventions.

References

Bambra, C., Egan, M., Thomas, S., Petticrew, M., and Whitehead, M. 2007. The psychosocial and health effects of workplace restructuring interventions reorganisation – 2. A systematic review of task. *Journal of Epidemiology and Community Health*, 61, 1028–37.

Biron, C., Burke, R.J. and Cooper C.L. To appear in 2013. *Creating Healthy Workplaces: Stress Reduction, Improved Well-being, and Organizational Effectiveness*. Farnham: Gower Publishing.

Biron, C., Cooper, C.L. and Gibbs, P. 2012. Stress interventions vs positive interventions: Apples and oranges? In *Oxford Handbook of Positive Organizational Scholarship*, edited by K.S. Cameron and G.M. Spreitzer. New York: Oxford University Press, 938–52.

Biron, C., Gatrell, C. and Cooper, C.L. 2010. Autopsy of a failure: Evaluating process and contextual issues in an organizational-level work stress intervention. *International Journal of Stress Management*, 17(2), 135–58.

Biron, C., Karanika-Murray, M. and Cooper, C.L. 2012a. *Improving Organizational Interventions for Stress and Well-being: Addressing Process and Context*. London: Psychology Press.

Biron, C., Karanika-Murray, M. and Cooper, C.L. 2012b. What works, for whom, in which context? Researching organizational interventions on psychosocial risks using realistic evaluation principles. In *Organizational Stress and Well-being Interventions: Addressing Process and Context*, edited by C. Biron, M. Karanika-Murray and C.L. Cooper. London: Psychology Press, Routledge, 163–84.

Biron, C., Karanika-Murray, M. and Cooper, C.L. 2012c. Organizational stress and well-being interventions: An overview. In *Improving Organizational Interventions for Stress and Well-being Interventions: Addressing Process and Context*, edited by C. Biron, M. Karanika-Murray and C.L. Cooper. London: Routledge, 1–18.

Blais, M.R., Lachance, L., Vallerand, R.J., Brière, N.M. and Riddle, A.S. 1993. L'inventaire des motivations au travail de Blais. *Revue québécoise de psychologie*, 14(3), 185–215.

Borritz, M., Bültmann, U., Rugulies, R., Christensen, K.B., Villadsen, E. and Kristensen, T.S. 2005. Psychosocial work characteristics as predictors for burnout: Findings from 3-year follow up of the PUMA Study. *Journal of Occupational and Environmental Medecine*, 47(10), 1015–25.

Brun, J.-P., Biron, C. and Ivers, H. 2007. Démarche stratégique de prévention des problèmes de santé mentale au travail, R-514. Québec, Canada: Institut de recherche Robert-Sauvé en santé et en sécurité du travail (http://www.irsst.qc.ca), 78.

Brun, J.-P., Biron, C., and St-Hilaire, F. 2009. *Guide pour une démarche stratégique de prévention des problèmes de santé psychologique au travail* (RG-618). Montreal Institut de recherche Robert-Sauvé en santé et en sécurité du travail.

Bureau de normalisation du Québec. 2008. Prévention, promotion et pratiques organisationnelles favorables à la santé en milieu de travail – Guide explicatif sur la norme BNQ 9700-800/2008. Québec: Bureau de normalisation du Québec, 46.

Cooper, C. and Dewe, P. 2008. Well-being, absenteeism, presenteeism, costs and challenges. *Occupational Medicine(London)*, 58(8), 522–4.

Cooper, C.L. and Cartwright, S. 1997. An intervention strategy for workplace stress. *Journal of Psychosomatic Research*, 43(1), 7–16.

Cox, T., Randall, R. and Griffiths, A. 2002. *Interventions to Control Stress at Work in Hospital Staff*. Nottingham: The Institute of Work, Health and Organisations, University of Nottingham.

Cox, T., Taris, T.W. and Nielsen, K. 2010. Organizational interventions: Issues and challenges. *Work and Stress*, 24(3), 217–18.

Cox, T., Karanika-Murray, M., Griffiths, A. and Houdmont, J. 2007. Evaluating organizational-level work stress interventions: Beyond traditional methods. *Work and Stress*, 21(4), 348–62.

Cox, T., Griffiths, A.J., Barlowe, C.A., Randall, R.J., Thomson, L.E. and Rial-Gonzalez, E. 2000. *Organisational Interventions for Work Stress: A Risk Management Approach*. Sudbury: HSE Books.

CSA. 2013. CAN/CSA-Z1003-13/BNQ 9700-803/2013 Psychological health and safety in the workplace – prevention, promotion, and guidance to staged implementation. CSA Group, Bureau de normalisation du Québec.

Dahl-Jørgensen, C., and Saksvik, P.Ø. 2005. The impact of two organizational interventions on the health of service sector workers. *International Journal of Health Services*, 35(3), 529–49.

Egan, M., Bambra, C., Thomas, S., Petticrew, M., Whitehead, M. and Thomson, H. 2007. The psychosocial and health effects of workplace reorganisation – 1. A systematic review of organisational-level interventions that aim to increase employee control. *Journal of Epidemiology and Community Health*, 61, 945–54.

Fernet, C., Guay, F. and Senecal, C. 2004. Adjusting to job demands: The role of work, self-determination and job control in predicting burnout. *Journal of Vocational Behavior*, 65, 39–56.

Foresight Mental Capital and Wellbeing Project 2008. *Final Project Report*. London: The Government Office for Science.

Giga, S., Faragher, B. and Cooper, C.L. 2003a. Identification of good practice in stress prevention/management. In *Beacons of Excellence in Stress Prevention*, edited by J. Jordan, E. Gurr, G. Tinline, S. Giga, B. Faragher and C.L. Cooper (Vol. HSE Research Report 133). Sudbury: HSE Books, 1–45.

Giga, S.I., Cooper, C.L. and Faragher, B. 2003b. The development of a framework for a comprehensive approach to stress management interventions at work. *International Journal of Stress Management*, 10(4), 280–296.

Giga, S.I., Noblet, A.J., Faragher, B. and Cooper, C.L. 2003c. The UK perspective: A review of research on organisational stress management interventions. *Australian Psychologist*, 38(2), 158–64.

Health and Safety Executive. 2003. Tackling stress in your organisation: A walkthrough. Retrieved in 2004 from http://www.hse.gov.uk/stress/walkthrough/index.htm.

Hesselink, J.K., Wiezer, N., Den Besten, H. and De Kleijn, E. 2012. The development of smart and practical small group interventions for work stress. In *Improving Organizational Interventions for Stress and Well-being: Address Process and Context Issues*, edited by C. Biron, M. Karanika-Murray and C.L. Cooper. London: Routledge, 258–82.

Israel, B.A., Baker, E.A., Goldenhar, L.M., Heaney, C.A. and Schurman, S.J. 1996. Occupational stress, safety and health: Conceptual framework and principles for effective prevention interventions. *Journal of Occupational Health Psychology*, 1(3), 261–86.

Johnson, J.V., Hall, E.M. and Theorell, T. 1989. Combined effects of job strain and social isolation on cardiovascular disease morbidity and mortality in a random sample of the Swedish male working population. *Scandinavian Journal of Work Environment and Health*, 15, 271–9.

Jordan, J., Gurr, E., Tinline, G., Giga, S., Faragher, B. and Cooper, C. 2003. *Beacons of Excellence in Stress Prevention*. Manchester: Health and Safety Executive.

Karasek, R. and Theorell, T. 1990. *Healthy Work: Stress, Productivity and the Reconstruction of Working Life*. New York: Basic Books

Kompier, M. and Cooper, C.L. 1999. Stress prevention: European countries and European cases compared. In *Preventing Stress, Improving Productivity – European Case Studies in the Workplace*, edited by M. Kompier and C.L. Cooper. London: Routledge, 312–36.

Kompier, M.A.J., Cooper, C.L. and Geurts, S.A.E. 2000. A multiple case study approach to work stress prevention in Europe. *European Journal of Work and Organizational Psychology*, 9(3), 371.

LaMontagne, A.D., Keegel, T., Louie, A.M., Ostry, A. and Landbergis, P.A. 2007. A systematic review of the job-stress intervention evaluation literature, 1990–2005. *International Journal of Occupational and Environmental Health*, 13, 268–80.

Lohela, M., Björklund, C., Vingård, E., Hagberg, J. and Jensen I. 2009. Does a change in psychosocial work factors lead to a change in employee health? *Journal of Occupational and Environmental Medicine*, 51(2), 195–203.

MacKay, C.J., Cousins, R., Kelly, P.J., Lees, S. and McCaig, R.H. 2004. "Management Standards" and work-related stress in the UK: Policy background and science. *Work and Stress*, 18(2), 91.

Marmot, M., Bosma, H., Hemingway, H., Brunner, E. and Stansfeld, S. 1997. Contribution of job control and other risk factors to social variations in coronary heart disease incidence. *Lancet*, 350, 235–9.

Munir, F., Nielsen, K. and Carneiro, I.G. 2010. Transformational leadership and depressive symptoms: A prospective study. *Journal of Affective Disorders*, 120(1–3), 235–9.

Nielsen, K. and Randall, R. 2009. Managers' active support when implementing teams: the impact on employee well-being. *Applied Psychology: Health and Well-Being*, 1(3), 374–90.

Nielsen, K. and Randall, R. 2012. Opening the black box: A framework for evaluating organizational-level occupational health interventions. *European Journal of Work and Organizational Psychology*.

Nielsen, K., Randall, R. and Albertsen, K. 2007. Participants, appraisals of process issues and the effects of stress management interventions. *Journal of Organizational Behavior*, 28, 793–810.

Nielsen, K., Taris, T.W. and Cox, T. 2010a. The future of organizational interventions: Addressing the challenges of today's organizations. *Work & Stress*, 24(3), 219–33.

Nielsen, K., Randall, R., Holten, A.-L. and Gonzalez, E.R. 2010b. Conducting organizational-level occupational health interventions: What works? *Work and Stress*, 24(3), 234–59.

Noblet, A. and LaMontagne, A.D. 2009. The challenges of developing, implementing, and evaluating interventions. In *The Oxford Handbook of Organizational Wellbeing*, edited by S. Cartwright and C.L. Cooper. Oxford: Oxford University Press, 466–96.

Nyberg, A., Alfredsson, L., Theorell, T., Westerlund, H., Vahtera, J. and Kivimäki, M. 2009. Managerial leadership and ischaemic heart disease among employees: the Swedish WOLF study. *Occupational and Environmental Medecine*, 66(1), 51–5.

Nytrø, K., Saksvik, P.O., Mikkelsen, A., Bohle, P. and Quinlan, M. 2000. An appraisal of key factors in the implementation of occupational stress interventions. *Work and Stress*, 14(3), 213–25.

Randall, R. and Nielsen, K. 2012. Does the intervention fit? An explanatory model of intervention success and failure in complex organizational environments. In *Improving Organizational Interventions for Stress and Well-being: Addressing Process and Context*, edited by C. Biron, M. Karanika-Murray and C.L. Cooper. London: Routledge, 1–17.

Randall, R., Cox, T. and Griffiths, A. 2007. Participants' accounts of a stress management intervention. *Human Relations*, 60(8), 1181.

Randall, R., Griffiths, A. and Cox, T. 2005. Evaluating organizational stress-management interventions using adapted study designs. *European Journal of Work and Organization Psychology*, 14(1), 23–41.

Rinfret, N. and Sénécal, C. 1996. La préférence envers une organisation est-elle tributaire du profil motivationnel des étudiants? [Organizational preference: Does it affect the motivational profile of students?]. *Revue Québécoise de Psychologie*, 17(3), 25–41.

Rossi, P.H., Lipsey, M.W. and Freeman, H.E. 2000. *Evaluation – a Systematic Approach* (7th edn). Thousand Oaks: Sage Publications Inc.

Sainsbury Centre for Mental Health 2007. Mental health at work: Developing the business case. *Policy Paper 8*. London: Sainsbury Institute for Mental Health.

Saksvik, P.Ø., Nytrø, K., Dahl-Jorgensen, C. and Mikkelsen, A. 2002. A process evaluation of individual and organizational occupational stress and health interventions. *Work and Stress*, 16(1), 37–57.

Saksvik, P.Ø., Tvedt, S.D., Nytrø, K., Andersen G.R., Andersen T.K., Buvik, M.P. and Semmer, N.K. 2006. Job stress interventions and the organization of work. *Scandinavian Journal of Work and Environmental Health*, 32(6, special issue), 515–27.

Siegrist, J. 1996. Adverse health effects of high-effort/low-reward conditions. *Journal of Occupational Health Psychology*, 1, 27–41.

Vézina, M., Chénard, C., Bhérer, L., Bourbonnais, R., Brun, J.-P., Gourdeau, P., Guimont, C., Lippel, K., Marchand, A., St-Arnaud, L. and Stock, S. 2008. *Outil de caractérisation préliminaire d'un milieu de travail au regard de la santé psychologique au travail*. Québec: Institut national de santé publique du Québec.

Vézina, M., Cloutier, E., Stock, S., Lippel, K., Fortin, É., Delisle, A., St-Vincent, M., Funes, A., Duguay, P., Vézina, S. and Prud'homme, P. 2011. *Enquête québécoise sur des conditions de travail, d'emploi et de SST (EQCOTESST)*. Montréal: IRSST.

Yarker, J., Donaldson-Feilder, E., Lewis, R. and Flaxman, P.E. 2008. *Management Competencies for Preventing and Reducing Stress at Work: Identifying and Developing the Management Behaviours Necessary to Implement the HSE Management Standards: Phase 2*. London: HSE Books.

16 Merging Occupational Health, Safety and Health Promotion with Lean: An Integrated Systems Approach (the LeanHealth Project)

TERESE STENFORS-HAYES, HENNA HASSON, HANNA AUGUSTSSON, HELENA HVITFELDT FORSBERG and ULRICA VON THIELE SCHWARZ

Abstract

In this chapter, the authors describe the integration of occupational health, safety and health promotion into the existing structures for organizational quality and production improvement (e.g. Lean management) at a hospital in Sweden. They furthermore describe current research in the field of Lean management and integration of organizational systems. The chapter centres on the initial implementation of the integration at one department of the chosen hospital. The early phase of the project prevents any firm conclusions as to the overall effect of this initiative. However, thus far it seems that the initiatives fulfil many factors that have been described as prerequisites for successful organizational-level occupational health interventions.

Introduction

Organizational-level occupational health interventions are the preferred way of improving working conditions and tackling work-related problems such as employee stress, according to current European legislation (EU-OSHA 2010a), and can be defined as changes in the design, organization and management of work, in order to improve working conditions, employee health and well-being (EU-OSHA 2010a, Nielsen and Randall 2012). However, these interventions have been criticized because of the lack of a large and consistent body of evidence of their positive impact (Briner and Reynolds 1999, Richardson and Rothstein

2008, Nielsen and Randall 2012). One reason they do not achieve the expected outcomes has been suggested to be their lack of integration into other organizational functions and goals, such as production. Occupational health interventions often become sidetracks that fade out or are down-prioritized.

NEED FOR INTEGRATION OF SYSTEMS

Organizational-level occupational health interventions are often intertwined with organizational systems and processes (Semmer 2003), making it difficult to separate them (Mikkelsen 2005). Therefore, occupational health interventions often require changes to organizational factors (Semmer 2003, Bambra et al. 2007). In other words, an occupational health intervention needs to be viewed as part of the organizational processes, structures and systems. This would also require a change from viewing organizational-level occupational health interventions as time-limited projects to seeing them as ongoing activities (Landsbergis and Vivona-Vaughan 1995). This could be achieved by using existing systems and processes, rather than creating new ones. This may make it easier to initiate, implement and sustain these interventions.

As a first step towards integration, Baker et al. (1996) suggested that health promotion and occupational health and safety work should be more integrated, as such interventions often operate in isolation from one another. Occupational safety and health have mainly dealt with the physical hazards of the work environment, while worksite health promotions have primarily concentrated on individual lifestyle behaviours. They proposed that organizations should consider using comprehensive approaches for improving employee health, i.e. initiatives that address individual, psychosocial, environmental and organizational factors, as well as broader policy issues that affect occupational health. The need for integrative approaches has recently been enforced by others, including the National Institute of Occupational Safety and Health (NIOSH 2012).

Other authors have suggested further integration by merging employee health and safety work with quality improvement and production systems (Wilkinson and Dale 1999, 2002, Jørgensen et al. 2006, Zwetsloot 1995). The underlying reason for this is that many organizations today work systematically with work environment and safety issues as well as quality and production development. As the development of these different systems is often undertaken in response to different types of needs, within different parts of an organization and at different points in time, it is not unusual for one organization to have a number of different systems for addressing the various issues. However, this increases the risk of unnecessarily complex bureaucracy, separate or even conflicting procedures, and higher costs (EU-OSHA 2010b). Some prior studies of these types of integrated systems assert that they are a way of saving resources and increasing an organization's efficiency. Furthermore, integrated systems can entail that changes within the different systems are made with consideration for each other, with decisions within the system being made with thought to not only quality but also the work environment and employee health. The different systems also require the same support from the social and organizational processes. Strategically, integrated systems are regarded as being able to contribute to a greater focus on continuous improvement work, better collaboration across functional borders and better preparedness for future integration of new standards (Smith 2002). To our knowledge, there are currently no previous studies that describe organizations working with integrated systems for health promotion, occupational health and safety, and production and quality improvement systems. Thus, the present study is one of the first to integrate health promotion and occupational health and safety work into an existing production and quality improvement system.

The Case of the Hospital in Enköping

The country district hospital in Enköping, Sweden, employs around 500 people in 12 departments, including surgery (general surgery, gynaecology, urology, orthopedics), radiology, internal medicine, acute care, intensive care, rehabilitation, hospital-governed home care, and some geriatrics. It also has consultation units for eyes, dermatology and oncology as well as an emergency unit that is open around the clock. The hospital provides clinical education for those studying to be physicians, nurses and other health professionals.

The work with health promotion for staff at the hospital was given more attention in 2001, when the hospital conducted the project "The hospital as a health-promoting workplace." At that time an exercise room for staff was built, and a health coordinator was hired. Later, the position of health representative was also established. Today there are one or two health representatives at each department, with the primary responsibility of distributing information about health promotion. For approximately ten years now, employees have been allowed to exercise during work hours for an hour each week, workload permitting. The health representatives meet annually to draft a health plan containing a health profile, exercising possibilities, lectures and a summer party. The hospital also works with health promotion as part of "Health-promoting hospitals and healthcare organizations," based on the World Health Organization's (WHO) project, Health Promoting Hospitals (WHO 1997). This network's aim is to increase the health benefits for patients and their next of kin, employees, as well as for those in the surrounding community.

In Sweden, all companies with more than 10 employees are obliged to systematically work with issues regarding the work environment; thus, within all larger Swedish companies there is an organization dedicated to working structurally with work environment matters (SFS 1977: 1160). At the hospital in Enköping, the systematic work with work environment issues builds on annual safety inspections following a checklist based on Swedish work environment legislation. Possible risks noted during the safety inspections are entered into an action plan, which is then followed up. The work environment law stresses that the employer should work not only to avoid accidents and reduce the risk of ill health, but also to generally attain a good work environment. The collaboration between employer and employee is highlighted, and employees' possibilities to participate in the design of their own working situation, as well as in the work towards change and development, are stressed. The work environment law also stresses that working conditions that allow for personal and professional development, self-determination and responsibility should be striven for. This can be regarded as an example of how companies are encouraged to work with not only rehabilitation (or reactive efforts) and prevention, but also promotion.

Integration Ideas Forming

In 2009, the hospital contacted the research group and requested support in further developing its health promotion work. The problem was that the health-promoting activities were mainly related to individual health habits and thus mostly attracted those who were already active. These discussions led to a reasoning concerning what is included in health promotion, and resulted in the conclusion that other measures were also relevant in promoting employees' health. The discussions between the hospital's work group and the researchers resulted in the identification of two needs: health promotion measures should be relevant from a production perspective, and production and quality improvement should be conducted with thought to employees' health. This focus sheds new light on their production and quality improvement system which is inspired by Lean and especially Kaizen (which is a commonly used tool in Lean). Could the system also be used to work with health promotion? The hospital, in dialogue with the research group, decided that the health promotion work would be integrated with the production and quality improvement system. Hospital management approved of this idea, seeing the advantages of coordinating multiple systems, further developing the work with their production and quality improvement system, and grouping health issues more closely with production issues. A research project was begun to investigate how health promotion work could be integrated with the existing Kaizen work. In connection with this, it was also decided that occupational health and safety would be integrated with the production and quality improvement system, which entailed that preventive as well as promotive health work would be integrated with the quality improvement system. Before the case is further described, current research in Lean and Kaizen will be presented below.

Lean and Kaizen

Lean is based on the production philosophy that has its origin in the Toyota Production System (Womack et al. 1990). The definitions of Lean and the characteristics included in the concept vary (Pettersen 2009). However, it has been suggested that Lean operates on two different levels: the strategic level, whereby understanding value in order to maximize value for the customer is essential; and the operational level, where tools for eliminating waste are in focus (Hines et al. 2004). Lean was originally applied in industry but has later been implemented in other sectors as well, such as healthcare (Poksinska 2010).

One of the operational-level tools used in Lean is Kaizen (Radnor and Walley 2008). The word *Kaizen* is originally Japanese and means "continuous improvements." In the Lean literature, it is described as a system that allows staff and workplace leaders to cooperate in a continual strive for improvement (Jacobson et al. 2009). The focus is on providing a structure for small-scale, continual improvements of such character that decisions can be made directly at the workplace level. Thus, it is often low-cost and low-risk changes that are in focus, while larger organizational changes that are more difficult to implement are regarded as falling under the responsibility of management (Jacobson et al. 2009). The idea is that the system should encourage the quick identification of a problem that arises in the work process, an understanding of the reasons behind the problem, and the trying of solutions (Holden 2011). The intent is to have a focus on employees' engagement and possibility to themselves investigate and improve their own work. However, some research shows that it is not necessarily desirable to only implement Kaizen, as there is a risk that the quick results that can be achieved will not be maintained if Kaizen is not integrated into the organization's overall strategy and vision (Mazzocato et al. 2010, Radnor and Walley 2008).

LEAN AND EMPLOYEE EFFECTS

As Lean becomes increasingly common, the discussion has increased concerning how it affects the work environment and employee health. Since the whole point of Lean is to achieve changes in work, it is reasonable to assume that both the physical and psychosocial work environment will be affected if it is implemented (Hasle 2011). A repeatedly shown effect of Lean on working environments is that it often leads to increased work pace, workload and work intensification (Landsbergis et al. 1999, Hasle et al. 2012). This might in turn lead to work-related musculoskeletal disorders (Brännmark and Håkansson 2012, Landsbergis et al. 1999), fatigue, tension and stress among employees (Landsbergis et al. 1999). However, certain Lean practices such as teamwork, task support and participation in Lean implementation have been suggested to have positive effects, such as reduced stress among employees (Conti et al. 2006). Other positive effects are breadth of employee role, cognitive demands, skill utilization, social relations (Jackson and Mullarkey 2000) and improved job content (Seppälä and Klemola 2004). It is further suggested that negative consequences (e.g. work-related musculoskeletal disorders) can be reduced by complementing the implementation of Lean with an ergonomic intervention (Brännmark and Håkansson 2012). Thus, it seems that Lean does not necessarily have a negative impact on employees' work environment and health but also has the potential to promote their health. These mixed results indicate that different Lean concepts and tools might have varied impacts on work environment and health. Lean strategies that lead to increased work pace and workload seem to have a negative impact on employees' health, while working environment strategies that facilitate participation, support and teamwork seem to have a positive impact.

Kaizen builds on a participatory approach by encouraging employees to be actively involved in evaluating and improving work processes. A participatory approach has been recommended as an appropriate intervention strategy, and plays an important role in well-known organizational occupational health interventions (Nielsen et al. 2010b). Such an approach has, for example, been shown to have a positive impact on work-related stress, job characteristics, learning climate and management style (Mikkelsen et al. 2000). From this perspective, Kaizen can be hypothesized to have a positive impact on employee work environment and health. At the same time, some research shows that Kaizen can also entail a form of self-rationalization, whereby employees reduce waste through the Kaizen work in a way that could potentially contribute to worsening their own working situation in the long run (Sewell and Wilkinson 1992, Moldasch and Weber 1998). This highlights a yet unanswered question concerning whether it is the components of the participation and cooperation that in themselves have the potential to lead to positive effects, or if it is rather the content of the changes that the improvement work entails.

Kaizen note

Group:	Serial number:	Area: ☐ Service ☐ Staff and climate ☐ Quality ☐ Economy

1a. Describe the problem:	Written by:	Date:	⊕
2. Suggestion for problem:	Responsible:	Date:	⊕
3. Suggestion will be tested Possible new suggestion:	Responsible:	Tested and evaluated. Date:	⊕
4. Decided solution:	Responsible:	New solution introduced. Date:	⊕

1b. Expected results when the problem is solved:

4. Achieved results:

⊕ Problem described.

◔ Suggestion for solution is decided. Responsible person appointed. Date for start.

◕ Suggestion is tested/evaluated.

● Solution is documented.

Voting decision
Manager ....

Voting is performed when needed at point 2 or 3.

Figure 16.1 A Kaizen note, used at department level as part of the Kaizen work at the hospital in Enköping to document the improvement work process

Source: Developed by © KAIZENsupport. The note has been translated from Swedish into English.

LEAN AND KAIZEN AT THE HOSPITAL IN ENKÖPING

Since 2009, the hospital in Enköping has worked with what they call the "Kaizen quality concept" as part of their production and quality improvement system. At the hospital Kaizen is described as a problem-solving model that is a structured way of getting more employees to participate, and every department has a great deal of freedom to form its Kaizen work as it sees fit. However, the general work process is the same. At each department there are a number of employees who serve as Kaizen representatives, responsible for the work at their respective department. However, all employees at the different departments are engaged in identifying problems in their work, documenting them on Kaizen notes (see Figure 16.1), and attending Kaizen and work group meetings where they propose solutions, test them and follow up on the results. Twice every term, the Kaizen representatives meet with a central Kaizen coordinator to discuss matters and exchange their experience of the local Kaizen work, in order to get support and inspiration in their work. The concrete work with Kaizen is different at the various departments, but it is most common that the Kaizen notes are discussed at the regular workplace meetings either weekly or monthly. Some departments write about important changes in their newsletters, while others have set aside 10–20 minutes each week for special Kaizen meetings, held informally in front of the Kaizen board.

Integration of Health Promotion, Occupational Health and Safety, and Kaizen

The fundamental principle in the integration of the three systems of health promotion, occupational health and safety, and Kaizen, was that no new structures would be created but rather that the integration would build on the existing Kaizen work. The integration is also based on the employees' activity, just like the Kaizen work is otherwise. The integration of health promotion and Kaizen was launched at the beginning of 2012. Two practical changes relating to the Kaizen work were introduced:

1. health promotion-related activities and improvements are addressed on the Kaizen notes, thereby integrating health promotion with other production and quality improvement issues; and
2. proposed solutions to problems mentioned on the Kaizen notes, regardless of which area the problem/proposal concerned, are analysed based on what effects the proposed solution might have on the employees' health.

The analysis is noted under "Expected results" (see Figure 16.1). When proposed solutions are pilot-tested, any effects on the employees' health are followed up and noted under "Achieved results." In this way, the production and quality issues are integrated with the health issues.

The change also entails that the roles and responsibilities of the local Kaizen as well as health representatives change. Instead of holding separate meetings like in the past, now joint meetings are held for the Kaizen and health representatives within the project. For the health representatives the integration also means that they, instead of primarily functioning as providers of information on health and health promotion, are now to direct proposals and ideas in this area to the Kaizen notes, and help their co-workers in analyses of health consequences.

During the planning of the integration of health promotion and Kaizen, the hospital also decided to integrate their occupational health and safety work with Kaizen. This was launched in late 2011, when all departments were instructed to write suggestions related to occupational health and safety as Kaizen notes. Practically, this meant that the departments were expected to work with occupational health and safety continuously over the year, rather than conducting annual risk assessments, action plans and revisions.

As a starting point for the integration of health promotion and Kaizen, a workshop was arranged by management and the researchers. The managers, Kaizen representatives and health representatives of the units participating in the intervention attended the workshop, in order to build a common understanding of the intervention's aim and approach and to prepare what the integration may look like in each department. The workshop also aimed to contribute to a better understanding of what health promotion is, as it was previously regarded as simply physical exercise. During the project, the hospital's central Kaizen and health representatives, as well as the intervention units' managers and Kaizen and health representatives, receive support in their work with the change through coaching by a researcher certified in supporting improvement work within healthcare.[1] The coaching, which is needs-based, aims to support the improvement work by seizing on ideas, helping with any obstacles and providing tools and methods. The central Kaizen and health representatives, in turn, coach the departments' Kaizen and health representatives, who in turn have a coaching role in relation to their co-workers at the department.

Two months after the intervention start a follow-up workshop was held, again attended by the department managers, Kaizen representatives and health representatives from the intervention units. The aim was to give the various departments and managers the possibility to exchange experiences, discuss difficulties and exchange ideas for the continued work. The workshop also offered the possibility to receive support and answers to questions from the Kaizen and health representatives, as well as the researchers. Key people from the departments as well as centrally in the hospital are keeping a diary during the project to reflect on the work, note any problems or questions for upcoming follow-up meetings, and note what has worked well and what has been difficult.

It is the hospital that owns and manages the integration work. The researchers actively collaborate with management as well as the Kaizen and health representatives to support the integration of health promotion with Kaizen, and are responsible for the evaluation of the project using an interactive research approach (Svensson 2002). The project integrating health promotion and Kaizen, called LeanHealth, is a three-and-a-half-year funded research project.[2]

The Research Design: Process and Outcome Evaluations

A research project with a quasi-experimental design was planned in order to examine the effects of the integration of health promotion work and Kaizen. The integration is

---

1   *Coaching Healthcare Improvement Teams*; Dartmouth Medical School and The Dartmouth Institute.

2   Funded by AFA Insurance.

therefore being done first at half of the hospital's departments while the other half serve as control units, continuing to work as they have previously with health promotion issues and Kaizen. The various units at the hospital were matched based on type of unit (e.g. opening hours, acute or non-acute care), size and working processes involving Kaizen. The intervention units were then randomly chosen from the matched pairs. As all departments have a great deal of freedom in designing their Kaizen work themselves, in practice this work is different at the various departments. This entails that the six departments in the intervention group also manage their integration work differently, based on their own needs and conditions. Thus, the intervention will also be different at the different departments. The researchers follow the processes at the different departments in order to study what conditions enable or obstruct the integration. The project uses a multi-method approach, whereby different types of data are collected through interviews, questionnaires and observations. The project's effects on health, work environment, productivity and efficiency are studied using validated questionnaires. Furthermore, the respective frequencies of sickness absence and the use of the "exercise hour" are examined. In addition, the Kaizen notes from 2009 and forward (numbering approximately 2,000) from all departments are analysed from a health perspective regarding problem approach and consequence analysis, as well number of health-related proposals and decisions.

Implementation of the Integration at Department x

In this section the integration of occupational health, health promotion and Kaizen at Department x will be described. The results are based on observations from seminars and meetings during the first four months of the project, and interviews with a Kaizen representative, a health representative, an occupational health and safety representative and the manager of the department.

Department x has approximately 25 members of staff, including two Kaizen representatives and two health representatives. The department has a well-established Kaizen system. The Kaizen board is reviewed at short weekly meetings that all staff attends. At these meetings a Kaizen representative reads all the new suggestions, which are then delegated to a group or a person to look into, if they cannot be resolved on the spot. The department has developed a structure with smaller Kaizen groups that take on new suggestions, and some suggestions may be forwarded to other departments or higher management. At the monthly staff meeting, decisions are made concerning whether new routines or solutions that have been tried out should be fully implemented. If the group agrees to do so, all staff members need to sign the proposal to show that they have received the information. The health promotion work before the LeanHealth project consisted of information at the monthly meetings about health-related lectures, social activities such as the staff summer party, a fruit basket and the possibility to exercise during work hours for an hour each week, workload permitting. The department has an occupational health and safety representative who conducted the annual risk assessment with the department manager, and the result of the safety inspection was presented at the monthly staff meeting. After the department was selected (through matching and randomization) to be an intervention group, the two Kaizen representatives, the two health representatives and the department manager actively participated in the crafting of the project at their department.

PERCEIVED AIM OF THE PROJECT

Through the interviews it was noted that the department manager understood the aim of the project as extending the scope of Kaizen to include suggestions for improvements with an emphasis on employee health. She felt that the project would help them focus on employee health aspects in relation to all organizational development and production issues:

> I think it's more about partly using Kaizen as a tool for making improvements but then letting health come up on the agenda, that we discuss it and it becomes natural. But that's what we do with work environment and all the other issues too. So it's just an extension of the Kaizen concept, actually, I think. But this is focused on the staff, the employees.

When the interviewed Kaizen representative described her understanding of the project, the emphasis on health in relation to production improvement was again stressed: "I mean I think one should try to improve the health of [pause] or not improve the health maybe but have more of a focus on health when we work with production improvement and hopefully that'll then lead to better health too."

The interviewed health representative emphasized that the project widened the scope of their health promotion work to include aspects other than just physical exercise and other health-related behaviours. She foresaw a potential broadening of the concept, so that issues like mental health and work-related health would also receive attention.

The manager believed she and the Kaizen and health representatives had a shared understanding of the project, but that the rest of the staff may not all have been as clear about it: "I think the Kaizen representatives, the health representatives and I get it, surely. But I'm not sure the whole group does. There might be those who think it's more diffuse, it depends on how much you've gotten involved ... in the Kaizen reviews and how you've listened. I think that's different."

WORKING WITH THE INTEGRATED SYSTEM

To introduce the project to the work group the Kaizen representatives brought back a poster they had made at the introductory workshop illustrating different aspects of health, and hung it in the staff room. Later the same week, the poster was presented at a weekly meeting. This generated a great deal of positive curiosity about the project. The manager described how the project was introduced to the staff gradually, led by the Kaizen representatives: "In the beginning I think it was mostly the Kaizen representatives who did it. That we together, I myself along with them too, set that perspective but now we have gotten better at bringing it up in connection with reviewing the Kaizen notes."

After the integration was launched, suggestions at the Kaizen board that could be linked to staff health were marked with a green sticker. Suggestions linked to occupational health and safety were marked with an orange sticker. All suggestions, whether colour-coded or not, were analysed from a health perspective, which means that the team discussed the health outcome of the proposals. For example, one problem identified at Department x was that the phone hours for call-backs to patients was at a time during which the nurse was occupied with other tasks. This caused patients to experience poor accessibility and the nurse to experience stress. A Kaizen note was written on the problem. The note was discussed in the work group during a Kaizen meeting and a suggested solution was to change the phone time to an hour later, as the nurse had fewer tasks then. The suggestion was analysed from a quality, productivity and health perspective. Expected outcomes of the solution were that patients would have better accessibility to the care and that the nurse's work environment would be improved, leading to decreased stress.

The manager believed the department was well on its way to starting to regard employee health as a natural aspect of organizational development and production, but that they still had some work to do before everyone was on track and had fully adopted the same mindset. The interviewed Kaizen representative was also very positive about the integration of the systems, and found no problems with it. She considered Kaizen to be a very efficient tool for making suggestions and improvements, and found it very helpful for communication. The integration seemed uncomplicated to her; it did not involve any extra tasks since the structure was already in place, which made it easy to simply add a few things. Health had been on the agenda before, but with the integration it became easier and more natural to discuss it, even in relation to quality improvements:

> That ... one reason I didn't think this Kaizen health project was tricky was because we already work like this, we just have to think about health. It's not any more work at all, and I guess it's going to be that way for everything that's put into Kaizen, that we already have the structure and then you just add a little ... It will be easier to address health, it will be easier to bring it up and talk about it because it's not always obvious when you're working with production improvement. It hasn't always been obvious to think from a health perspective, even if it follows along so that it's easier, you have an extra reason to bring it up.

The manager described the engagement among the staff as high, and said that everyone was engaged in Kaizen in one way or another. She felt that the general positive attitude towards Kaizen would spill over to health, and stressed that most employees seemed to find nothing strange about the integration of health promotion and occupational health with Kaizen. The researcher's observations of a Kaizen meeting seemed to confirm the department manager's view. It seemed that the staff was involved and positive about working with Kaizen, as well as the project. The different identified problems and suggestions were well received, and the staff was active in discussing how to solve the problems and deciding who should be in charge of the Kaizen note.

The department manager, health representative and occupational health and safety representative all suggested that there was an overlap between the areas of health promotion and occupational health. This overlap made it trickier to categorize suggestions and possible consequences. However, the occupational health and safety representative perceived the overlap as positive, as it gave more weight to suggestions and issues within these fields (Suntliv.nu 2012). This overlap between the fields that was identified at the department had never been discussed before the project started.

ROLES

The project meant that the health representatives could now take a more active role (together with the rest of the staff) in suggesting improvements through Kaizen, rather than merely passing on information about health promotion. The interviewed health representative, however, did not feel her role had changed much yet due to the integration.

She still provided information and monitored health issues. However, she acknowledged that there had been a change of behaviour among the staff, which she attributed to the Kaizen representatives and the fact that everyone was starting to think differently about these issues. The project also meant that health issues had become more of an issue for everyone, as everyone could make suggestions.

The interviewed Kaizen representative described her role as making sure things "kept rolling along," explaining that she often asked the department manager to give pep talks, and remind everyone how important it was and why they were doing it. She did not feel her role had changed much, but found that it had begun to encompass more and more:

> It's taken on more branches, so there are more things that happen when you discover how much you can actually work with it. Both health and other things. But as a role, you could say that … but it's also about growing in your role and getting more and more responsibility, or it takes on more significance in some way the whole time, the Kaizen work, so I guess your role also grows.

Neither the Kaizen representative nor the health representative who were interviewed felt the project had thus far led to increased collaboration between these two roles, as most of the project work was done by the Kaizen representatives. This is evident in the observations of the Kaizen meetings. The Kaizen representatives lead the Kaizen meetings, and are responsible for the work with the Kaizen notes in the same way as before the project. This means that the health representatives have a less pronounced role in the process of integrating health promotion with Kaizen. This might be assumed to be a natural step because health promotion is being integrated into the Kaizen system, and not the other way around.

The manager understood her role as being supportive and positive:

> [My most important task is] being supportive to the whole group and the Kaizen representatives, and being positive and showing that it's an important issue. Then I also actively participate, then as manager you have to have a more reserved role too, it's something you have to have a feel for … I shouldn't just come in with the solutions like that, and it can easily happen that you do that. Instead it's important to have discussions.

NEEDS

When discussing what support may be needed in the future, the importance of having key persons as well as management that holds everything together and supports their work was emphasized by the manager, and the health and Kaizen representatives in the interviews. The health representative felt that this would help attract and keep people's interest, and this was important for her to be able to keep doing what she did:

> Then you know people get interested too and then you have the energy to keep going. Because after all it does demand a number of changes, but at the same time there's less moaning and groaning when something's to be changed because then everybody's been in on the decision that it should be changed. So it's important to stress the positive.

The Kaizen representative expressed a need for further information to confirm that she was doing the right thing, as she expected the project to entail a heavier workload:

The only thing I've had a hard time keeping track of is what I'm supposed to do now. But maybe that's because I thought it seemed very simple, and then you start wondering were we supposed to do more? Have I missed something? That's what I was confused about in the beginning, is this what LeanHealth is? Are we there or not? That's it … I mean you want to know that … that we're so to say living up to the project. If you get what I mean? That [pause] we're doing what we're supposed to be doing, or however you might put it. That it's sufficient and that we've found the right work method.

The manager does not share these concerns, however, but rather believes the expectations on the department in the project are clear: "Yes, I feel that the Kaizen and health representatives, that they know what's going on and that means we've gotten the support we need."

For the near future, the Kaizen representative emphasized a need to come up with health-promoting activities. She felt a responsibility to identify these as Kaizen notes: "I mean I feel like if I don't do it then probably nobody else will really do it, or it can happen that somebody actually does it, but even so I feel a responsibility, that I also need to perform and produce a little note."

RESULTS SO FAR

Looking forward in the interview, the Kaizen representative foresaw that health would be a more central issue, closely tied to production improvement. The health representative believed the integration would make people reflect more on their work practices. The department manager expected that the integration with Kaizen would lead to their getting better at continuously following up the action plans on occupational health issues. The health representative felt the staff had become better at identifying occupational health and health promotion aspects on Kaizen notes. Both the Kaizen and the health representatives believed the project had already led to changes. However, this was not always easy to specify: "I can't put my finger on it because I don't remember the notes, but you bring it up and think about it and that there have been changes.{NOTE:0148}" The Kaizen representative, however, also mentioned that the analysis of the suggestions easily fell into one simple explanation of "less stress," rather than being explored on a more detailed or deeper level. The manager also suggested that it sometimes seemed a bit awkward to add a health perspective: "Sometimes it can be a little forced if you're always supposed to have a health perspective on everything, because if you're going to improve work duties or the way we do things then the whole time the natural health perspective can be 'less stress'."

Based on information given in the interviews and from observations, changes that have been implemented at the department since the project's start in early 2012 include:

- Incidents with patients whom the staff have perceived as threatening have led to two staff members always being present at the department.
- Putting the necessary equipment together during emergency CPR was perceived as a stressful task; therefore this step is now done beforehand so the equipment is ready to use.
- More health information was requested, and after a discussion among staff two lectures were held on sleep and nutrition. This has led to some improvements in eating habits among staff, such as healthier breakfasts.

- It has been decided that half of the biannual planning day will focus on health-related issues.
- A health promotion perspective on an after-work activity has been suggested, as having fun together outside work was mentioned as creating a better work atmosphere.
- Shut doors at the department are seen as leading to inaccessibility; thus a suggestion for doors with windows has been proposed (but has not yet been realized). Doors with windows would both increase accessibility and decrease the risk of threats and violence towards staff.

The department is now also providing simplified Kaizen notes for patients in the waiting room. These suggestions are then integrated into the staff Kaizen board for further discussion. This initiative was suggested to the department by a member of hospital management, and shows that the department continues to be an early adopter. Other departments have now been inspired by the integration work taking place at Department x, and have adopted their use of colour-coded stickers on the Kaizen board to identify health promotion and occupational health and safety suggestions.

Discussion

One reason organizational-level occupational health interventions do not attain expected outcomes has been suggested to be a lack of integration of this type of intervention into other organizational process and structures. In this chapter, we describe how an integration of occupational health, safety and health promotion with a quality and productivity system can be conducted.

Recently, the logic of integrating health promotion and occupational health and safety work has been highlighted (NIOSH 2012). Others have highlighted the need to integrate occupational health and safety with quality and production systems (EU-OSHA 2010b). At the hospital in Enköping, all three of these – previously separate – systems were integrated:

1. occupational health and safety;
2. health promotion; and
3. the Lean production and quality improvement system.

Thus, this project adds to the literature describing integration between occupational health and safety and health promotion on the one hand and occupational health and safety and quality and production systems on the other. Based on the results of the interviews with the staff at one department, the joint integration of occupational health and safety and health promotion into Kaizen seems to be a rational approach. Those interviewed acknowledged that it felt natural to do it this way. They even described difficulties in separating the different areas, as they are all intertwined in practice.

Lean is becoming a widespread production system in all sectors of work. Given that it has been associated with negative effects on employee health (Brännmark and Håkansson 2012, Landsbergis et al. 1999), there has been scepticism towards Lean among occupational health practitioners and researchers. However, there is also evidence indicating that its effect may vary depending on which components are implemented. Kaizen uses a participatory approach. It has many attributes in common with the socio-technical approach, which among other things strongly emphasizes the involvement and active participation of staff in the improvement of work. This approach has historically had a strong influence on Scandinavian organizations. Employee participation is important not only because participation in itself is related to improved employee health; it is also essential to the success of any intervention. Occupational health interventions on an organizational level are assumed to have the best chance of succeeding when they are structured and follow a participatory intervention process in all phases of the intervention (Nielsen et al. 2010a, 2010b). We know that employees often actively craft their jobs through their daily activities to fit their own personal goals at work. It is therefore important that employees be part of developing ongoing intervention activities and content, in contrast to top-down approaches (Nielsen et al. 2007, 2010a). Using employees' job-crafting strategies and day-to-day autonomy to help them share the responsibility for ensuring a healthy organization may be a way to facilitate a successful organizational-level intervention (Nielsen et al. 2010a). At Department x, all staff members have become involved in analyzing Kaizen notes from a health perspective. However, the implementation of the other step of the integration, in which health-related problems are suggested as Kaizen notes, has been somewhat slower and the interviewed Kaizen representative suggested that she needed to take on more of a role-modelling role for this to happen. For her, personally, her increased participation in health-related work issues was clear.

Another important aspect of Kaizen is that it is designed for working continuously with small-scale changes. This could be called the golden standard of many quality improvement systems, but it is less common within occupational health interventions. Observations in this study provide an example of the potential benefits of this approach in this area, e.g. how a challenging and stressful situation for one occupational group could be improved through a minor change in routine. Also, this approach means that each department will form its own sub-interventions, based on its own needs. This is likely to increase the fit between the organization and the intervention. However, it also has implications for research, as interventions will differ between departments. One way of handling this may be to approach such projects as natural experiences, whereby not only the outcome but also the variation in intervention and process is studied.

Conclusion

In occupational health interventions, an apprehension is always that the intervention will become sidetracked, fade off or be down-prioritized as other issues take the upper hand over time. As the intervention presented in this chapter took place only six months prior to the writing of this chapter, its sustainability is not clear. However, it is promising that the staff representatives perceived the intervention as "natural," and pointed out that it did not add anything to their workload. Based on the interviews and observations, employees and key stakeholders seemed to be highly motivated to take part in the integration at Department x. They seemed to share common goals for the intervention and understand the process and potential benefits. Nevertheless, change is not a discrete event. All parts of the integration do not happen from Day One; instead, it seems to evolve bit by bit. Thus, the engagement and involvement of the key stakeholders in the department is crucial.

In this chapter we have presented the integrated approach developed and implemented at a Swedish hospital. The early phase of the project prevents any firm conclusions as to the overall effect of this initiative. However, we have described how one of the departments has understood and approached the integration, and thus far it seems that the initiatives fulfil many factors that have been described as prerequisites for successful organizational-level occupational health interventions.

References

Baker, E., Israel, B. and Schurman, S. 1996. The integrated model: implications for worksite health promotion and occupational health and safety practice. *Health Education Quarterly*, 23(2), 175–90.

Bambra, C., Egan, M., Thomas, S., Petticrew, M. and Whitehead, M. 2007. The psychosocial and health effects of workplace reorganisation. 2. A systematic review of task restructuring interventions. *Journal of Epidemiology and Community Health*, 61(12), 1028–37.

Brännmark, M. and Håkansson, M. 2012. Lean production and work-related musculoskeletal disorders: Overviews of international and Swedish studies. *Work: A Journal of Prevention, Assessment and Rehabilitation*, 41, 2321–8.

Briner, R.B. and Reynolds, S. 1999. The costs, benefits, and limitations of organizational level stress interventions. *Journal of Organizational Behavior*, 20(5), 647–64.

Conti, R., Angelis, J., Cooper, C., Faragher, B. and Gill, C. 2006. The effects of lean production on worker job stress. *International Journal of Operations & Production Management*, 26(9), 1013–38.

EU-OSHA (European Agency for Safety and Health at Work) 2010a. *European Survey of Enterprises on New and Emerging Risks: Managing Safety and Health at Work*. Luxembourg: Publications Office of the European Union.

EU-OSHA (European Agency for Safety and Health at Work) 2010b. *Mainstreaming OSH into Business Management*. Luxembourg: European Agency for Safety and Health at Work.

Hasle, P. 2011. Lean production: An evaluation of the possibilities for an employee supportive lean practice. *Human Factors and Ergonomics in Manufacturing & Service Industries*.

Hasle, P., Bojesen, A., Jensen, P.L. and Bramming, P. 2012. Lean and the working environment: A review of the literature. *International Journal of Operations & Production Management*, 32(7), 829–49.

Hines, P., Holweg, M. & Rich, N. 2004. Learning to evolve: A review of contemporary lean thinking. *International Journal of Operations & Production Management*, 24(10), 994–1011.

Holden, R.J. 2011. Lean thinking in emergency departments: A critical review. *Annals of Emergency Medicine*, 57(3), 265–78.

Jackson, P.R. and Mullarkey, S. 2000. Lean production teams and health in garment manufacture. *Journal of Occupational Health Psychology*, 5(2), 231.

Jacobson, G.H., Mccoin, N.S., Lescallette, R., Russ, S. and Slovis, C.M. 2009. Kaizen: A method of process improvement in the emergency department. *Academic Emergency Medicine*, 16(12), 1341–9.

Jørgensen, T.H., Remmen, A. and Mellado, M.D. 2006. Integrated management systems: Three different levels of integration. *Journal of Cleaner Production*, 14(8), 713–22.

Landsbergis, P.A. and Vivona-Vaughan, E. 1995. Evaluation of an occupational stress intervention in a public agency. *Journal of Organizational Behavior*, 16(1), 29–48.

Landsbergis, P.A., Cahill, J. and Schnall, P. 1999. The impact of lean production and related new systems of work organization on worker health. *Journal of Occupational Health Psychology*, 4(2), 108.

Mazzocato, P., Savage, C., Brommels, M., Aronsson, H. and Thor, J. 2010. Lean thinking in healthcare: A realist review of the literature. *Quality and Safety in Health Care*, 19(5), 376–82.

Mikkelsen, A. 2005. Methodological challenges in the study of organizational interventions in flexible organizations. In *Anthology for Kjell Grønhaug in Celebration of his 70th Birthday*, edited by A.M. Fuglseth and I.A. Kleppe. Bergen: Fagbokforlaget, 150–178.

Mikkelsen, A., Saksvik, P.Ø. and Landsbergis, P. 2000. The impact of a participatory organizational intervention on job stress in community health care institutions. *Work & Stress*, 14(2), 156–70.

Moldasch, M. and Weber, W.G. 1998. The "three waves" of industrial group work: Historical reflections on current research on group work. *Human Relations*, 51(3), 347–88.

Nielsen, K. and Randall, R. 2012. Opening the black box: Presenting a model for evaluating organizational-level interventions. *European Journal of Work and Organizational Psychology*.

Nielsen, K., Randall, R. and Albertsen, K. 2007. Participants' appraisals of process issues and the effects of stress management interventions. *Journal of Organizational Behavior*, 28(6), 793–810.

Nielsen, K., Taris, T.W. and Cox, T. 2010a. The future of organizational interventions: Addressing the challenges of today's organizations. *Work & Stress*, 24(3), 219–33.

Nielsen, K., Randall, R., Holten, A.L. and González, E.R. 2010b. Conducting organizational-level occupational health interventions: What works? *Work & Stress*, 24(3), 234–59.

NIOSH 2012. Research Compendium: The NIOSH Total Worker HealthTM Program: Seminal Research Papers 2012. Washington, DC: U.S: Department of Health and Human Services, Public Health Service, Centers for Disease Control and Prevention, National Institute for Occupational Safety and Health, DHHS (NIOSH).

Pettersen, J. 2009. Defining lean production: Some conceptual and practical issues. *The TQM Journal*, 21(2), 127–42.

Poksinska, B. 2010. The current state of Lean implementation in health care: Literature review. *Quality Management in Healthcare*, 19(4), 319.

Radnor, Z. and Walley, P. 2008. Learning to walk before we try to run: Adapting Lean for the public sector. *Public Money and Management*, 28(1), 13–20.

Richardson, K.M. and Rothstein, H.R. 2008. Effects of occupational stress management intervention programs: A meta-analysis. *Journal of Occupational Health Psychology*, 13(1), 69.

Semmer, N.K. 2003. Job stress interventions and organization of work. In *Handbook of Occupational Health Psychology*, edited by J.C. Quick and L.E. Tetrick. Washington, DC: American Psychological Association, 325–53.

Seppälä, P. and Klemola, S. 2004. How do employees perceive their organization and job when companies adopt principles of lean production? *Human Factors and Ergonomics in Manufacturing & Service Industries*, 14(2), 157–80.

Sewell, G. and Wilkinson, B. 1992. "Someone to watch over me": Surveillance, discipline and the just-in-time labour process. *Sociology*, 26(2), 271–89.

SFS 1977:1160 Arbetsmiljölag. Ändrad t.o.m. SFS 2011:741 ed. Stockholm: Arbetsmarknadsdepartementet.

Smith, D. 2002. *IMS: Implementing and Operating*. London: BSI British Standards Institution.

Suntliv.nu. 2012. *Flyt i arbetet främjar hälsan. Ständig jakt på problem- och förbättringar*. Retrieved on 29 August 2012 from www.suntliv.nu.

Svensson, L. 2002. Bakgrund och utgångspunkter. In *Interaktiv forskning – för utveckling av teori och praktik*, edited by G.B.L. Svensson, P.-E. Ellström and Ö. Widegren. Stockholm: Arbetslivsinstitutet, 173–207.

WHO (World Health Organization) 1997. *The Vienna Recommendations on Health Promoting Hospitals*. Copenhagen: World Health Organization.

Wilkinson, G. and Dale, B.G. 1999. Integrated management systems: An examination of the concept and theory. *The TQM Magazine*, 11(2), 95–104.

Wilkinson, G. and Dale, B.G. 2002. An examination of the ISO 9001: 2000 standard and its influence on the integration of management systems. *Production Planning & Control*, 13(3), 284–97.

Womack, J.P., Jones, D.T. and Roos, D. 1990. *The Machine that Changed the World: The Story of Lean Production*. New York: Rawson Associates.

Zwetsloot, G.I.J.M. 1995. Improving cleaner production by integration into the management of quality, environment and working conditions. *Journal of Cleaner Production*, 3(1), 61–6.

17 eHealth Interventions for Organizations: Potential Benefits and Implementation Challenges

HENNA HASSON, ULRICA VON THIELE SCHWARZ, KARIN VILLAUME and DAN HASSON

Abstract

Workplace leaders in many industrialized countries are required by law to assess the psychosocial work environment and occupational health on a regular basis. However, leaders often report that they lack knowledge about how to utilize and act upon the results from these assessments to improve occupational health. This may illustrate a gap between science and practice. Scientifically, there is well-established knowledge about psychosocial factors that can cause work-related ill health. At the same time, interpreting the results from occupational health surveys and knowing what do and how to focus interventions based on the results is not easy for many leaders. Moreover, the number of evidence-based interventions for occupational health promotion is increasing. This can make it more difficult to know which of the methods would be the most suitable one. In this chapter, we discuss how an Internet-based approach can be used for working with occupational health and making the scientific evidence easily available for workplace leaders.

Introduction

ASSESSMENT OF THE PSYCHOSOCIAL WORK ENVIRONMENT AND OCCUPATIONAL HEALTH

Assessments of the psychosocial work environment and occupational health, often called risk assessments, are increasingly delivered via the Internet. Web-based risk assessment surveys have offered solutions to some of the limitations of traditional assessment methods (Cook et al. 2000). One important limitation with traditional pen-and-paper assessment methods is that the results are often presented several months after the assessment

(Rhodes et al. 2003). During this period, the organization may have been subjected to a wide variety of changes, which can make the results outdated. Improvement efforts based on these results can therefore be of limited value, or perhaps even detrimental, since wrong interventions may negatively impact the department or the company. Interactive web-based surveys can instead provide instant feedback, directly after the assessment is finalized, both to the individuals and to the managers. Other advantages of using web-based surveys include that a large number of people can be reached at a lower cost than traditional methods (Griffiths et al. 2006). Furthermore, questionnaires can be adaptive, so that respondents only see the questions that apply to them, making them more relevant and time efficient. Also, the administration processes are convenient in terms of easy storage (Rhodes et al. 2003, van Gelder et al. 2010). It is also easier to monitor work groups' response rates. Several studies have indicated that participants are more open and honest when answering web-based surveys compared to paper-and-pen versions and that social desirability decreases (Davis 1999, van Gelder et al. 2010, Rhodes et al. 2003). Disadvantages with web-based surveys are that the risk of data hacking and that not everyone has access to the Internet (Rhodes et al. 2003).

Results from a risk assessment can often be extensive and difficult to understand. As a consequence, leaders will spend a disproportionate amount of time on analysing the results, or, in the worst case, avoid getting into them at all. Furthermore, the more complicated the results displayed, the greater the risk of misinterpretations of the results. This may lead to bad priorities between different areas in need of improvement. Web-based survey tools provide a possibility of automated analyses of large amounts of data with instant standardized and tailored result feedback (Lustria et al. 2009). This could offer guidance and support for leaders in selecting interventions to improve areas identified in a work environment survey.

eHEALTH INTERVENTIONS

Internet-based interventions have mainly focused on *individual level* health promotion (Wantland et al. 2004). There are few examples of web-based interventions that leaders could use for promoting health and the psychosocial work environment at group or organizational level. Several of the individual level web-based interventions such as programmes for stress reduction (Kawakami et al. 2006, Hasson et al. 2005), physical activity (Spittaels et al. 2007), nutrition (Oenema et al. 2001), smoking cessation (Strecher et al. 2008) and alcohol consumption (Kypri et al. 2004) have shown positive results. Users of health promotion programmes have also been found to prefer web-based versions when compared with more traditional deliveries (Cook et al. 2007, Mangunkusumo et al. 2007). This opens up opportunities to provide eHealth interventions also for group and organizational level well-being. Some web-based programmes have used workplaces as a way to reach participants, such as programmes for improving dietary behaviours (Block et al. 2004), increasing physical activity (Hager et al. 2002, Marshall et al. 2003) and reducing alcohol consumption (Matano et al. 2007). However, these programmes have had individual level focus rather than work group or organization level focus. Consequently, there is limited knowledge of how leaders can make full benefit of group and organization level web-based occupational health interventions.

In summary, there are few eHealth tools that target groups and organizations rather than individuals. The web-based tools that focus on organizations and groups have mainly offered risk assessment surveys, but not included any interventions to occupational health improvement. The present chapter describes an ongoing project of an interactive web-based tool i.e. HealthWatch that includes both risk assessment surveys and evidence-based, practical interventions for leaders at all organizational levels as well as the human resource (HR) department. The project aims at improving organizational, group and individual well-being by improving the psychosocial work environment, optimizing work ability and job satisfaction as well as preventing ill health, sickness presenteeism and long-term sick leave by offering a practical and interactive web-based tool.

The Web-based Tool: A Case Report

Like many other programmes, the precursor of the tool started as an intervention targeting individuals, in the so-called Health-IT study. The prospective randomized, controlled research trial was conducted in 2002 on companies from the information technology and media sector (Hasson et al. 2005). The aim was to develop a web-based brief screening tool for the psychosocial work environment with possibilities for unlimited assessments and instant feedback. The intervention also provided tools for employees' health promotion and stress management. The study found that the web-based tool exhibited beneficial effects in improving employees' perceptions of their ability to manage stress, improved sleep quality, increased mental energy and concentration ability in the intervention group compared to the control group. In addition, significant improvements in employees' biomarkers were found at the end of the six-month intervention (Hasson et al. 2005).

Since the Health-IT study, the tool HealthWatch has been publically available free of charge for individuals as well as for smaller groups and has currently approximately 15,000 users. One fundamental principal of the tool is that it should be user-driven. Therefore, it is continuously developed based on comments from users and advancements in the technological fields. One such development was to make the tool available for company use and, today, it is used by approximately 35 companies. The company users have also identified a need for practical interventions that helps leaders improve occupational health in their work groups. Today, the tool includes an interactive self-learning tool for managers to conduct interventions aiming at improving the psychosocial work environment for the group as well as tools for leadership development. As the tool was implemented at the company level, users also expressed a need for guidance on how to implement the system in an organizational setting. This led to the current project, which is described in this chapter. Funding for development of the web-based tool for group and organizational use was received from the European Social Fund (ESF) in Stockholm. They see the web-based tool as a possible way of strengthening employees' positions in working life.[1] The tool currently consists of surveys for assessment of the psychosocial work environment and occupational health interventions that can be used at individual, group and organization levels. The content of the tool is presented in Table 17.1.

---

1   The ESF-council supports measures to prevent and combat unemployment and to promote employees' training.

Table 17.1 The content of the web-based tool

	Individual level (employees and leaders)	Group, department, organization level
Assessment	*Brief survey* 15 seconds, real-time monitoring of well-being, health, stress and work conditions; responded regularly – from daily up to every second week. *Extensive survey* 15 minutes screening of the most common public health disorders and psychosocial work environment. Conducted 1–4 times annually (frequency decided by the organization).	
Interventions	*Instant feedback* on individual's own ratings of the brief (graphs illustrating current ratings and comparisons with same socioeconomic group), development over time and extensive survey (written feedback on answers and a *referral* to the occupational health care provider when needed). *Self-help exercises* Approximately 20 health promotion and stress management (scientifically proven) exercises with elements of cognitive behavioural therapy, classical and innovative relaxation techniques, structured problem solving, conflict management, skills training, body awareness, etc. *Diary* for notes and expressive writing used with optional frequency. *Information about stress, health and well-being*	Regular, real-time *feedback on work group's well-being*, health, stress and psychosocial work environment (from *brief survey*) Real-time feedback on *work group's perceptions of psychosocial work environment and health variables 1–4 times per year* (from *extensive survey*). *Self- and organizational development exercises* Approximately 30 exercises and tools to improve leadership skills with regards to organizational health promotion and systematic psychosocial work environment development. *Information about stress, health and well-being*

THE ASSESSMENTS AND INTERVENTIONS AT THE INDIVIDUAL LEVEL

The brief survey can be used by individuals, e.g. both leaders and employees, as often as they wish to survey their own health and well-being. It takes approximately 15–30 seconds to fill in and consists of 11 questions concerning health, sleep, concentration, energy, control, social life, stress, work efficiency, work load, work satisfaction and work climate. A possibility to activate a reminder at an interval of choice is provided for the individuals. The reminders are delivered via email with encrypted automatic login links so that participants can access the brief survey tool as easily and effortlessly as possible. For individuals, immediate tailored feedback on the brief survey is given in the form of graphs using different colours as indications for healthy level (green), improvement needed (yellow) and unhealthy level (red). An individual can also compare his/her own results with aggregated mean values of all respondents in the database, and those with the same socioeconomic profile. Self-help exercises targeting the different surveyed areas are provided and are accessed by clicking on each specific bar in the results graph. Thus, based on the results from the brief survey, the system provides the individual with simple and time-efficient tools for health promotion and stress management. The self-help exercises for individuals include techniques for relaxation, improving sleep quality, cognitive reframing, time management, emotional control and self-awareness, strengthening self-esteem, life reflection and dissociation. Most of the exercises are presented in three formats to fit different pedagogic preferences. These can be viewed in html, downloaded as a PDF or experienced as an animation with both picture and sound. The tool also includes a diary with multiple utilities. The users can for instance use it to conduct expressive writing or add other health relevant measures such as blood pressure, hormone levels and physical activities (biofeedback). In the reading view of the diary, the user can access past ratings on the brief survey, notes and other values. The tool also provides popular scientific information about stress, health and well-being in terms of brief summaries of medical and health-related news.

The extensive survey is used to get an overview of the health-related variables and the psychosocial work environment. The extensive survey takes approximately 15–18 minutes and is usually distributed one to four times a year. The frequency is decided by the companies and/or work groups. Individuals receive instant and tailored feedback on some of their health-related answers after filling in the survey. They can also receive an automated referral to the occupational healthcare provider if the results indicate needs such as symptoms of long-term stress or musculoskeletal pain problems. The decision concerning if and how to involve the occupational healthcare provider is decided by each company.

THE INTERVENTIONS FOR LEADERS AND HR DEPARTMENT

The brief survey offers work group leaders, HR departments and organizational leaders an opportunity to follow trends in well-being and the psychosocial work environment at a group level. It is recommended that that each work group and organization collectively decides a minimum frequency for filling in the brief survey so that group level results can be generated continuously. Reminders are set and provided at group level. This means that a leader can set the minimum frequency of responses in his/her group and reminders are automated for those who have not logged on to the system during the past week or two weeks. It is also recommended that the group should use the tool at least every 14th day and preferably once a week. If individuals utilize the tool more frequently, they will yield a more correct mean value to the group results. For instance, if the questionnaire is answered only once a week, situational circumstances such as a night of poor sleep will bias the results. If it is filled in three times per week, this situational bias will decrease. All leaders are advised to regularly (at least once a month) display and discuss the results of the brief surveys at group meetings in order to highlight employee health and work environment as key aspects of the agenda.

The results of the *extensive survey* are presented at group level to the group leader via an automatic structured display of results. In order to reduce a stressful task for many leaders, i.e. to interpret results and prioritize, a short summary of positive health- and work-related results and three to four improvement/focus areas is provided. It is called "The 5-minute report" and the idea is that a leader within five minutes should know the strengths of the groups and what kind of improvements are needed. The surveys are fully dynamic and can be tailored according to the organizational needs and additional focus areas or questions can be included. For ethical and personal integrity reasons, results of the brief and extensive surveys are not presented for groups consisting of less than 10 individuals or if the response rate is less than 50 per cent. There is an exception however. If a group of seven to eight individuals insists on viewing the group feedback and everyone signs a consent form, this blockage can be removed. In this case all of the group members need to answer the surveys in order for the feedback to be given.

All leaders are also provided with approximately 30 interactive self-learning educational exercises. These cover a broad area of issues relating to leadership, work and health, including conflict management, work climate, employee development, crises management, problem solving, goal setting, communication and information, recruitment, monitoring and feedback, strategies for change, time management, the work environment act, rehabilitation and stress management. The exercises are based on organizational behavioural analysis which is the application of learning theories (applied behaviour analysis) on the organizational level (Wilder et al. 2009). The dual aim of the exercises is to promote both employee work engagement and organizational productivity through positive reinforcement strategies.

THE UNDERLYING LOGIC OF THE TOOL

The development of the tool was guided by five principles: time efficiency; change from reactive to systematic and proactive; adaptive system; to follow trends rather than capturing snapshots; and multiple level self-help and educational exercises. These principles are briefly described in the following section.

Time efficient

The main idea of the previous Health-IT study was that a stress management and health promotion programme should be time efficient in order to be utilized by the users. The idea from the beginning was that stressed individuals do not want to spend time on stress management and stress management should not be stressful. The hypothesis was that even highly stressed individuals have the time to take their medicine. The time to pour a glass of water and swallow a pill was estimated to take 30–40 seconds, and that was the basic time frame that the intervention was built upon. Another underlying theory concerns neural plasticity (Concise Medical Dictionary 2010), which hypothesizes that frequent repetition of a behaviour for introducing habits is more sustainable than less frequently repeated behaviours. According to this idea, it is more beneficial for health promotion and health behaviours to spend 30–60 seconds per day to become aware of health, stress and well-being than one hour monthly on meditating about it. This is the basis of the brief survey in which the user is also "rewarded" with instant feedback. The interventions also need to be time efficient and, therefore, the time frame is displayed for all exercises. The time it takes to go through an exercise ranges from a few seconds to an hour or more, with others taking 15 minutes to learn but then only a few seconds to apply later on.

From reactive to systematic and proactive

As the denotation "risk assessment" suggests, many assessment of psychosocial work conditions and work-related health are focused on risk prevention. Even though it is possible to use this tool that way, the main idea is to use it for maintenance, promotion and improvement of health and well-being. This is in line with the recent suggestions by Nielsen and Randall (2013) who state that more work is needed to test and describe interventions that include positive aspects of work and employee development. In our case, this is achieved by encouraging an ongoing, systematic assessment and monitoring of psychosocial work conditions and employee health as part of a participatory, continuous improvement effort. The frequent assessment also allows for monitoring progress as actions are taken. This can be used both to assess the effectiveness of the actions taken, and for encouraging striving towards goals.

The brief and extensive surveys also function as an early warning system so that potential risks can be handled at an early stage. Although this apparently is prevention rather than promotion, it is set aside from many traditional risk assessments by its instant feedback and regularity of measurements. These features facilitate early detection, as leaders and HR departments can monitor the development of departments/groups in real time. If there are unusual changes in the patterns, leaders and HR can detect this and take action. Or, if the patterns reflect a positive trend, these should work as positive reinforcements for the group. Also, the early warning system acts as an ice breaker between HR and managers, as it offers an opportunity for coaching and collaborative problem solving. The tool also finds individuals with signs of ill health, and automatically offers them the opportunity to be contacted by the occupational healthcare provider. If a person responds positively to the offer she/he should be contacted within 24 hours. The request to the occupational healthcare provider (which can be customized for each organization) is sent by the individual via the system.

Adaptive

Another underlying idea for the tool is that it should be adaptive to the needs of individuals, groups, departments and organizations. This is addressed in several ways. The adaptive features for the individuals have been described above. For the leaders and management, the exercises were built using the concept "From quick fix to fundamental change … and something in between." The idea was to offer exercises and tools from the most basic quick-fixes that take minutes to work through, to programmes that take months to complete. This means that leaders with limited amounts of time are able to use some tools whilst managers aiming at long-term personal and organizational development can find tools for that as well. The tool is also adaptive in the sense that it is continuously modified based on the needs and preferences of the users. Each organization has the possibility to add and omit questions in the extensive survey. For instance, a work group with specific problems or positive goals can be offered additional questions for evaluation purposes. The need for adaptive tools that take the different local needs of the work groups into consideration has been highlighted (Brisson et al. 2006). Also, previous research of web-based programmes have shown that tailoring messages for each individual or for work groups is more effective than presenting generic information in terms of engaging individuals, building their self-efficacy and improving health behaviours (Lustria et al. 2009).

Following trends rather than capturing snapshots

The tool emphasizes the importance of following trends over time instead of only measuring and reacting upon cross-sectional results. The current state when filling out the questionnaire will most probably influence the ratings of the last month (Gorin and Stone 2001, Holte et al. 2003, Hasson 2005, Hasson and Arnetz 2005). In our experience, measurements of occupational health and well-being, such as questionnaires, are like snapshots with a camera. They capture the moment and are determined by mood and the situation. One will see different results depending on weather; better results when it is sunny, and worse should it rain or snow. With this in mind, looking at trends gives a more accurate picture of the situation since our health and well-being is dynamic. Figure 17.1 shows the trends for employee self-rated health in four departments in a company that have used the tool regularly for 18 months. Self-rated health is a simple, yet powerful, measure that has been linked to future morbidity and mortality (Idler and Benyamini 1997, Bailis et al. 2003) as well as functional decline, disability and utilization

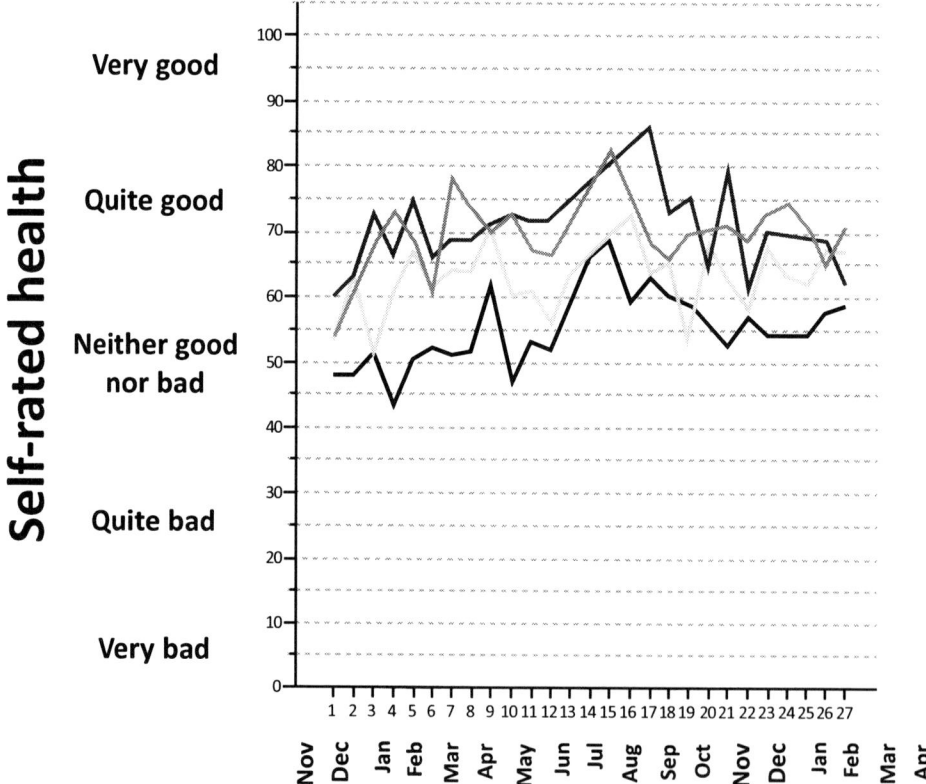

Figure 17.1 Mean values on self-rated health for employees at four departments during 18 months

of healthcare (Farmer and Ferraro 1997, Goldman et al. 2004, Bailis et al. 2003). Lower levels are therefore an important early warning for increased risk for ill health. From Figure 17.1, it becomes clear that there are huge fluctuations in employees' self-rated health over time. A cross-sectional survey, therefore, is highly vulnerable for the timing of the survey. However, monitoring trends makes it possible to intervene only when there is a true need rather than temporary fluctuations. This means that repeated assessments make it possible to use resources in a more efficient and cost-effective way and to make sure that the interventions are relevant to treat an existing problem. However, this also means that the survey needs to be quick and easy to complete.

Self-help and educational exercises at different levels

The exercises are provided for leaders at all levels and the HR department so that they all can work with continuous improvement of the psychosocial work environment and employee health promotion at group level. This is expected to be a way to involve all key actors in different levels of an organization. It also offers an opportunity to tailor the tool to the needs of these key actors. Previous studies have suggested that tailoring of organizational-level interventions should include adapting interventions to meet the requirements of specific individual employees. A lack of tailoring to individual needs has been cited as a potential problem with organizational interventions. In the light of this, a combination of multi-level interventions, e.g. individual and organizational, has been suggested as an optimal strategy (Lamontagne et al. 2007).

THEORETICAL OUTCOMES OF THE TOOL

A programme theory including short- and long-term theoretical effects of the tool is shown in Table 17.2 on the next page. The first column presents the immediate behavioural impacts of using the tool. The second and third column presents the theoretical impacts and final outcomes after using the tool according to the description on the first column.

The implementation of the tool

The participating organizations in the ongoing project were primarily recruited via an occupational pension plan company and four of them were contacted directly by the research group. Nineteen organizations were contacted and 10 of these enrolled in the study. The main reasons for organizations to decline participation was lack of time and commitment from HR or management, inappropriate timing or that the organization was currently focused on other in-house projects.

The implementation of the tool in the 10 organizations was initiated during the spring of 2011. The implementation process was conducted in a structured way. The process started with a presentation of the tool and the project by the last author to representatives of the HR department and company management. Thereafter, a similar presentation was held for department and group leaders. Only departments and work groups interested in participating were included. Thus, the programme was not always implemented in all workplaces within the participating organizations. From the researchers' perspective this

Table 17.2   A programme theory of the web-based tool

Behavioural impacts	Impacts	Outcomes
Individual level	*Individual level*	*Individual level*
Frequent monitoring of stress, health and well-being.	Improved general knowledge and awareness of health and well-being as well as own current and previous health, well-being, psychosocial work environment and public health disorders.	Improved concentration ability (i.e. reduced stress levels), sleep quality, mental energy, social life and job satisfaction.
Active use of interventions and exercises to improve health (e.g. active stress management and health promotion).		Reduced work related health problems.
Health behaviour changes (e.g. increased physical activity, healthy behaviour at work and at home).	Improved knowledge of and ability for how to work with issues related to health, stress and relaxation.	
Seeking care from occupational healthcare provider.	Increased contact with occupational healthcare providers.	
Applying new skills in different situations.	Increased control over own health and increased motivation for active use of health promoting activities.	
Frequent documentation and/ or expression of thoughts and feelings.		
Active search for knowledge within health, lifestyle and medicine.	Increased knowledge about how health variables affect each other over time.	
Increased discussions at work group about health, well-being and PWE.	Insights and better under-standing of factors underlying health and well-being.	
Leaders	*Leaders*	
Frequent monitoring of work group's/department's/ organization's stress, health and PWE ratings.	Improved general knowledge and awareness of stress, health and wellbeing.	
Active use of interventions to improve work group's/unit's/ organization's stress, health and PWE ratings.	Increased knowledge and awareness of work group's/ unit's/organization's current and previous stress, health and PWE	
Active role in improving employees' health and PWE (for instance giving feedback on the surveys and taking actions based on the results).	Improved knowledge of how to work with issues related to employees' health and PWE.	
	Improved sense of control over work group's/unit's/ organization's health, wellbeing and PWE.	
Applying new skills in different situations.		
	Increased motivation and ability for active use of health promoting activities.	

Table 17.2 Continued

Behavioural impacts	Impacts	Outcomes
		Group/organization level
		Improved psychosocial work environment.
		Reduction in sickness presenteeism and long-term sick leave.

was based on theories showing more successful implementation when the participants can influence and negotiate in the design and implementation of interventions (Nytrø et al. 2000, Saksvik et al. 2002). It was also common that the organizations wanted to test the tool on a small scale before implementing it on a larger scale. When a decision to participate was formed also by the group/department leaders, an inspirational implementation seminar was arranged for the employees. The implementation seminar usually started with one or more top-level managers (or HR) introducing the aim of this particular project in relation to the organization and that adherence was very important. Thereafter, one or two representatives from the project management (usually the last author) presented the background, the previous research findings from using the tool and demonstrated the tool. In addition, employees were given opportunities to ask questions. The goal of the seminar was to inform about the tool and boost willingness to participate. During the seminar, invitations to the tool were sent out via email so that an account could be opened directly after the seminar. In some cases, the participants were asked to open an account and to fill in the survey immediately after the seminar and then meet again after 30 minutes for reflections and questions. This strategy seems to yield the best response rates, in some cases up to 100 per cent. The project management was available for answering any questions or resolving any technical issues. In other instances, participants would open an account and fill in the survey whenever they wished within a certain period of time (decided by the group, managers and/or HR), usually within two to three weeks after the seminar.

PARTICIPANTS' EXPERIENCES AND USE OF THE TOOL

During the project, information was continuously collected from the users via questionnaires and interviews in order to investigate their experiences of the tool and of the project. In addition, measures that are automatically produced by the system such as logins have been analysed. The interviews were conducted by an external project assistant and the project managers were not involved in that process. The interviews were conducted with a selection of employees, HR professionals, workplace- and organization leaders. The focus was on enabling and hindering factors for using and implementing the tool as well as own reflections for successful use of the tool in the organization or group. Some preliminary findings from the interviews are presented in this chapter. These findings are based on a total of 57 interviews with employees (n=38), work group and department leaders (n=11) and HR representatives (n=7). The interviews were conducted in each of the participating organizations approximately six months after the implementation of the programme. Since the development of the organizational level exercises was done in parallel with the implementation, these exercises were not fully implemented at the time for the interviews. Hence, the interviews focused on the survey, feedback and individual exercise as well as implementation issues. The baseline survey, e.g. the first extensive survey that is conducted directly after the implementation seminar, includes some items measuring the participants' views of the programme. These items were adapted from a previously validated questionnaire, the Intervention Process Measure (Randall et al. 2009).

At the baseline, a majority of the employees and leaders looked forward to starting the intervention. Approximately 55 per cent of the employees and 65 per cent of the leaders who responded to the baseline questionnaire had high expectations that the intervention would improve their health (Figure 17.2).

Most of the interview respondents perceived that their immediate leader had been positive about implementing the tool. This was also shown in the results from the baseline questionnaire (Figure 17.3). However, employees reported in the interviews that the interest from organization management was less evident. In fact, some said that the organization management had not given any signals at all concerning the tool. Many of the respondents did not know whether the management supported the project and used the tools at all. The lack of support from senior management might have a "trickle down" effect on middle managers (Saksvik et al. 2002). This type of trickle down effect was not seen in the questionnaire results, but it is possible that such effect could occur on a longer run if no clear management signals are given to the work group leaders and employees. Figure 17.3 also shows how employees and leaders at the baseline questionnaire rated the level of information received from their immediate leader. It has been shown that the level of information and communication plays an important role in the effects of interventions (Jimmieson et al. 2004). It has also been found that open communication helps employees to understand the intentions behind occupational health interventions thus improving employee commitment to and participation in the intervention (Nytrø et al. 2000, Weick et al. 2005).

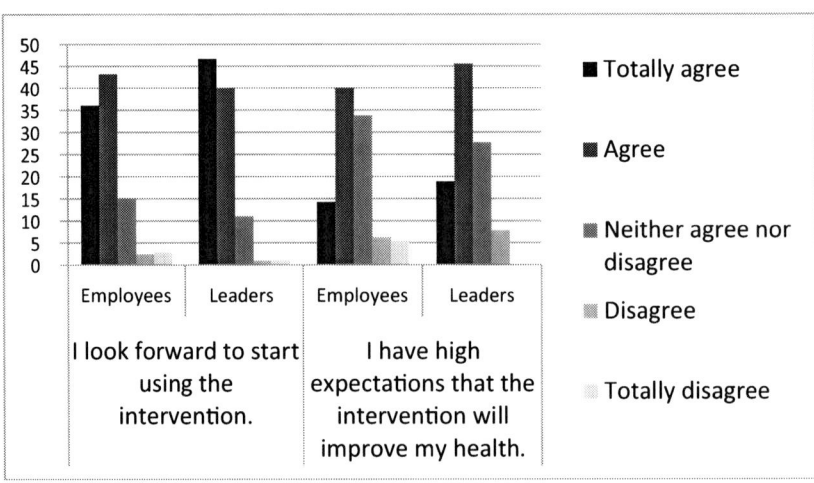

Figure 17.2 Employees' and leaders' ratings of expectations concerning the tool at baseline

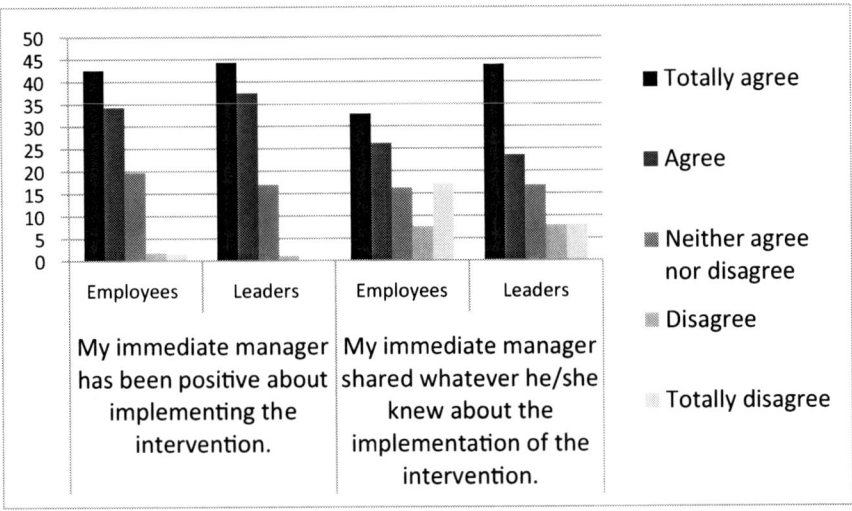

Figure 17.3 Employees' and leaders' ratings of their leader's attitude and information sharing at baseline

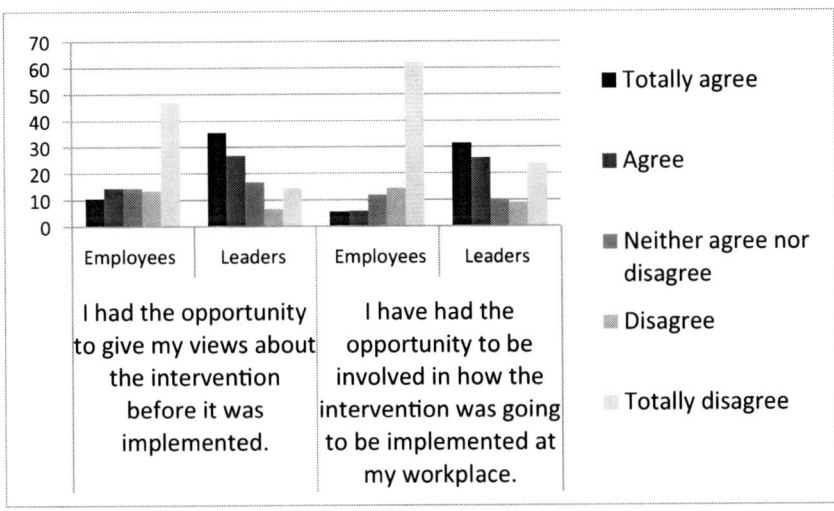

Figure 17.4 Employees' and leaders' ratings of opportunities for participation in the implementation at baseline

Another aspect important for commitment is opportunities for participation in programme planning and implementation (Nytrø et al. 2000, Saksvik et al. 2002). The interviews revealed that employees wished that they would have had more to say about the decision and implementation of the programme at their workplaces. The baseline questionnaire ratings (Figure 17.4) also shows that a quite low proportion of employees perceived that they had had the opportunity to give their views about the intervention before it was implemented as well as to be involved in how the intervention was implemented at their workplace. A larger proportion of leaders perceived opportunities for involvement in implementation as compared to the employees at the baseline questionnaire. According to several authors, a participatory approach is a desirable strategy for implementing occupational health programmes (Nielsen et al. 2010). Lines (2004) found that the amount of participation was negatively related to resistance of change, and positively related to achievement of goals and organizational commitment. Similarly, Nielsen et al. (2007) found that high levels of participation in change were associated with low levels of behavioural stress symptoms and higher job satisfaction post-intervention. Nielsen et al. (2010) emphasizes the importance of establishing a steering group composed of both employers and employees as a first step to ensure participation. It is possible that this type of participatory activities would have positive effects on employees' commitment to the present intervention in later stages of the programme implementation. Currently, the implementation was to a high extent depending on the leaders' commitment, which makes the use of the tool vulnerable to, for instance, change of leaders.

Use of the tool

By the end of April 2012, a total of 1,149 individuals in the participating organizations were invited to use the tool and of these 906 individuals had created a personal account (corresponding to 79 per cent). The lowest participation rate in an organization was 64 per cent and the highest 91 per cent. When studying the automatic measures of participants' use of the programme, we found that participants had logged in on average four times per month. The mean amount of logins per month was constant during the first 14 months of the project. The participants had been logged-in on average 43 minutes per month. There were some variations between the participating companies and departments. The mean login time per organization ranged between 33 and 52 minutes. Previous studies have pointed out that participants' limited exposure to Internet interventions has been the major concern when delivering web-based interventions (Evers et al. 2003, Buller et al. 2005). Some studies indicate that participants tend to leave the website before completing the intervention (Brouwer et al. 2008, Danaher et al. 2006). Furthermore, few participants visited an intervention more than once (Andersson et al. 2002, Ström et al. 2000, Pressler et al. 2010). In the current study, high percentages (almost 80 per cent) of the invited individuals entered the web-based tool. The compliance rate was also quite stable over time since the amount of logins did not change much over time. On the other hand, it is uncertain whether four logins per month is frequent enough to improve health and well-being. This will be investigated in forthcoming articles where health results over time can be studied in relation to participants' use of the tool.

The employees and leaders described that the tool they used most often was the brief survey, which also could be seen from the automatic measures. Some interview respondents described that they used the brief survey as often as daily, but more often it was used at least once a week. Most commonly, the decision about how often they responded was made collectively in the work group in order to receive work group results continuously.

The preliminary findings of the interviews showed that, in general, the employees perceived the content of the programme as positive. They felt that the focus on solutions and positive aspects of work was appealing. Many employees initially perceived the project as a signal that the organization cared about them, which was appreciated. Employees explained that it was crucial for their use of the tool that it was time efficient and simple to use. These were the main factors contributing to the initial use of the web-based programme. Here are two citations that describe the employees' perceptions of the tool: "to me it is very simple to just receive an email, then I'm continuously reminded about it" and "I thought it was good that the tool was brought up. It was an indication that the employer cares about the employees and not only the results."

The interview results indicate that employees thought it was highly important for them that their leader took the results from the tool into consideration, gave feedback on the results and acted upon it. This is in line with prior research showing that provision of feedback on work progress can lead to increased intervention activity among participants (Eklöf and Hagberg 2006). Small groups, which due to ethical considerations were not allowed feedback on group level results, expressed frustration over this. This highlights the difficulty of balancing between the need to protect individuals from being identified (which is possible if groups are too small) and to provide group feedback.

Participants' perception of the benefits of the tool is one crucial determinant for their participation and use of a health programme (Saksvik et al. 2002). Employees in the current project had some different opinions concerning who would benefit most from using the tool. Some employees described that they believed the tool would have positive effects on them as individuals, but had difficulties seeing how the tool would contribute to well-being of the work group or the organization. Others expressed the opposite view: they did not feel a need for the tool at the individual level personally. However, they used the brief survey so that the workplace leader would have results at group level and so that group interventions could be implemented based on those results. These individuals expressed that they had good health and well-being and didn't need any tools for additional improvements. This is interesting since most of the prior studies have indicated that health promotion tools in general target best individuals that are already healthy and have healthy lifestyles (Saksvik et al. 2002). It will be interesting to follow the participants during the entire project time and study how their perceptions of the programme benefits for themselves versus the organization develop over a longer time period and how that affects their use of the programme.

Workplace leaders were also positive about the project and the tool. They had a strong confidence in the tool and they saw it as an innovation within the work environment area. The basic ideas and the work process of the programme were in line with how they wanted to work with work environment issues. They described how important it was to move from a reactive to a more proactive type of work environment. Leaders also expressed a need for a mobile version of the tool so that it would be easier for them to use it outside the office. They emphasized the importance of following trends of employee well-being rather than measuring at one time point and having quick results and feedback on the measurements. They also felt that the tool provided a hands-on support in issues relating to their work as leaders rather than being an additional work task. More concretely, the programme was helpful in raising issues and being a support in discussions both in groups and with individual employees. The results were also used when discussing with the organization managers:

after the first survey I tried to tell the management that this is how it is for my unit. The workload has been very high, several employees feel bad and have worked too much and we won't be able to deliver all that has been planned for us. The tool can help me embed why I have to prioritize certain tasks, for instance because people don't feel so good. That way I definitely think that this tool can reduce the stress, by using it actively.

Leaders varied greatly in their reported levels of knowledge and confidence on how to deal with work environment issues in general and with the tool more specifically. This contributed to different attitudes towards their own role in the process of working with the tool. Some of the leaders expressed that the tool facilitates their work as work group leader. These leaders often checked the group's results on the brief survey, gave feedback based on the results to the group and discussed the results continuously with the employees. They experienced that the brief survey gave concrete information about the group's well-being. Without the brief survey, a leader just needs to "guess" how the group is feeling. This group of leaders realized how central a role they had in the implementation of this type of tool. One leader summarized his opinions in the following way:

to not just say that we have a tool and it's now up to you to use it, but using it as a tool to follow up. Because it's like all other work, if you don't follow up on it there's a risk it might fizzle out. So it's absolutely critical that we as managers use it.

Other leaders did not have the interest and competence to make optimal use of the tool. These leaders still showed interest in the tool and used it to improve their individual health, but not as group leaders in order to improve the group's well-being. These leaders often felt that the responsibility lay higher up in the organization. They also felt guilty as they realized that they were not using the tool as they should be. These leaders requested more support from organizational leaders and project management concerning how to use the tool as a leader. This shows the different stages the leaders were at when it comes to using a tool like this. Not all leaders felt that they had reasonable interest, knowledge or organizational support to use the tool to monitor and develop his/her work group's work environment and employees' health. Prior studies show that it is usually middle managers that are subsequently responsible for communicating and implementing change (Guth and Macmillan 1986). Therefore, middle managers play a crucial role in many organizational-level occupational health interventions (Randall et al. 2007). Dahl-Jørgensen and Saksvik (2005) reported that middle managers resisted change by restricting the time spent on interventions by employees.

Interview results: Experienced benefits

The interviews showed that employees and leaders perceived that the use of the tool had increased their monitoring and awareness of stress, health, well-being and psychosocial work environment. Some of the respondents also used the exercises to adopt a more active health promotion strategy. The respondents also perceived that the tool contributed to increased knowledge about stress and health. They also reported that they had reflected more on their own health and stress. The respondents also reported that they had a more relaxed attitude towards short-term stress and ill health since they understood from the tool that this did not have to be harmful if combined with sufficient recovery. They reported that use of the tool contributed to some health behaviour changes such as better skills to prioritize work tasks with consideration to own health as well as increased contact with occupational healthcare providers. The respondents also considered that the six-month period was too short to judge whether the tool had had some effects on the outcomes, i.e. their health and well-being. One employee described this as follows:

> I thought it was good for your own sake to reflect a bit, because often, you never think about how you really feel, because life goes by pretty quickly and it's good to stop and think about it every now and then. And then I think it's incredibly good that the leaders get to see how the employees are doing.

Possible effects on the outcome measures, e.g. employees' health, well-being, productivity, psychosocial work environment, ill health, presenteeism and sick leave, are assessed with validated questionnaires at baseline and follow-ups (six and 12 months). These evaluations are currently underway and not presented in this chapter.

Conclusion

The chapter presented an eHealth tool that could facilitate leaders' work for improving groups' and organizations' occupational health. The tool consists of surveys to continuously measure employees' well-being, but also interventions to act upon the results of the surveys. The goal is to offer concrete interventions that leaders at different levels in organizations can use to improve their leadership in relation to employees' psychosocial work environment and occupational health. The initial results showed that the workplace leaders and employees had positive perceptions of the web-based tool. It suited their mindset of modern occupational health work with proactive and time efficient focus. However, the employees differed concerning their opinions on whether the tool would mainly have impact on individuals' or on groups' well-being. Some employees described that they believed the tool would have positive effects for them as individuals, but had difficulties in perceiving how the tool would contribute to the well-being of the work group or the organization. Others expressed the opposite view: they did not feel a need for the tool at the individual level. However, they used the tool to contribute the work group survey results so that the workplace leader would have results at the group level and so that group interventions could be implemented based on those results. Employees' perceptions of the tool's impact on individual rather than organizational well-being might be highly dependent on how the work group leaders are using the tool. From employees' point of view, one of the main factors contributing to employees' active use of the tool was their immediate leader's role in giving feedback on group survey results on a regular basis. Workplace leaders reported different levels of knowledge and interest on occupational health issues, which in turn contributed to different leader actions and roles in relation to the tool. Some leaders took an active role and continuous dialogue with employees concerning the tool and the results obtained. Others reported lack of knowledge on how to work with the tool and requested more support. This further reflects the role and activities of leaders at higher organizational level. In this study, they were reported to have a quite passive role with few signals to middle leaders and employees concerning the use of the tool as organizational intervention.

Employees and leaders perceived that the use of the tool had contributed to some of the immediate individual level behavioural impacts that were presented in the programme theory (Table 17.1). They reported more frequent monitoring of their stress, health, well-being and psychosocial work environment. Some of the respondents had also switched to a more active health promotion strategy by using the interventions provided in the tool. The respondents also perceived increased knowledge and awareness about stress and health.

In sum, the initial results of the evaluation of this web-based tool showed that it is promising in meeting many of the expectations on what web-based tools can deliver, e.g. being proactive, interactive, time-lined and time-efficient. However, it is important that the characteristics of the intervention are not confused with issues relating to implementation. A web-based tool, like any other, needs to be implemented in order to have an effect. A reflection from this study is that the implementation of a web-based programme may be as challenging as occupational health interventions that are delivered in more traditional ways. Particularly, getting upper management involved in occupational health issues is central. In order for this to happen, occupational health issues need to be a more strategic question, e.g., their role in relation to the aim of the organization needs to be clarified. Otherwise, there is a risk that the use of the tools will stand the risk of being down-prioritized in favour of other issues.

References

Andersson, G., Stromgren, T., Strom, L. and Lyttkens, L. 2002. Randomized controlled trial of internet-based cognitive behavior therapy for distress associated with tinnitus. *Psychosomatic Medicine*, 64(5), 810–16.

Bailis, D.S., Segall, A. and Chipperfield, J.G. 2003. Two views of self-rated general health status. *Soc Sci Med*, 56(2), 203–17.

Block, G., Block, T., Wakimoto, P. and Block, C. 2004. Demonstration of an e-mailed worksite nutrition intervention program. *Preventing Chronic Disease*, 1:A06.

Brisson, C., Cantin, V., Larocque, B., Vézina, M., Vinet, A., Trudel, L. and Bourbonnais, R. 2006. Intervention research on work organization and health: Research design and preliminary results on mental health. *Canadian Journal of Community Mental Health (Revue canadienne de santé mentale communautaire)*, 25(2), 241–59.

Brouwer, W., Oenema, A., Crutzen, R., De Nooijer, J., De Vries, N.K. and Brug, J. 2008. An exploration of factors related to dissemination of and exposure to internet-delivered behavior change interventions aimed at adults: A Delphi study approach. *Journal of Medical Internet Research*, 10(2), e10.

Buller, D.B., Buller, M.K. and Kane, I. 2005. Web-based strategies to disseminate a sun safety curriculum to public elementary schools and state-licensed child-care facilities. *Health Psychology*, 24(5), 470.

Concise Medical Dictionary 2010. Oxford Reference Online. Vol. 2012. Oxford: Oxford University Press. Available at: http://www.oxfordreference.com/views/ENTRY.html?subview=Main&entry=t60. e13601.

Cook, C., Heath, F. and Thompson, R.L. 2000. A meta-analysis of response rates in web-or internet-based surveys. *Educational and Psychological Measurement*, 60(6), 821–36.

Cook, R.F., Billings, D.W., Hersch, R.K., Back, A.S. and Hendrickson, A. 2007. A field test of a web-based workplace health promotion program to improve dietary practices, reduce stress, and increase physical activity: Randomized controlled trial. *Journal of Medical Internet Research*, 9(2), e17.

Dahl-Jørgensen, C. and Saksvik, P.O. 2005. The impact of two organizational interventions on the health of service sector workers. *International Journal of Health Services*, 35(3), 529–49.

Danaher, B.G., Boles, S.M., Akers, L., Gordon, J.S. and Severson, H.H. 2006. Defining participant exposure measures in web-based health behavior change programs. *Journal of Medical Internet Research*, 8(3), e15.

Davis, R.N. 1999. Web-based administration of a personality questionnaire: Comparison with traditional methods. *Behavior Research Methods*, 31(4), 572–7.

Eklöf, M. and Hagberg, M. 2006. Are simple feedback interventions involving workplace data associated with better working environment and health? A cluster randomized controlled study among Swedish VDU workers. *Applied Ergonomics*, 37(2), 201–10.

Evers, K.E., Prochaska, J.M., Prochaska, J.O., Driskell, M.M., Cummins, C.O. and Velicer, W.F. 2003. Strengths and weaknesses of health behavior change programs on the Internet. *Journal of Health Psychology*, 8(1), 63–70.

Farmer, M.M. and Ferraro, K.F. 1997. Distress and perceived health: Mechanisms of health decline. *Journal of Health Social Behavior*, 38(3), 298–311.

Goldman, N., Glei, D.A. and Chang, M.C. 2004. The role of clinical risk factors in understanding self-rated health. *Ann Epidemiol*, 14(1), 49–57.

Gorin, A.A. and Stone, A.A. 2001. Recall biases and cognitive errors in retrospective self-reports: A call for momentary assessments. In *Handbook of Health Psychology*, edited by A. Baum. London and Mahwah, NJ: Lawrence Erlbaum Associates, 405–13.

Griffiths, F., Lindenmeyer, A., Powell, J., Lowe, P. and Thorogood, M. 2006. Why are health care interventions delivered over the internet? A systematic review of the published literature. *Journal of Medical Internet Research*, 8(2), e10.

Guth, W.D. and Macmillan, I.C. 1986. Strategy implementation versus middle manager self-interest. *Strategic Management Journal*, 7(4), 313–27.

Hager, R., Hardy, A. and Aldana, S. 2002. Evaluation of an internet, stage-based physical activity intervention. *Health Education Journal*, 33, 329–35.

Hasson, D. 2005. *Stress Management Interventions and Predictors of Long-term Health: Prospectively Controlled Studies on Long-term Pain Patients and a Healthy Sample from IT and Media Companies*. Acta Universitatis Upsaliensis : Univ.-bibl., Uppsala.

Hasson, D., Anderberg, U.M., Theorell, T. and Arnetz, B.B. 2005. Psychophysiological effects of a web-based stress management system: A prospective, randomized controlled intervention study of IT and media workers. *BMC Public Health*, 5(1), 78.

Hasson, D. and Arnetz, B.B. 2005. Validation and findings comparing VAS vs. Likert scales for psychosocial measurements. *The International Electronic Journal of Health Education*, 8, 178–92.

Holte, K.A., Vasseljen, O. and Westgaard, R.H. 2003. Exploring perceived tension as a response to psychosocial work stress. *Scandinavian Journal of Work, Environment & Health*, 29(2), 124–33.

Idler, E.L. and Benyamini, Y. 1997. Self-rated health and mortality: a review of twenty-seven community studies. *Journal of Health and Social Behavior*, 38(1), 21–37.

Jimmieson, N.L., Terry, D.J. and Callan, V.J. 2004. A longitudinal study of employee adaptation to organizational change: The role of change-related information and change-related self-efficacy. *Journal of Occupational Health Psychology*, 9(1), 11.

Kawakami, N., Takao, S., Kobayashi, Y. and Tsutsumi, A. 2006. Effects of web-based supervisor training on job stressors and psychological distress among workers: A workplace-based randomized controlled trial. *Journal of Uccupational Health*, 48(1), 28–34.

Kypri, K., Saunders, J.B., Williams, S.M., McGee, R.O., Langley, J.D., Cashell-Smith, M.L. and Gallagher, S.J. 2004. Web-based screening and brief intervention for hazardous drinking: A double-blind randomized controlled trial. *Addiction*, 99(11), 1410–17.

Lamontagne, A.D., Keegel, T., Louie, A.M., Ostry, A. and Landsbergis, P.A. 2007. A systematic review of the job-stress intervention evaluation literature, 1990–2005. *International Journal of Occupational and Environmental Health*, 13(3), 268–80.

Lines, R. 2004. Influence of participation in strategic change: Resistance, organizational commitment and change goal achievement. *Journal of Change Management*, 4(3), 193–215.

Lustria, M.L., Cortese, J., Noar, S.M. and Glueckauf, R.L. 2009. Computer-tailored health interventions delivered over the Web: Review and analysis of key components. *Patient Education and Counseling*, 74(2), 156–73.

Mangunkusumo, R.T., Brug, J., Duisterhout, J.S., de Koning, H.J. and Raat, H. 2007. Feasibility, acceptability, and quality of internet-administered adolescent health promotion in a preventive-care setting. *Health Education Research*, 22(1), 1.

Marshall, A., Leslie, E., Bauman, A., Marcus, B. and Owen, N. 2003. Print versus website physical activity programs: A randomized trial. *American Journal of Preventative Medicine*, 25, 88–94.

Matano, R., Koopman, C., Wanat, S., Winzelberg, A., Whitsell, S., Westrup, D., Futa, K., Clayton, J., Mussman, L. and Taylor, C. 2007. A pilot study of an interactive web site in the workplace for reducing alcohol consumption. *Journal of Substance Abuse Treatment*, 32, 71–80.

Nielsen, K., Randall, R. and Albertsen, K. 2007. Participants' appraisals of process issues and the effects of stress management interventions. *Journal of Organizational Behavior*, 28(6), 793–810.

Nielsen, K. and Randall, R. 2013. Opening the black box: A framework for evaluating organizational-level occupational health interventions. *European Journal of Work and Organizational Psychology*, 27(3).

Nielsen, K., Randall, R., Holten, A. and Rial Gonzale, E. 2010. Conducting organizational-level occupational health interventions: What works? *Work & Stress*, 24(3), 234–59.

Nytrø, K., Saksvik, P., Mikkelsen, A., Bohle, P. and Quinlan, M. 2000. An appraisal of key factors in the implementation of occupational stress interventions. *Work & Stress*, 14(3), 213–25.

Oenema, A., Brug, J. and Lechner, L. 2001. Web-based tailored nutrition education: Results of a randomized controlled trial. *Health Education Research*, 16(6), 647–60.

Pressler, A., Knebel, U., Esch, S., Kölbl, D., Esefeld, K., Scherr, J., Haller, B. and Schmidt-Trucksäss, A. 2010. An internet-delivered exercise intervention for workplace health promotion in overweight sedentary employees: A randomized trial. *Preventive Medicine*, 51(3–4), 234–39.

Randall, R., Cox, T. and Griffiths, A. 2007. Participants' accounts of a stress management intervention. *Human Relations*, 60(8), 1181.

Randall, R., Nielsen, K. and Tvedt, S.D. 2009. The development of five scales to measure employees' appraisals of organizational-level stress management interventions. *Work & Stress*, 23(1), 1–23.

Rhodes, S., Bowie, D. and Hergenrather, K. 2003. Collecting behavioural data using the world wide web: Considerations for researchers. *Journal of Epidemiology and Community Health*, 57(1), 68–73.

Saksvik, P., Nytrø, K., Dahl-Jørgensen, C. and Mikkelsen, A. 2002. A process evaluation of individual and organizational occupational stress and health interventions. *Work & Stress*, 16(1), 37–57.

Spittaels, H., De Bourdeaudhuij, I. and Vandelanotte, C. 2007. Evaluation of a website-delivered computer-tailored intervention for increasing physical activity in the general population. *Preventive Medicine*, 44(3), 209–17.

Strecher, V.J., McClure, J.B., Alexander, G.L., Chakraborty, B., Nair, V.N., Konkel, J.M., Greene, S.M., Collins, L.M., Carlier, C.C., Wiese, C.J., Little, R.J., Pomerleau, C.S. and Pomerleau, O.F. 2008. Web-based smoking-cessation programs: Results of a randomized trial. *American Journal of Preventive Medicine*, 34(5), 373–81.

Ström, L., Pettersson, R. and Andersson, G. 2000. A controlled trial of self-help treatment of recurrent headache conducted via the internet. *Journal of Consulting and Clinical Psychology*, 68, 722–7.

van Gelder, M.M.H.J., Bretveld, R.W. and Roeleveld, N. 2010. Web-based questionnaires: The future in epidemiology? *American Journal of Epidemiology*, 172(11), 1292–8.

Wantland, D.J., Portillo, C.J., Holzemer, W.L., Slaughter, R. and McGhee, E.M. 2004. The effectiveness of web-based vs. non-web-based interventions: A meta-analysis of behavioral change outcomes. *Journal of Medical Internet Research*, 6(4), e40.

Weick, K.E., Sutcliffe, K.M. and Obstfeld, D. 2005. Organizing and the process of sensemaking. *Organization Science*, 16, 409–21.

Wilder, D.A., Austin, J. and Casella, S. 2009. Applying behavior analysis in organizations: Organizational behavior management. *Psychological Services*, 6(3), 202.

Conclusion: Positive vs. Stress Interventions – Does it Really Matter?

CAROLINE BIRON

This volume is an attempt to highlight that healthy workplaces can be developed using different paths. Contributors draw on the literatures from both positive psychology organizational scholarship (POS), as well as on the more "traditional" occupational stress prevention. POS can be defined as Positive "the study of that which is positive, flourishing, and life-giving in organizations" (Cameron and Caza 2004: 731). It puts emphasis on generative dynamics that make organizations, organizational units, and organizational members flourish and thrive. Interventions presented include both individual-level, that is interventions attempting to improve individuals' capacity to cope with stress, and organizational-level, or interventions aiming to reduce exposure to the sources of stress in the work environment. Interventions have various targets, from leadership training, to improving physical health, work–life balance, or developing participative work climate and workers' engagement. In the last section of the book, contributors highlight the importance of paying attention to the implementation process and the factors that can hinder and facilitate the intervention. Overall, in this volume, we wish delve into how interventions emerging from different approaches, with different scopes and targets, can contribute to creating healthier workplaces.

As concluding remarks, we would like to insist on one particular point: the differences and similarities between the more positive orientated approaches (i.e. aiming to promote human excellence, flourishing people, well-being, satisfaction) and the intervention emerging from a risk management approach (i.e. aiming to reduce exposure to psychosocial job constraints such as high job demands, lack of job control or social support, in order to prevent stress). The literature on work-related stress has been dominated by studies attempting to demonstrate a causal relationship between exposure to various types of sources of stress, and various consequences of this exposure on employee health and organizational performance. Many debates still exist regarding the very definition of stress, what causes it, what are the consequences, and whether we should even use the term "stress." Indeed, some researchers argue that the term is ambiguous (Briner and Reynolds 1999) and practitioners and managers often find it threatening or negative. The same confusion exists in the positive psychology field, where the term well-being is used to describe life satisfaction, wealth, happiness and various other terms. In both fields, researchers attempt to clarify and to present models that will allow measuring, and defining the constructs so that they can be used to document interventions effectiveness (Jayawickreme et al. 2012). Nevertheless, many agree that psychosocial risks such as the lack of job control, of social support, of recognition for one's efforts and work overload are, in the long term, strong predictors of poor physical and psychological health (Semmer 2006) and other effects of strain on workers (psychological distress, exhaustion, psychosomatic complaints, health problems) and on organizations (increased absenteeism, presenteeism, productivity losses). From a theoretical perspective, aiming to reduce exposure to psychosocial risks or any attempt to modify aspects of work that are creating stress should have positive effects on workers' health and well-being. These changes should, at least theoretically, be associated with decreased strain reactions, which in turn should be linked with decreased adverse consequences on individuals and organizations. There is a growing body of research demonstrating the relevance of work- and organizational-level interventions to reduce strain and ill-health by reducing exposure to psychosocial constraints (see for example Bambra et al. 2007, Biron, Ivers, Brun and Cooper 2011, Bourbonnais et al. 2006a, Bourbonnais et al. 2006b, Egan et al. 2007, Gilbert-Ouimet et al. 2012, Nielsen, Randall and Alberten 2007). However, many deplore the scarcity of solid theoretically sound intervention research, the lack of methodological rigour, the overemphasis on individual-level interventions instead of on organizational-level ones, as well as the lack of attention paid to how and why interventions produce or fail to produce certain effects (Biron et al. 2012b, Cox et al. 2007, 2010).

In addition to these criticism, there has been a call for a less disease-oriented approach to psychology and to organizational behaviour (Gable and Haidt 2005, Seligman and Csikszentmihalyi 2000). This is particularly evident in the establishment of several new scholarly movements within the past 10 years that emphasize the study of more positively orientated phenomena. In 1999, at an APA inaugural address, Martin Seligman argued that psychology had been focused strictly on understanding what makes people ill, as opposed to what makes people flourish. Indeed, human excellence and flourishing has been overlooked in the stress literature and in psychology in general. This call for more attention to be paid to positive characteristics of human beings has been answered with much enthusiasm by both researchers and practitioners. Just to illustrate this point, a research on PsycInfo using the word "happiness" for publications in the decade before Seligman's call for more attention to positive psychology got 567 hits. The same search got 2,303 hits for the last decade (2002–2012), representing an increase of 75 per cent. As a contrast, the key words "occupational stress" got 3,993 hits (between 1989 and 1999) vs. 6,452 hits (between 2002 and 2012), representing an increase of 38 per cent. The *Journal of Happiness Studies*, which was founded in 2000 and is at the moment edited by Antonella Delle Fave, has an impact factor of 1.875 (Journal Citation Report 2011).

Despite this sharp increase in the interest for positive, flourishing human excellence, there has been some criticism (mainly among academics) of the positive movement, especially in terms of what it implies for organizations. For example, some have argued that it equates to putting old wine in new bottles, that it forces to workers to have structured fun (Warren 2003), that it runs the risk of stigmatizing workers who are struggling with negative emotions (Fineman 2006), and that interventions are mainly focused in individual issues instead of organizational issues (Hackman, 2009) . For example, a recent systematic study on positive interventions which included over 1,500 participants in 15 studies, interventions related to loving-kindness meditation, enhancing resilience, psychological capital, gratitude, solution-focused coaching (Meyers, van Woerkom, and Bakker 2012). Although the results of the review showed that these interventions were generally effective to improve a range of outcomes such as well-being and performance, the focus of these interventions is on changing individuals instead of changing pathogenic characteristics of the workplace.

We do not aim to heat up the debate on the points and counterpoints of positive organizational behaviour, but for the purpose of this book, we argue that both approaches can be integrated. We posit this book's focus on interventions, whether they emerge out of one paradigm or the other. We also posit, in line with Biron et al. (2012a), that in terms of *actual changes* made in the workplaces or changes made to help workers cope with their workplaces, both the positive approach and the stress reduction one aim towards the same goal: *creating healthier workplaces*. The actual content of the changes made are rather similar, but the path taken to attain the goal differs. For example, appreciative inquiry (AI) aims to improve the workplace and to set positive goals for teams, as Lewis describes in Chapter 12. This approach uses different means, but its ultimate goal is not very different than what Jauvin et al. describe in Chapter 14. Jauvin et al. indicate that their interventions followed structured steps, namely:

1. identification of psychosocial constraints;
2. identification of solutions to act on these constraints;
3. prioritizing solutions;
4. dissemination of information;
5. feedback loop.

This problem-solving approach is very similar to the one described by Biron et al. in Chapter 15, or any problem-solving approach. Appreciative inquiry is defined by as being different to "deficiencies-based" models or the problem-solving approach in that is it defined as an asset-based approach. Instead of identifying problems, AI aims to identify what is appreciated by workers, what is working well, and to envision how things might be in the future. Although AI is said to be fundamentally different to problem-solving approaches, we argue here that the actual changes made in the workplaces are likely to be very similar, and that some elements of the process are even likely to be the same. For example, some of the facilitators and barriers are likely to be similar. To illustrate this point, let's use one of the barriers identified by Jauvin et al. in this volume and by Biron et al. regarding the importance of management support to implement interventions. Whether a positive approach or a problem-solving approach is retained, both are going to require management support, such as financial support to implement the proposed changes, and to free up employees for the time needed to plan, develop and implement the interventions. In chapters 9 and 10, Osatuke et al. and Leiter describe an intervention based on workers' engagement and respect in the workplace. Leiter highlights that this organizational intervention was founded on positive psychology, but it followed the same 10 criteria for effective intervention programmes as the ones identified by Kompier et al. (2000), which were developed for stress interventions (e.g. using a stepwise, systematic approach, having clear tasks and responsibilities, using a participative approach, responsible management, and monitoring the intervention). Thus, we argue that in terms of intervention process, although positive interventions and stress interventions can differ, there are many similarities and the two paradigms are not mutually exclusive. There are, however, pros and cons to using either approach. For example, stress can be seen as a threatening concept by managers, who might fear they are opening a Pandora's box. In that sense, a more positively oriented approach might increase managers' and employees' readiness to change. Yet, it is well known that change cannot be forced on someone who does not want to change: there must be a felt need for change. In that sense, a risk assessment of psychosocial constraints and psychological health (a stress approach) might raise the alarm and trigger the desire to improve some work-related aspects (for a more thorough discussion on the pros and cons of stress and positive approaches, see Biron et al. 2012a).

In sum, in this volume, we aimed to bring together researchers and practitioners who are involved in interventions, either at the individual level or at the organizational level, and who use either the positive approach or the psychosocial risks and stress reduction approach. We conclude by arguing for a more integrated approach encompassing a comprehensive approach to both preventing stress and promoting health and well-being. As Biron et al. (2012a) argue, conceptually, both approaches ultimately aim to bring about changes in the workplace. Although each approach has its own set of paradigmatic assumptions, the concrete changes in the workplaces are not conceptually distinct when arising from the positive (promoting health) and the negative (preventing stress). Like the popular expression, we posit that "All roads lead to Rome" or, in other words, regardless of which approach is chosen, what really matters is very much like what Semmer suggests, and we feel this is rather well put and would like to conclude with this:

> *Let me follow up on the remarks made by several authors about the potential danger of talking about stress interventions (or similar terminology). This is not simply a problem of terminology, but also a conceptual issue. It highlights the fact that interventions are unlikely to succeed unless they become an integral part of everyday operations, rather than a "health project" ... The health focus should not detract from acknowledging that the issues we are dealing with overlap to a considerable extent with issues discussed long ago in terms of fostering internal motivation and personal development ... in terms of aligning technical and social systems to each other ... or, more generally, in terms of quality of working life ... What it boils down to in the end is, in my view, something like this: We are talking about work that people like, that motivates them, that gives them a sense of meaning and fulfillment in, participation in social life, a sense of accomplishment, etc. It may not mainly be about health. It may "simply" be about: good work. (Semmer 2012: xii)*

References

Bambra, C., Egan, M., Thomas, S., Petticrew, M., and Whitehead, M. (2007). The psychosocial and health effects of workplace restructuring interventions reorganisation: A systematic review of task. *Journal of Epidemiology and Community Health*, 61, 1028–37.

Biron, C., Cooper, C.L. and Gibbs, P. (2012a). Stress interventions vs positive interventions: Apples and oranges? In *Oxford Handbook of Positive Organizational Scholarship*, edited by K.S. Cameron and G.M. Spreitzer. New York, NY: Oxford University Press, 938–52.

Biron, C., Ivers, H., Brun, J-P. and Cooper, C.L. (2011). *The More the Merrier? A Dose-response Analysis of an Observational Study of Organizational-level Stress Interventions*. Paper presented at the Work, Stress, & Health Conference, Orlando, Florida.

Biron, C., Karanika-Murray, M. and Cooper, C.L. (2012b). Organizational stress and well-being interventions: An overview. In *Improving Organizational Interventions for Stress and Well-being Interventions: Addressing Process and Context*, edited by C. Biron, M. Karanika-Murray and C.L. Cooper. London: Routledge, 1–17.

Bourbonnais, R., Brisson, C., Vinet, A., Vézina, M., Abdous, B., and Gaudet, B. (2006a). Effectiveness of a participative intervention on psychosocial work factors to prevent mental health problems in a hospital setting. *Journal of Occupational & Environmental Medicine*, 63, 335–42.

Bourbonnais, R., Brisson, C., Vinet, A., Vezina, M., and Lower, A. (2006). Development and implementation of a participative intervention to improve the psychosocial work environment and mental health in an acute care hospital. *Occupational and Environmental Medicine*, 63, 326–34.

Gilbert-Ouimet, M., Brisson, C., Vézina, M., Trudel, L., Bourbonnais, R., Masse, B., Baril-Gingras, G. and Dionne, C.E. (2011). Intervention study on psychosocial work factors and mental health and musculoskeletal outcomes. *HealthcarePapers*, 11(Sp), 47–66.

Briner, R.B. and Reynolds, S. 1999. The costs, benefits, and limitations of organizational level stress interventions. *Journal of Organizational Behavior*, 20, 647–64.

Cameron, K.S. and Caza, A. (2004). Introduction: Contributions to the discipline of positive organizational scholarship. *American Behavioral Scientist*, 47, 731–9.

Cox, T., Taris, T.W. and Nielsen, K. (2010). Organizational interventions: Issues and challenges. *Work & Stress*, 24(3), 217–18.

Cox, T., Karanika-Murray, M., Griffiths, A. and Houdmont, J. (2007). Evaluating organizational-level work stress interventions: Beyond traditional methods. *Work & Stress*, 21(4), 348–62.

Egan, M., Bambra, C., Thomas, S., Petticrew, M., Whitehead, M. and Thomson, H. (2007). The psychosocial and health effects of workplace reorganisation: A systematic review of organisational-level interventions that aim to increase employee control. *Journal of Epidemiology and Community Health*, (61), 945–54.

Fineman, S. (2006). On being positive: Concerns and counterpoints. *Academy of Management Review*, 31(2), 270–291.

Hackman, J.R. (2009). The perils of positivity. *Journal of Organizational Behavior*, 30, 309–19.

Gable, S.L. and Haidt, J. (2005). What (and why) is positive psychology? *Review of General Psychology*, 9(2), 103–10.

Jayawickreme, E., Forgeard, M.J.C. and Seligman, M.E.P. (2012). The engine of well-being. *Review of General Psychology*, 16(4), 327–42.

Journal Citation Reports® Social Sciences Edition. 2011. Thomson Reuters.

Kompier, Michiel A.J., Cooper, Cary L. and Geurts, Sabine A.E. (2000). A multiple case study approach to work stress prevention in Europe. *European Journal of Work & Organizational Psychology*, 9(3), 371.

Meyers, M.C., van Woerkom, M. and Bakker, A.B. (2012). The added value of the positive: A literature review of positive psychology interventions in organizations. *European Journal of Work and Organizational Psychology*, 22(5), 618–32.

Nielsen, K., Randall, R. and Albertsen, K. (2007). Participants, appraisals of process issues and the effects of stress management interventions. *Journal of Organizational Behavior*, 28, 793–810.

Seligman, M.E.P. and Csikszentmihalyi, M. (2000). Positive psychology: An introduction. *American Psychologist*, 55, 5–14.

Semmer, N.K. (2006). Job stress interventions and the organization of work. *Scandinavian Journal of Work and Environmental Health*, 32(6, special issue), 515–27.

Semmer, N.K. (2012). Foreword. In *Improving Organizational Interventions for Stress and Well-being: Addressing Process and Context*, edited by C. Biron, M. Karanika-Murray and C.L. Cooper. London: Routledge.

Warren, S. (2003). Humour as a management tool? The irony of structuring fun in organizations. In *The Ironic Eye*, edited by U. Johannson and J. Woodilla. Copenhagen: Copenhagen Business Press.

Index

Note: Figures indexed in italic with bold page numbers and Tables indexed in bold with bold page numbers.

20–item strain symptoms checklist   73
24–hour access   77
24–hour shifts, working   162

ABLE   67, 72, 76
   coaching programme   75–8, 81–5
   treatment group   78, 81–2
   workshops   72, 83
absences   1, 4, 9–10, 25, 30–31, 113, 115,
      119, 135, 207, 216, 252
   costs   113, 115–17, 120, 122, 135
   long-term   119
   short-term   119
abusive leaders   206, 208, 210, 217
Academy of Management   5
ACBS   60
acceptance and action questionnaire   57
acceptance commitment therapy, origins
    of   52–3
accidents   1, 4, 25, 30–31, 135, 283
   occupational   265
   workplace   135–6
account managers   94
accountants   78
achieved results   286–7 *see also* expected
   results
achieving balance in life and employment
    see ABLE
Achor, S.   226–7, 230
ACSM   110
ACT   10–14, 23–5, 39–43, 51–62, 91–104,
   109–17, 138–42, 169–76, 206–8,
   246–50, 265–8, 270–77, 287–95,
   307–8, 312–16
   for creating healthy workplaces   57–60

   groups   58
   interventions   52, 57–61
   models   54
   practitioners   53
   protocols   58
   theorists   54
   treatment groups   59
   workshops   58–9
   worksite interventions   60
action plans   11, 23, 61, 176, 248, 262–3,
   271–2, 275, 283, 288, 293
   evidence-based   277
   first   132
   models of   274–5
active management   208
active participation of employees   246, 250
Adams, G.A.   68–70
Adelson, B.M.   42
administrative assistants   78
Administrative Sciences Association of
    Canada   5
Adrianson, L.   195
advancement reports   275
AES   148–9, 153, 155, 160, 178, 194
   longitudinal data   149
   psychological safety measure quartiles
    160
AFA Insurance   288
aggregated reports   272
AI Practitioner   232
Åkesson, I.   113
Aldana, S.G.   101, 113
Alexy, B.B.   111
All Employee Survey *see* AES
AMA   110

American College of Cardiology 101
American College of Sports Medicine *see*
    ACSM
American Express 97
American Federation of Government
    Employees 155
American Heart Association 92
American Psychological Association *see* APA
ANACT 136
Anagnos, J. 8
anasakti, concept of 43
Anderson, N. 9, 92, 150, 152, 209
Andersson, G. 312
annual cost of absenteeism to small businesses
    (per employee) **136**
Antoniou, A. 1–2
APA 5, 184, 224, 322
appraisals, health risk 10, 92, 96, 101–2
appreciative inquiry *see* AI
appreciative leadership 223, 232–3
appropriation of processes 247
Argyris, C. 4, 176
Armour, S. 141
Arnetz, B. 306
Arnold, K.A. 185, 206, 210
'Art of Healthy Cooking' (educational
    seminar subject) 95
artefact production 38
Arterburn, D. 92
Asakura, T. 189
Ashby, G.F. 226
Ashforth, B.E. 210
assessment surveys 171
Association for Contextual Behavioral
    Science *see* ACBS
ataraxia, concept of 43
Atlanta area 91, 93, 95, 103
Augustsson, Hanna 281
Aviva Family Finances Report 139
Avolio, B.J. 206–7, 209, 213–14, 229, 233

Baer, R.A. 56
Baicker, K 92
Bailis, D.S. 306–7
Baker, S. 282
Bakker, A.B. 152, 170, 186, 196, 322
Balogun, J. 5

Bamberger, P. 122, 213
Bambra, C. 3, 52, 269, 282, 322
Bandura, A. 15
Barbier, R. 251
Baril-Gingras, G. 240, 249, 252
Barling, J. 2, 6, 8, 69, 185, 205–13, 215–18
Barr-Anderson, D.J. 111
BART 141
Bartone, P.T. 73
baseline 59, 97–8, 102–3, 115–16, 228,
    310–11, 315
    assessments 98, 171
    biometric assessments 97–8
    civility perceptions 149
    participants 103
    questionnaire 310, 312
    questionnaire ratings 311
    surveys 101, 310
Bass, B.M. 206–7, 209
Bassi, Marta 37
Baumeister, R.F. 38, 114
BBC 141
BC Biomedical Laboratories 166
Beehr, T.A. 70
Behavioral Risk Factor Surveillance Survey
    99–101
Bejerot, E. 113
Belgian service vouchers ('Titres-services')
    140
Belgium 130, 142
Bellavia, G.M. 68
Bellis, M. 223
Belton, L.W. 147, 149
Benedict, M.A. 92
benefits 141, 299, 313
    to employers 92
    exponential 164
    health 98, 119, 283
    long-term 226
    social 134, 142
    sport and culture (Finland) 142
Benjamin, Stacey A. 91
Benmehdi, Hassan 135
Benyamini, Y. 306
Benzer 153
Berchtold, N.C. 120
Beresford, Brent 51

Berger, M.L. 113, 115
Berne, E. 151
Bilsky, W. 38
Biron, Caroline 1, 3–4, 6, 9, 12–13, 240–41, 251, 253–4, 261–3, 265–7, 269, 271–2, 275, 321–4
Biswas-Diener, R. 228
Blais, M.R. 270
Block, G. 99, 300
blood pressure 1, 3, 7, 71, 85, 91, 101, 228, 303
    control of 100
    diastolic 98, 101
    digital meters 98
    high 68
    levels 92
    lowering of 97
    systolic 112
blue-collar employees 184
Blumenthal, J.A. 119–20
BMI 91, 99–100
body mass index see BMI
Bohlmeijer, E.T. 52
Bonanno, G.A. 55
Bond, F.W. 52, 55, 57–60, 71, 251
Bono, J.E. 209–10
Bookhart, Bettie 156
Borloo, Jean-Louis 139
Bormans, L. 231
Borritz, M. 266
Bosma, H.A. 40
Bourbonnais, Renée 239–40, 244, 250, 322
Boyd, R. 37, 39
Brännmark, M. 285, 294
Brazil 133, 135, 141
    and the influence of the Brazilian World Food Programme 135
    and the introduction of meal vouchers 133
    provides increasing commercial competition 6
    and the use of transport vouchers 130, 141
Brazilian World Food Programme see PAT
Brewerton, P. 229
brief description of the projects: common methods 242

Briner, R.B. 3, 281, 321
Brinkborg, H. 59
Brisson, C. 244, 305
British Academy of Management 5
British government 139
British researchers 239
Brooks, A. 115, 229
Brouwer, W.B. 115, 312
Brun, J. 239, 262, 264–8, 270, 276, 322
Buckingham, M. 229
Buddhism 43, 56
Buffett National Wellness Survey 7
Bulgaria 133–4, 137
    beneficiaries of 133
    parliament of 133
Buller, D.B. 312
Bültmann, U. 188
Bunce, D. 52, 57–8, 71, 251
Bureau de normalisation du Québec 265
Burke, Roland J. 1–4, 6, 262
Burns, J.M. 209
Burton, J. 184
Bycio, P. 209
Byosiere, P.B. 68

Calegari, P. 38
call centre employees 58–9
Cameron, Kim 4, 12, 224–5, 233, 321
Campbell, J.D. 70, 110
Canada 77, 166, 176, 211, 239–41, 244, 263, 265
    and CREW activities within hospitals 171
    and the display of data from health care units 177
    identifying the organizational factors behind mental health problems at work 240
    research conducted in 2010 7
    and Union Gas 5
cancers 30
Caplan, R.D. 224
Carameli, K. 152
Carayon, P. 69–70
cardiovascular risks 92–3, 97–8, 101, 103–4
    early 97
    elevated 95

reduction of 91
Carnethon, M. 92, 101
Cartwright, S. 6, 8, 12, 69–70, 75, 183–4, 187, 264
Carver, C.S. 71
Casablanca 135
case reports 113–16, 301
case studies 91, 104
  dealing with treatment for leadership 212
  enhancing development and performance through leadership 212–13
  investing in transformational leadership 211
  means, standard deviations and inter-correlations among study variables **80**
  means, standard deviations and inter-correlations among the study variables **79**
  Ministry of Quebec Public Service 267
  repeated measures means, standard deviations, and t-tests for outcomes **81**
  summary of the well-being means and standard errors for the ABLE treatment participants and the control group at Time 1 and Time 2 **82**
  workshop intervention programme 72
  workshop intervention results 74
  worksite wellness programme for reducing cardiovascular risks 91
Cash, Maureen 147
Caspersen, C.J. 110
Cassirer, N. 138
Caulfield, N. 3, 69
Caza, A. 321
CBI 61
CBOCs 148
CBT 8, 59, 302
CCs 24
CDPP 30
CDs 60
census surveys 148
Center for Disease Control and Prevention 92, 99, 101
Center for Epidemiologic Studies Depression 189

CEOs 94, 149, 163, 273, 276–7
CESU vouchers 129, 139–40
Chambaud, L. 249
Chamberlain, T. 69, 73
Chan, R.S.Y. 44, 111
Chan Osilla, K. 92
Chang, D. 119
'changing VHA culture' (CREW slogan) 150
character strengths 228–30
Cheque Emploi Services Universel *see* CESU
Cherns, A.B 4
childcare vouchers 129, 138–9
Chilean mine disasters 2
Chinese mine disasters 2, 6
choices 3, 26, 96, 130, 140, 143, 227, 232
  conscious 227
  healthier food 26, 96
  and the role of the French government 140
  unethical 3
  in the workplace 130
cholesterol 7, 96–7
  blood 112
  controlling of 96
  total 98, 100
  total serum 91
Chronic Care Model (Wagner) 99
civil behaviours 151–3
civil relationships 180
civility 15, 145, 147, 149–53, 155–61, 163–4, 169, 175–80
  improving 173, 178
  levels of 155, 163, 169, 174, 176
  measure of 153
  references to 177
  relationship of 178
  respect and engagement in the workplace *see* CREW
  team 177
civility scale 153
Clayton, Michael 156
Clifton, D. 229
co-workers 148, 151–3, 159–61, 175, 184–5, 193, 196, 210, 288
Coffman, C. 229
cognitive-behavioural intervention *see* CBI
cognitive behavioural therapy *see* CBT

cognitive defusion 54
Cohen, D. 110, 230
Colcombe, S. 119
Colditz, G. 109
Collaborating Centres for Occupational
Health see CCs
Colquitt, J.A. 209
**comments from participants 12
months after the start of the
intervention 274**
common sense model of health behaviour
and health outcomes for a worksite
wellness programme to reduce
cardiovascular risks **97**
Commonwealth Director of Public
Prosecutions see CDPP
communication plans 262, 266–7, 273,
276
community based outpatient centres see
CBOCs
Commuter Checks (USA) 130
companions 156–7, 172–3
CREW 154, 173
mentoring the facilitators 173
NCOD 154–5, 157–8
working with facilitators and
coordinators to interpret the surveys
176
competitive advantages 4, 6, 162, 196, 267
components of a successful worksite
programme and design and
implementation issues from the
pilot cardiovascular risk reduction
programme **102**
**conceptual model delineating how
physical exercise during work
hours can affect productivity
and sickness absence at the
employee and workplace levels
118**
conferences 5, 156, 172, 261
Academy of Management 5
Administrative Sciences Association of
Canada 5
British Academy of Management 5
European Academy of Occupational
Health Psychology 5, 261

European Academy of Work and
Organizational Psychology 5
'conflict interventions' 69–70
'Confucian perspective' 44
Congress of the United States 165
Congressional Budget Office 148
Conn, V.S. 92, 111–13
Conrad, P. 111
content of the web-based tool 302
control groups 9, 58–9, 71, 75–6, 78, 81–4,
98, 184, 211–13, 242, 301
Cook, J.V. 299–300
Cooper, C.L. 1–2, 6, 8, 30, 68–70, 75, 85,
183–6, 261–2, 264, 322
Cooperrider, D.L. 150, 231
coordinators 156, 162, 172–6
core processes 54, 56–9
corporate social responsibility see CSR
**correlations of engagement with
social constructs 179**
Cortina, L.M. 15, 176
Cortinovis, I. 42
Cotman, C.W. 120
Cottington, E.M. 184
counselling 8, 12, 212
employee assistance 26
feedback 212
for leadership training 212
private 102
psychological 8
sessions 212–13
voluntary 29
counsellors 59
Cox, T. 9–10, 32, 61–2, 261–2, 264–5, 322
Crawford, E.R. 152
creating healthy workplaces 4, 51–2, 57,
183, 186, 193, 196, 223
Cree, T. 205
CREW 147–66, 169, 171–8, 180
activities 159, 172, 175–6
approaches of 147, 150, 152, 172
behaviours 150, 163
'changing VHA culture' (CREW slogan)
150
Civility Scale **153**, 176, 178
civility scores 153, 155, 176–7
community 157, 163

companions 154, 156–7, 172–3
concepts 147, 151, 155
coordinators 152, 154–8, 162–4
data 166
definition of civility resting on intrinsic
    qualities 175
developers 149
employee participation 1, 14, 92, 95,
    101–2, 150, 154–5, 183–5, 196, 251,
    263–4, 274, 295
experience 156
experiments 164
facilitation 162, 172
facilitators 151, 156–7
groups 152, 155, 162, 178
impact of 155, 161, 171
implementation of 147, 150, 155–6,
    172–3, 177–8
initiatives 151–2, 154, 158, 173, 176
intervention approach 151
intervention outcomes 158
intervention processes 152
legitimizing facilitation 172
marketing of 155
meetings 150–51, 153, 158–9, 162–3,
    173–4
models 151
outcomes 160–61
participants 150, 155, 159, 164, 173
philosophy of 157, 163–4
post-intervention surveys 161
principles of 150
processes 150, 155, 158–9, 161–3,
    165–6, 169, 171–8, 180
profile of civility scores, before and after
    177
programmes 165, 174
qualities of 169, 178
sessions 171–4
sites 153
success of 157, 163
testimonials from employee participants
    (also displayed on CREW webpage,
    NCOD Internet site) **155**
Toolkit 171
uses a principle-driven rather than
    protocol-driven style 153

webpage 155
and work engagement 178
workplace interventions 150–52, 155,
    171, 178, 180
Cronbach's alpha for Time 1 and Time 2
    73, 188–9
cross-functional teams 195
CSA 265
Csikszentmihalyi, M. 4, 39–40, 42, 150, 322
CSR 130, 141–2
cultural differentiation 37
culture 4, 7, 24, 37–40, 101, 147, 153, 158,
    175, 187, 196, 215, 249, 263, 267
    flourishing 226
    local 154
    national 11
    positive 6
    workplace 25, 173
Cummins, R.A. 44
Cumulative Fatigue Symptoms Index 189
Curtis, J. 78
Cushway, D. 71
*cycle of poor nutrition and low national
    productivity* **134**
Czabala, C. 240–41
Czech Republic 133

da Silva, Lula 133
Dahl, J. 52, 59
Dahl-Jørgensen, C. 251, 275, 314
‹Daily Hassles Scale› 81
Dale, B.G. 282
Dalrymple, K.L. 52
DALYs 30
Damasio, A. 224
Danaher, B.G. 312
Daniels, K. 14
Davidson, M.J. 3
Davis, L.E. 4, 300
Day, Arla 67
Dayton VA Medical Center 156
de Geus, E.J.C. 120
de Los Angeles Moreno Uriegas, Maria 131
Deci, E. 40
*Decreased rates of informal and formal
    complaints to the VA Office of
    Resolution Management regarding*

Equal Employment Opportunities at a VA medical centre since it started participating in CREW **157**
DeJoy, D.M.   249
Delle Fave, A.   37–44, 322
Demerouti, E.   40, 186, 196
demographic characteristics of participants   190
Denison, D.R.   187
Denmark   240, 263
dental nurses   114
department managers   212, 288–93
departments   13, 113, 159, 170, 244, 281, 283, 287–91, 293–6, 300–302, 305–7, 312
   hierarchical   267
   hospital's   289
   HR   114, 303, 305, 307
   Kaizen representatives working at their respective   287
   university research   185
   various   287–9
depression   1, 3, 30, 52, 77, 109, 184, 188–9, 191–3, 210, 226
   doctor-diagnosed   188
   increasses in   68
   in job dissatisfaction   12
   in studies   188
DeRango, K.   68
Deschenaux, J.   141
development of interventions   11, 246
developments, organizational   4, 170, 233, 274–5, 290–91, 305
Dewa, C.S.   31
Dewe, P.   68, 85, 261
Dickinson, H.O.   101
DiClemente, C.C.   51
Diener, E.   44, 228, 231
disability-adjusted life years *see* DALYs
Dishman, R.K.   111
distress   8, 52, 54–5, 59, 183, 267, 273
   employee   51–2
   measure of   270
   psychological   2, 51, 55, 58, 184, 188, 240, 267, 270, 321
   reducing   61
   reporting less   59

   symptoms of   52, 59
Dollard, M.F.   7
Donaldson-Fielder, E.J.   58
Donohoe, Meghan   205, 266
Douglas, Roblin   91
Drago, R.W.   2
Druce, C.   135
Dubin's theory of segmentation   43
Ducki, A.   240, 249, 254
Durup, M.   43
Dussault, J.   244
Duxbury, L.   68
DVDs   95, 154
Dvir, T.   209, 212, 214
Dyrenforth, S.R.   147

Eastern Europe   41
Eckman, K.   174
'Ecocheques' (environmental or green vouchers)   130, 142
ECQOTESST survey   240
Edington, D.W.   112
Edmondson, A.   160
educational seminar topics
   'art of healthy cooking'   95
   'prehypertension'   95
   'reading nutrition labels'   95
EEOs   157
effects of participatory intervention   184
Egan, M.   52, 101, 269–70, 322
Ehrenreich, B.   228
eight-item civility scale (VHA process)   176
Eklöf, M.   313
emergency CPR   293
employee assistance programmes   8, 69
employee distress   51–2
employee participation   1, 14, 92, 95, 101–2, 150, 154–5, 183–5, 196, 251, 263–4, 274, 295
employee safety   166
employee surveys   153
employee well-being   3, 6, 12, 69–70, 76, 83, 85, 120, 185, 205–6, 208, 210, 217, 313
   enhanced   210
   outcomes   81
   reduced   67

sustaining 170
employees 4–8, 57–61, 70–78, 91–5, 110–15,
    120–22, 129–43, 183–5, 193–5,
    205–12, 214–18, 250–53, 262–76,
    283–5, 306–16
  attitudes 208–10, 216
  blue-collar 184
  call centre 58–9
  co-finance 141
  disciplining of 163
  disgruntled 2
  distressed 58
  engagement 5, 94, 101–2, 149, 152, 163
  front-line 176
  health of 7, 91, 208, 269
  healthy 4, 112
  helping 70, 111
  improving health of 13, 93, 183, 193,
    196, 282, 301, 308
  individual 10, 12, 120, 150, 153, 185,
    207, 307, 313
  Japanese white-collar 187
  low income 142
  outcomes 67, 85, 214–15
  participating 84, 95
  public health service 59
  punishing 206
  representatives 29, 170, 242, 246, 248–9
  scepticism of 171
  stress levels 68, 185, 266, 281
  well-being 196, 205, 218
  white-collar 187, 196
employees' and leaders' ratings of expectations
    concerning the tool at baseline **310**
employees' and leaders' ratings of opportunities
    for participation in the implementation
    at baseline **311**
employees' and leaders' ratings of their leader's
    attitude and information sharing at
    baseline **311**
'employer of choice' 270, 273, 276–7
employers 5, 24–5, 28, 30–33, 78, 91–5,
    97, 101–4, 119–20, 130–33, 136–7,
    139–43, 261–2, 276–7, 283
  and their desire of employees to add
    value 115
  mid-size 92

responsibilities to employees 142
  and wellness programmes 7, 92, 132
Engbers, L.J. 92, 110
enhancing development and performance
    through leadership 212–13
enhancing self-awareness of current
    leadership styles 214
Enköping, Sweden 283, 286–7, 294
Environment of Care Newsletter 154
environments 29, 160–61, 264, 285
  physical 23, 29
  present respective 271
  stressful 186
epidemiological studies 119
equal employment opportunities *see* EEOs
Eriksen, H.R. 111–12
Ernst & Young 137
ESF 301
estimated worksite wellness
    programme costs per
    participant (amortized on 60
    baseline participants) 103
EU 29–30, 143
EU-OSHA 29, 31, 281–2, 294
Eurest Lunchtime Report 135
Eurobarometer 2007 136
Europe 5, 32, 136, 138, 184
European Academy of Occupational Health
    Psychology 261
European Academy of Work and
    Organizational Psychology 5
European Agency for Health and Safety at
    Work *see* EU-OSHA
European Commission 142
European Council 137
European Ecolabel 142
European Journal of Work and Organizational
    Psychology 5
European Network for Workplace Health 120
European psychology and management
    conferences 5
European Social Fund *see* ESF
European Union 129
evaluation reports 97
evaluations 9–11, 13–14, 25, 33, 91–4,
    97–8, 100–104, 112, 177–8, 214–16,
    254, 261–2, 268–9, 288, 315–16

measures 98–100
    process and outcome 288
    programme 94, 97
events, stressful 231
Evers, K.E. 312
experimental design 98
exponential benefits 164
external mediation 247, 252

F/V servings 99
facilitators 152–3, 156–8, 162, 171–6, 180,
        248, 252, 323
    duties 162
    first-time 172
    recruitment of 158
    role of 172–3
    skilful 162
    training of 152, 154, 173
factors 1, 3, 5–6, 11, 32, 111, 117, 119–21,
        188, 225, 239–41, 243–7, 250–52,
        254–6, 273–4
    adverse 244
    complementary 137
    critical 183, 253, 273
    distinguishing 231
    environmental 193
    hindering 239, 241, 245, 250, 309
    non-occupational 12
    non-work 118
    obstructing 240
    organizational 6, 240, 282
    psychosocial 98, 299
    work-related 120
**factors which facilitated or hindered
    the intervention process 246**
Fagard, R.H. 101
Falkenberg, L.E. 113, 117, 119–20
families 1, 3, 7, 23, 25, 27, 31, 41–3, 67–8,
        76, 92–3, 133–4, 137, 139, 164
FAO 131, 135
FAQ sheets 154
fast food 96
fatigue 119, 188–9, 191–3, 285
    decreased 118–19
    measuring 189
    used to examine the health condition of
        participants 188

Fave, Antonella Delle 37
Fernet, C. 270
Ferraro, K.F. 307
Ferrie, J.E. 29
Fineman, S. 322
Finland 130, 142
Finnish hospital personnel 188
fiscal arguments 149
fiscal recovery 136
five-point Likert scale 73–4, 78
Flaxman, P.E. 58–60
focus groups 232, 244–5, 264
Folkman, S. 69, 71
Food at Work 131
food insecurity 135
FOOD programme 142n3
food safety 96
food services 131, 161, 163
food vouchers 133–4
Ford, G.A. 101
formal presentations 255
Forman, E.M. 52
Forsberg, Helena Hvitfeldt 281
four avenues of influence **26**
*four critical factors for the success of
        interventions and concrete indicators
        characterizing these factors* **273**
France 129, 135, 137, 139–40, 143
Francis, Lori 67–8, 70, 75, 84, 216
Franzini, L. 68
Fredrickson, Barbara 4, 226, 228
Freeman, H.E. 113
Frese, M. 67
Friedenreich, C.M. 109
Friedman, S.D. 2
Frone, M. 68
Frost, Robert 51
Fujigaki, Y. 188
Fukui, S. 188
*fulfilling workplace: the organization's role in
        achieving individual and organizational
        health* 1–2

Gable, S.L. 322
Gabriel, P. 30
Gagnon, C. 252
Gallup research 229

Galper, D. 109
Galt, V. 5
Ganster, D.C. 57, 69
Garcia, E. 15
Gélinas, A. 252
Gemson, D.H. 101
Giga, S.I. 9, 71, 261–3, 265–6
Gignac, S. 244
Gilbert-Ouimet, M. 240, 244, 322
Gilbreath, B. 205
GIROST researchers 241, 241n2
Glaser, S.R. 187, 194
global financial and mental health impact of work-related stress 30
Global Plan of Action on Workers' Health 23
Godshalk, V. 210
Goetzel, R.Z. 30, 91, 93, 101
Goldenhar, L.M. 10, 242, 251, 255
Goldman, N. 307
Gollwitzer, P.M. 40, 61
Gorin, A.A. 306
Gosta Esping-Andersen 140
Gottman, J.M. 227
Graham-Rowe, E. 111
Grandey, A.A. 206
Grawitch, M.J. 183–4
Great Britain 129–30, 140–41, 251
Greece 43, 132
Greenhaus, J.H. 2, 68
Griffiths, A. 9–10, 300
Grosch, J.W. 92
Grossman, P. 56
group facilitation 157, 171
group feedback 304, 313
group leaders 303, 307, 314
group meetings 151, 156, 163, 227, 303
group members 152–3, 158–9, 162, 185, 304
group participants 158
group processes 173–4, 194–5
group training 205, 211, 213
groups 38, 74, 80–84, 116, 151–5, 158–9, 161, 165, 171, 173–6, 226–7, 251, 289–90, 300–305, 312–15
  employer 93, 95, 98
  functioning 155, 180

participating 152, 155, 162
small 60, 313
Gundersen, M. 240–41, 248, 251
Guth, W.D. 314

Hackman, J.R. 4, 224, 322
Hagberg, J. 313
Hager, R. 300
Haidt, J. 322
Hain, C. 68, 70
Håkansson, M. 285, 294
Hall, E.M. 242
Hammer, T.H. 68, 169
Hansen, T.M. 195
Harris, R. 60
Harvard University study 7
Harvey, P. 206
Haskell, W.L. 110
Haslam, C. 109, 112
Hasle, P. 285
Hasson, Dan 240, 244, 254, 299–301, 306
Hasson, Henna 109, 281, 299
Hater, J.J. 209
Haworth, J.T. 40, 42
Hayes, S.C. 52–4, 56–60, 122
health behaviours 29, 97–8, 304
health benefits 98, 119, 283
health care 6, 61, 78, 169
  corporations 212
  costs 92, 115
  large public dental organizations 113
  primary 27
  professions 113
  units 177
health conditions 37, 183, 188
health education 92, 94, 100, 104
health effects 29, 112, 115
  employee 116
  positive 117
health improvements 112–13, 116
health information 29, 293
health interventions
  organizational-level occupational 281–2, 294, 314
  organizational occupational 285
  successful organizational-level occupational 281, 296

web-based occupational 300
health issues 5, 24, 132, 287, 292
    grouping 284
    important employee 102
    mental 53
    monitored 292
    occupational 293, 315–16
    veteran-related 148
health perspectives 289–91, 293, 295
‹health-promoting hospitals› 283
health promotion 6, 14, 25, 29, 94, 104,
        110, 130, 267, 281–4, 287–9, 291–4,
        301–4, 308
    hospital work 284, 288–90
    integration of 287–8, 291
    organizational 302
    programmes 300, 304
    programs 92
    strategies 314, 316
health representatives 283, 287–93
health risk appraisals 92, 96, 101–2
health risks 96, 104, 231
health services researchers 100–101
health status 23, 31, 98–100
    experience of 98
    improved perceived 112
    and productivity 97
healthcare 31, 147–8, 164, 284, 288, 307
    costs 31
    organizations 165, 283
    private sector organizations 166
    quality of 148
    services 31
    systems 147–8
    total budgets 132
    workplace 153, 166
HealthWatch 301
HealthWorks resources 93
healthy organizations 4, 6
healthy people 6, 92
healthy workplaces 4, 24–5, 28–9, 33, 51–2,
        57, 61–2, 130, 183–6, 193, 196, 223,
        225, 231, 321
    creation of 4, 51–2, 57, 183, 186, 193,
        196, 223
    initiatives 24–5
    practices 184

third wave behavioural-cognitive
        approach 51
Heaney, C. 184
Heany, C. 253
Heaphy, E. 227
Hein, C. 138
Hellmann, P.S. 185, 210
Hemp, P. 239
Henrich, J. 38
Hepburn, C.G. 69
Herbert, J.D. 52
Hesselink, J.K. 264
Heuvel, M. 41
Hewlett, S.A. 2
Hexagon Model of Psychological Flexibility **55**
Hibbard, J.H. 99
Hicks, J.A. 39
Higgins, C. 68
Hill, S. 40
Hinduism 43
Hines, P. 284
HIV/AIDS programmes 29
Hjelmquist, E. 195
HMOs 101, 103–4
Hodson, R. 70
Hoffman, D.A. 4
Hogan, R. 2
Holden, R.J. 285
Holloway, E. 3
Holte, K.A. 306
home services vouchers 139
Hope Hailey, V. 5
hospital centre 244, 253
    experimental 244
    short-term 241
hospital's work groups 284
Howell, J.M. 206, 209
HPA (hypothalamic-pituitary-adrenal)
        system 231
HR 5, 7–8, 78, 94, 113, 121, 142, 148, 156,
        253, 272–3, 276, 301, 305, 309
HR departments 114, 303, 305, 307
HRA *see* health risk appraisal
HRC 148–9
Hu, F.B. 68, 70
Hughes, M.C. 101
human excellence 224, 321–2

human resources   7, 113, 142, 156, 166,
      253, 276, 301 *see also* HR
   chief of   156
   CREW group   159
   departments   265, 272
   management practices   8
   managers   78, 94
human resources committee *see* HRC
human resources managers   78
Hungary   143
Hurley, J.   185
Hurrell, J.J.   6, 8, 67–9, 71, 85

Ibarra, H.   195
IBOPE   133
ICOH   24
ICOSI   133, 135
identifying risks   242
Idler, E.L.   306
IG *see* intervention groups
IHD   185
ill health   61, 112–13, 119, 283, 305, 307,
      314–15
   accidents caused by   30
   preventing employee   6
   work-related   110, 299
*illustration of connection between
      psychological safety and civility*   **160**
ILO   24, 129, 131, 134, 138
*impact of ABLE intervention on life
      satisfaction*   **82**
impact of ABLE on hassles   **83**
impact of ABLE on negative mood   **83**
impact of ABLE on perceived stress   **82**
implementation seminars   309–10
improvements in health   71, 96, 109,
      111–12, 118, 186, 231, 290
*improving organizational interventions for
      stress and well-being: addressing process
      and context*   1–2
income tax   136, 140
increasing workplace civility   3
India   6, 29, 41
individual employees   10, 12, 120, 150,
      153, 185, 207, 307, 313
individual health   6, 84, 314

habits   284
various   1
individual-level interventions   8–9, 111, 322
individuals, stressed   304
Industrial Revolution   223
*influences on health according to the WHO
      'Healthy Workplace' model*   **32**
information   37–9, 72, 77, 84–5, 96,
      99–100, 154, 159, 194–5, 197, 243–4,
      264–5, 276, 289, 291–3
   cost–benefit   85
   cultural   38–9
   demographic   187
   dissemination of   255, 264, 266, 323
   distributing of   283
   exchanges of   38–9, 42
   family status   73
   flows   190–91, 194–5, 197, 229
   generic   306
   health   29, 293
   levels of   310
   new   39, 227
   popular scientific   303
   qualitative   9
   sharing of   169
   supplying of   185
   symbolic   37
   technology   148, 301
   transfer of   251, 255
   transmission of   38, 266
   withholding of   206
information-sharing mechanisms   255
Inhofe, Kyle   156
Innovation Promotion Program *see* IPP
insecurity   29
INSPQ   264
Instituto Brasileiro de Opinião Pública e
      Estatística *see* IBOPE
integrated model of development   **45**
integration   6, 8, 39, 41, 92, 184, 214,
      281–2, 287–91, 293–6
   dynamic   46
   emerging   6
   of health promotion   287
   joint   294
   of monitoring   176

processes 41
of systems 282
work 288–9, 294
work and social 41
work–family 3
interactive web-based surveys 300
inter-correlations 79–81, 191
International Labour Organization *see* ILO
internet-based interventions 300, 312
intervention groups
  activities of 247, 252
  creation of 243, 251
  initial 84
  joint 250
  members of 247
  modes of consultation 247
intervention projects 3, 5, 7, 10, 113, 241,
    243–5, 248–53, 255–6, 261, 265, 272,
    277
  complex 180
  managed 170
  marketing of 267
intervention research
  longitudinal 68
  projects 239, 241–2, 245
  theoretically sound 322
interventions 3–5, 8–14, 58–61, 68–72,
    74, 109–12, 114–17, 169–70, 176–8,
    239–56, 261–77, 294–5, 300–305,
    310–12, 321–4
  activities 5, 10, 265, 272–3, 277, 295
  applied 3
  cognitive-behavioural 61
  complex 171, 266
  conducting of 301
  conflict 69–70
  development of 11, 246
  dietary 112
  early 23, 51–2, 68, 70
  effectiveness of 13, 23, 166, 256, 321
  eHealth 299–300
  emotion-focused 58
  evaluation of 14, 170
  evidence-based 299
  experimental leadership 212
  failed 13

groups 76, 112, 114, 184, 242–3, 247–8,
    251, 253–4, 289, 301
  implanting 262
  implementation of 6, 11, 241, 248–50,
    252, 255–6, 263, 272, 309
  individual-focused 8–9, 52, 111, 322
  internet-based 300, 312
  leadership 213–15
  maintenance of 273
  multi-level 307
  multi-modal 13
  OHS 112
  organizational-level 1, 5, 8–9, 11, 13,
    15, 52, 295, 307, 322
  organizational stress management 62
  outcomes 10, 13
  physical activity 111–13
  policy-level 33
  positive 12, 322–3
  primary 8, 69, 76
  processes 4, 9–10, 13, 157, 239, 243,
    245–6, 248–56, 262–4, 267, 272–3,
    275, 277, 323
  psychosocial 8, 61
  research 169
  secondary 76
  short-term 71
  small-scale 264
  strategic 272
  studies 169, 214, 240
  sustainable 4
  system-wide 172
  tertiary 52, 70
  tertiary stress prevention 52
  training 212–14
  units 288–9
  web-based 300, 312
  work-related 112
  work scheduling 8
  work-stress life-coaching 71
  workplace-based 8, 112
  worksite 52
  worksite health-promotion 115
  worksite stress management 58–9
IOHA 24
IPP 58

IPSOS 132
Isaac, F.W. 7
ischemic heart disease *see* IHD
Isen, Alice 226
Israel 264
Israel Defense Forces 212
IT-based communication 195
Italy 41, 143
Ivancevich, J.M. 52
Ivanov, I. 23
Ivanova, Lyudmila 133
Ivers, H. 322
Iwasaki, Y. 70

Jablonka, E. 37
Jacobson, G.H. 284
Japan 183–4, 188–9
Japanese corporations 187
Japanese government 183
Japanese white-collar employees 187
Japanese working people 188
Jason, L.A. 44
Jauvin, Nathalie 5, 239, 323
Jayawickreme, E. 321
Jex,m S.M. 69–70
Jimmieson, N.L. 310
Job Demand/Resources Models 179, 186
job insecurity 29
job stress 52, 67–70, 72, 76, 84–5, 112, 184, 210, 242, 264
    identifying of 71
    interventions 85
    lessening of 70
    reducing work-related 184
    and work–life balance 67–8
job stressors 184
Johnson, D. 75, 166, 226, 242, 266
The Joint Commission *see* TJC
Jordan, J. 262, 264–7, 271, 276
Jørgensen, T.H. 282
Journal of Applied Behavioral Science 5
Journal of Happiness Studies 322
Journal of Occupational Health Psychology 5
journals 5, 14, 166
    academic and practitioner 5, 14
    AI Practitioner 232

European Journal of Work and Organizational Psychology 5
Journal of Applied Behavioral Science 5
Journal of Happiness Studies 322
Journal of Occupational Health Psychology 5
peer-reviewed scientific 166
Work and Stress 5
Jreige, S. 69–70
Jung, D.I. 209

Kabat-Zinn, J. 56
Kahn, R.L. 68
Kaiser Foundation Health Plan 91
Kaiser Permanente Georgia *see* KPGA
Kaiser Permanente Georgia's Research Department 104
Kaiser Permanente Georgia's Wellness and Health Promotion Department and Marketing Department 104
Kaiser Permanente's HealthWorks programme 93
Kaizen 284–95
    board of directors 287, 289–90, 294
    central coordinator of 287
    concept of 287, 290
    definition of 284
    groups 289
    health project 291
    and the holding of joint meetings 287, 291–2
    note, used at department level as part of the Kaizen work at the hospital in Enköping to document the improvement work process **286**
    notes 287–90, 292–5
    representatives 287–90, 292
    system 289, 292
    work 284–7, 289, 292
Karanika-Murray, M. 13
Karasek, R.A. 230, 240, 242, 266
Karasek's work stress models 240
Karlsson, M.L. 120
Kashdan, T.B. 55
Kashihara, K. 119
Kasser, T 70
Katzenbach, J.R. 4

Kawai, N. 185
Kawakami, N. 188, 300
Kegan, R. 40
Keller, S.D. 209
Kelley, G.A. 101
Kelloway, Kevin 68–9, 71, 75, 84, 185, 205–8,
     210, 212–18, 266
Kenny, D.T. 68
King, L.,A. 39
Kinoshita, E. 189
Kivimäki, M. 30
Klemola, S. 285
Kline, C.E. 119
Kluckhohn, F. 38
Kobayashi, Y. 184, 240, 254
Koenig, H.G. 44
Koh, H. 92, 209
Kompier, M. 11, 85, 170–71, 177–8, 180,
     240, 248, 250–51, 255, 262, 265, 275,
     323
Koopman, C. 99
Koopmanschap, M.A. 115
Kortum, Evelyn 23–4, 265
Kossek, E. 67
Kotter,J. 5
Kottke, T.E. 93
Koukoulaki, T. 30
KPGA 91, 93–5, 98, 100–101, 103
   account managers 94
   manager of wellness 94
   nurses 96
   project teams 94, 97
   research programme 97, 100, 102
   research staff 94, 101–2
   teams 94, 97, 101–2
KPGA Institutional Review Board 93, 97
Kraemer, W.J. 110
Kramer, A.F. 119
Kristensen, T.S. 68, 249, 253
Kubzansky, L.D. 230
Kunnen, E.S. 40
Kuoppala, J. 92, 210
Kusy, M. 3
Kypri, K. 300

lab technologists 78
label reading 96

Lachman, M.E. 231
Lamb, M.J. 37
Lambourne, K. 119
LaMontagne, A.D. 9–11, 52, 267, 269, 307
Landsbergis, M P. 184, 282, 285, 294
Langan-Fox, J. 6
Latham, G.P. 70, 185, 189, 193
Lavie, C.J. 101
Lawler, E.R. 4
Lazarus, R.S. 69, 71
Le Blanc, P. 43
leadership 6–8, 15, 28, 32, 97, 148, 184–5,
     205–7, 210, 212–13, 216–17, 232,
     246, 249, 266
   abusive 206, 217
   appreciative 223, 232–3
   behaviours 209–18, 229
   development 213–14, 218, 301
   employer group's 101
   interventions 213–15
   managerial 185
   organizational 205
   passive 206, 210, 217
   positive 207, 232–3
   skills 206, 302
   styles 185, 205, 207–8, 211–14, 216,
      266
   training 205–6, 211–15, 218, 321
   transformational 211
leadership behaviours 214, 216, 218
   abusive 218
   bad 206
   changed 214
   current 214
   display transformational 206
   effective 211
   improved 211
   reducing negative 218
   teaching transformational 212
   transformational 214
leadership engagement **27**
Lean 281, 284–5, 287–9, 293–4
   concepts of 285
   implementation of 285
   literature 284
   management 281
   practices 285

strategies 285
LeanHealth Project *see* Lean
Lee, R.T. 210
Lefevre, J. 40
legislation 6, 14, 25, 31, 131, 141
   current European 281
   OHS 31
   passed by the UK, the Netherlands,
      Sweden and Norway 14
   supporting workplace interventions 14
   Swedish work environment 283
   transit benefits 141
   UK 13
leisure 43, 45, 92, 98–100, 142–3
   particular 43
   regular 100
   vigorous 101
     wellness programmes and 142
Leiter, Michael 2, 43, 67, 152, 159, 161,
      163–4, 166, 169, 171, 176, 178, 180,
      218, 323
Leka, S. 14, 32–3, 61–2
Lencioni, P. 4, 6
Lenneman, J. 113
LePine, J.A. 194
Lescarbeau, R. 249
Lester, D. 2
Leung, K. 38, 44
levels 1–3, 8, 10, 12, 14, 37, 40–43, 57–9,
      158–9, 174–6, 178, 180, 274–5, 307,
      315
   biological 39
   compensation 119
   decreased 116
   departmental 271
   employee's 58
   global 30, 33
   increasing 2, 6, 12, 109
   of management 11, 246, 248, 275
   measured 58
   mental health 57
   physical activity 101, 111
   recommended 99
   social 39, 43–4
   team 271
   tertiary 8–9, 52

unhealthy 303
Leventhal, J. 98
Levi, L. 32
Lewis, Sarah 223, 230–33, 323
Lewis-Payton, Rica 155
library assistants 78
life domains 42–5
   independent 43
   multiple 44–5
   relevant 44
   single 44
lifestyle 91, 95, 97–101, 308
   attitudes 101
   behaviours 12, 92
   changes 96, 99
   data 95, 102
   healthy 91, 97–8, 100, 265, 313
   problems 229
   recommendations 100
   sedentary 121
Liimatainen, M.R. 30
Likert, R.J. 4, 183
Likert scale, five-point 73–4, 78
Lim, D. 51, 209
Lincoln, A. 148
Lindfors, Petra 109
Lingley-Pottie, P. 76
link between poverty and sub-standard working
     conditions in developing countries **31**
Linley, P.A. 41, 229
Linnan, L. 92
Livingstone, H. 67–8, 71
Locke, E.A. 70, 185, 189, 193
Loehr, J. 2
Lohela, M. 266
long-term benefits 226
longitudinal studies 196
Longman, Philip 148
Lourijsen, E. 251
Lowe, G. 6, 193, 209
Lu, L. 43
Lubit, R.H. 3
Luce, C.B. 2
Lustria, M.L. 300, 306
Lutalo, M. 29
Luthans, F. 230

Lutz, Aleda E. 156

M&A 183, 187, 194
  activities 187
  companies undergoing 194
  putting participants under stress 187
Macdonald, W. 32
Macik-Frey, M. 169
Mackay, C. 14, 265
Macmillan, I.C. 314
Mactavish, J. 70
Maes, S. 242
Mafioso barons 229
maintenance of interventions 273
Maitland, A. 3
managed care organizations see MCOs
management conferences 5 see also
  conferences
management practices 8, 170, 270–71
management support 11, 323
Mangunkusumo, R.T. 300
Marcus, B.H. 109, 111
Marmot, M.G. 270
Marshall, A.L. 110–11, 300
Maslach, C. 15, 73, 176
Maslach Burnout Inventory – General Scale
  176
Maslach Burnout Inventory – General Survey
  73
Masley, S. 119
Maslow, A.H. 151
Massimini, M. 38–9
Matano, R. 300
Matlin, M.W. 227
Mattice, Tom 156
Mattke, S. 115
Mazzocato, P. 285
Mazzon, A. 133, 135
McAndrew, L.M. 98
McCracken, L.M. 52
McEachan, R. 112
McGrath, Patrick 67, 76, 195
McKee, M.C. 210
McNemar's test 100
MCOs 93, 99
MCS-12 measures variation in emotional
  functioning 99

mean values on self-rated health for employees
  at four departments during 18 months
  306
means and SDs of work demand and
  work control items 191
mechanisms explaining physical exercise
  effects on organizational output
  117–20
Medical Centers 154–6, 163
medicines 304, 308
Mein, G.K. 121
Member States 23–4, 30, 137, 140
Ménard, Julie 51
Mental Component Summary see MCS
mental content 53, 56
mental health levels 57
mental health problems 30, 33, 53,
  239–41, 256, 265–6
  interventions to prevent 239
  preventing of 239
mental models 9–10, 263–4
mergers and/or acquisitions see M&A
merging occupational health 281–96
  safety and health promotion with Lean
  281
Merrill, R.M. 101
Merritt, R.K. 110
Meshoulam, I. 213
Messenger, J.C. 190
Metcalfe, C. 29
Meterko, M. 153
Meyers, J.P. 70, 98, 322
Mickelson 135
Micklethwait, J 223
middle managers 10, 121, 275, 310, 314
Midpoint gathering 173
Mikkelsen, A. 71, 184, 196, 240–41, 248,
  251, 255, 282, 285
Milani, R. 101
Mileråd, E. 113
Miller, P. 109, 112
Mills, P.R. 113
mine disasters 2, 4
Ministries of Health 24, 183
Ministries of Labour 24
Ministries of Tourism (Switzerland, France,
  Hungary and Italy) 143

Ministry of Education, Culture, Sports,
    Science and Technology (Japan) 196
Ministry of Employment, Labour and Social
    Dialogue (Belgium) 140
Ministry of Quebec Public Service 267
MLQs 211, 214
Mohr, L.B. 98, 153
Moldasch, M. 285
Monday–Friday shifts 73
monitoring of work groups 300
Moore, S.C. 232
Moos, R.H. 70
Morocco 135
Morris, M.W. 38
motivated workgroups 155
motivation status quo measure see MSQ
Moyle, P. 210
MSQ 189
Mullarkey, S. 285
Mullen, J.E. 206, 210, 214–16
multifactor leadership questionnaires see
    MLQs
**multiple regression on engagement
    179**
Munir, F. 266
Muñoz Sastre, M.T. 44
Murillo, J. 133, 136–7
Murphy, L.R. 69

NAPOC 133
Natali, E. 33
National Association of Services for
    Individuals 140
National Center for Organization
    Development see NCOD
National Employment Office see ONEM
National Institute for Occupational Safety
    and Health see NIOSH
National Institute of Public Health see
    INSPQ
National Leadership Board see NLB
National Public Opinion Center see NAPOC
National Statistics Institute of Romania
    133
National Worksite Health Promotion Survey 92
NCOD 148–9, 153–8
    assessment of current progress 158

companions 154–8, 172–3
coordinators 154, 156, 162, 172–6
CREW logistics 154
internet sites 155
logistic and operations coordinators 154
staff members 154, 156, 158
Netherlands 6, 14, 139, 188, 250, 263
New Directions in Organizational Psychology
    and Behavioral Medicine 1–2
New Zealand Health Authority 166
Newson, L. 37
Nicholson, S. 115, 117
Nielsen, K. 9–10, 13, 70, 210, 240–41, 251,
    253–4, 261, 263–6, 269, 272–6, 281–2,
    285, 295, 312
NIOSH 5, 265, 282, 294
Nistico, H. 44
Niven, N. 75
NLB 148–9
Noblet, A. 10, 266, 269
non-communicable diseases 29, 32
non-participative climates 193
non-smoking policies 26, 253
Nordic countries 138
Norway 14, 263
Norwegian Postal Service 248
Nova Scotia 77
nurses 4, 41, 165, 283, 290–91
    dental 114
    registered 78, 98, 164
nutrition 68, 70, 96, 131–2, 134–5, 293, 300
    addressing issues concerning 95
    balanced 142
    counselling 7
    education programmes 72, 92
    good 96, 132, 134
    importance of 131–7
Nuwayhid, I.A. 31
Nyberg, A. 185, 266
Nyman, J.A. 115
Nytrø, K. 263, 276, 309–11

Öberg, T. 113
obesity 1–3, 92, 132
occupational health 5, 10, 24, 31, 33, 184,
    264–5, 277, 281–5, 287–91, 293–6,
    299–303, 305–6, 308, 314–16

care providers   302–3, 305, 308, 315
  identifying of   293
  improvements in   301
  integration of   281, 289, 294
  interventions   10, 277, 282, 295, 301,
    310, 316
  practitioners   294
  programmes   312
  psychology   5
  research   31, 187
  surveys   299
  traditional   24
occupational health and safety see OHS
occupational safety   132, 282
occupational stress   69, 184, 261–3, 322
OCS   187
OD   4, 170–71, 231, 233, 274–5, 290–91,
    302, 305
OECD   129, 139
Oenema, A.   300
Oetzel, J.G.   70
Offermann, L.R.   185, 210
Office of Resolution Management   157
OHS   24, 112, 264, 287
  interventions   112
  legislation   31
Oklahoma City   156
Oldham, G.R.   4
Oldman, G.R.   224
ONEM   140
oppressive work cultures   226
Oram, R.   141
organization management   310
organizational changes   6–7, 171
organizational climate
  assessments   149
  favourable nature of   193
  measurement of   187–8, 196
organizational commitment   70, 93, 102,
    118, 120, 178, 185, 211, 230, 312
organizational context   9, 13, 85, 110, 122,
    147, 263
organizational culture (see also culture)   4–5,
    9, 26, 155, 185, 187, 233, 246, 262
  changing of   150
  conservative   241
  oppressive   226

organizational culture scale see OCS
organizational development see OD
organizational effectiveness   6, 160, 183,
    185, 193, 196, 208–9, 262
organizational environments   166
organizational health   1–2, 4, 6, 9, 11, 24,
    148–50, 164
  initiatives   157, 164
  management programme   28
organizational initiatives   2, 116, 120, 149,
    164
organizational interventions   4, 8–9, 14,
    62, 68, 169, 253–5, 262, 266–7, 307,
    315, 323
  complex   266
  evaluating of   170
  preventive   266
  successful   4, 180, 255
organizational leaders   7, 147, 180, 205,
    218, 256, 303, 314
  local   157
  risk investing time, energy and
    credibility into activities   170
organizational learning   172, 174
organizational-level interventions   1, 5,
    8–9, 11, 13, 15, 52, 295, 307, 322
organizational outcomes   161, 206
organizational outputs   109, 113, 117, 120
organizational performances   6, 13–14,
    185, 189, 321
  and the effects of toxic organizations
    on   2–3
  enhancing of   184
  and organizational intervention
    covering well-being and successful
    performance   4
organizational practices   84, 93, 225
organizational processes   209, 282, 294
organizational psychology   9
organizational scholarships   224
organizational well-being   3, 12, 315
organizations   6–9, 84–5, 112–14, 120–22,
    147–9, 172–8, 194–7, 209–18, 223–7,
    230–33, 245–9, 251–6, 267–71, 281–3,
    312–16
  besieged   226
  chaotic   148

complex 171
contracting with external consultants
175
downsizing of 185
employer 32
flourishing 223, 225
governmental 59
health maintenance 92–3, 100
healthcare 165, 283
healthy 4–6, 164, 183, 295
international 129
large 3, 223, 268, 275
local government 59
managed care 93
national prevention 265
non-governmental 24
non-profit 27
participant 277
public sector 169
relevant intergovernmental 24
well-performing healthcare 166
origins of acceptance commitment therapy
52–3
Osatuke, Katerine 147, 149, 152–4, 159,
161, 163, 169, 171–2, 176, 178, 180,
323
O'Shea, R.M. 113
Oskamp, S. 40
Öst, L. 52
Ostry, A. 29
Oxenburgh, M. 30
Oxfam 135
Ozeki, C. 67
Ozminkowski, R.J. 91, 115

PANAS 74, 78
Panter, C. 28
paramedics 78
Parasuraman, S. 67
Parker, S. 43, 188
Parks, K.M. 113, 229–30
Parsonage, M. 239
participative approaches 170, 174, 323
participative climate 183, 185–9, 191–4,
196–7
buffers 193
dimension of 193–4

enhancing of 193, 196
expected 186
favourable perception of 186
higher 184, 191
items of 188, 197
subscale of 189–91
used as an individual-level variable 188
participative culture 184–5
participative leadership 185
participatory intervention 184, 245
approaches to 184
effects of 184
research projects 241
studies 184
passive leadership 206, 210, 217
Passmore, J. 8
PAT 133, 135
Pate, R.R. 110
Patient Protection and Affordable Care Act
92
Pauly, M. 113, 115, 117
PCS 99
Pearson, C.M. 150, 152
Pearson correlations of change in lifestyle
and biometric measures 100
Peiró, J.M. 69
Pelletier, K.R. 92, 112–13
Perez, A.P. 101
Pernambuco (Brazil) 135
personal health resources 23, 25
personality 119, 217, 229
based explanations 151
styles 159
traits 229
types 159
pertinent sources of information
and main results of the
intervention effects 244
Peters, T.J. 4
Peterson, L.R. 228–30
Pettersen, J. 284
physical activity 70, 91–2, 96, 98–100,
109–12, 116–17, 130, 142, 300, 303
daily 110
effects of 111, 117
implementing of 110
increasing 300, 308

interventions 111–13
levels 101, 111
programmes 92, 113
self-rated 112
short bouts of 111
workplace-based 111–13
physical component summary *see* PCS
physical exercise 109–11, 113–14, 116–22,
    288, 290
aerobic 110
condition 114, 116
effects of 109, 115, 117, 119
implementing of 109
intervention groups 116
intervention projects 113
interventions 109–22
levels 116
mandatory 114, 116
regular 119, 121
workplace-based 109, 120, 122
physical health 1, 3, 33, 119, 183, 230,
    242, 269
decreased 68
diminished 2
improving 321
of individuals 4
Piccolo, R.F. 208–9
Pillai, R. 209
pilot projects 252, 267–8, 270, 276
Pinder, C.C. 189
Pinto, M.B. 195
Ployhart, R.E. 209
Podsakoff, N.P. 209–10
Pohjonen, T. 111–12
Poksinska, B. 284
politicians 31
Pollard, T.M. 29
Pollyanna argument 227–8
Pond, S.B. 71
POS 321
positive and negative affectivity scale *see*
    PANAS
positive psychology 150, 223–4, 226, 228,
    232–3, 321–3
post-industrial society 38, 42–3
post-traumatics stress disorder *see* PTSD
PowerPoint presentations 154

pre-tax salary deductions 141
preferential taxation 131
'Prehypertension' (educational seminar
    subject) 95
Pressler, A. 312
prevention of stress 129, 324
prevention strategies 265
preventive interventions 240, 243, 253,
    255, 262–3, 267
implementation of 239, 241
success of 256, 266
Pribram, K.H. 39
PricewaterhouseCoopers 132
PRIMA-EF European-wide stakeholder
    survey 32
primary interventions 8, 69, 76
Probst, T.M. 185
process and contextual factors 9, 245, 248,
    256
process and outcome evaluations 288
process evaluation model 9
processes, appropriation of 247
Prochaska, J.O. 51
production 121, 131, 136–7, 226, 277,
    282–4, 287, 291
improvements 281, 290–91, 293
lean 294
levels of 115
perspectives 284
sustainable 142
systems 282, 294
profile of civility scores before and after CREW
    177
programme activities 96, 101–2
contract 103
directed 94
formal 95
justified orienting worksite wellness 97
selected 95
tailoring to 96, 101–2
worksite wellness 91, 95
programme theory of the web-based
    tool 308
programmes 29, 61–2, 67–8, 75–8, 83–5,
    92, 94–5, 101, 132–3, 149–50, 153–5,
    162–3, 170, 300–301, 310–13
activities 93–5, 101–2, 104

designs of 91, 93–4, 101, 103
implementation of 103, 312
participation in 91, 93–5
stress management 8, 52, 69, 71, 92
theory of 307–8, 316
vouchers (for employees) 129
web-based 300, 305, 313, 316
project management 175–6, 266, 309, 314
project teams 195, 270, 272–3, 276
projects 3, 170, 175, 208, 217, 241–5,
248–50, 252–6, 263–7, 272–6, 283,
287–96, 307, 309–10, 312–13
competing 263
in-house 307
intervention 3, 5, 7, 10, 113, 241, 243–5,
248–53, 255–6, 261, 265, 272, 277
long-term 169
organizational 173
particular 309
pilot 252, 267–8, 270, 276
research 171, 243–4, 248–9, 252–5, 284,
288
stress intervention 265, 277
time-limited 282
Pronk, N.P. 93, 112–13
providers 92, 287
external 133
health-care 5
occupational healthcare 303, 305, 308,
315
service 131, 138
Titre-services 140
PsyCap states 230
psychological distress 2, 51, 55, 58, 184,
188, 240, 267, 270, 321
psychological flexibility
development of 54, 61
Hexagon Model of **55**
levels of 58
model 57, 60
processes 58
role of 57
*Psychological Health and Safety in the
Workplace* (Canada) 265
psychological selection results 39
psychological well-being 188, 210, 230–31,
267

Psychologically Healthy Workplace Program
(APA) 184
psychology 52, 173, 322
psychology organizational scholarship *see*
POS
psychosocial risk management 265
psychosocial risks 31, 33, 52, 262, 264–5,
267–71, 275, 321–2, 324
diagnosis of 262–3
including 24
managing of 6
psychosocial working environment 29
PTSD 2, 148, 224
public health 24, 109–10, 264
agendas 24
approach 24, 32
disorders 302, 308
publications 166, 244, 322
Environment of Care Newsletter 154
European Network for Workplace Health
120
Expert Consultation on Worker's Feeding
131
Food at Work 131
*The Fulfilling Workplace: The Organization's
Role in Achieving Individual and
Organizational Health* 1–2
*Improving Organizational Interventions
for Stress and Well-being: Addressing
Process and Context* 1–2
*New Directions in Organizational Psychology
and Behavioral Medicine* 1–2
Risky Business 1–2
*WHO Global Plan of Action on Workers'
Health* 23
Puetz, T.W. 119
Purvanova, R.K. 209
PWE ratings 308

QLW 262
quality of life at work *see* QLW
Quebec 239–41, 244, 261–2, 265, 267–70
government 262
survey 240
workers 270
QWL 268, 270, 276
approach 268, 275–6

intervention 269
project 272

Rabat 135
Racette, S.B. 101
Radar, Kae 232
Ramadan 135
Randall, R. 9–10, 13, 263–4, 266, 269,
    272–6, 281–2, 305, 310, 314, 322
**range, means, SDs and correlations of
    the variables used in this study
    (N=625) 192**
**range, number of items, means, SDs
    and means per items of the
    subscale of participative climate
    (N=625) 191**
Ranta, R. 111–12
Rantanen, J. 31
Rappaport, J. 193
Ratzan, S.C. 7
'Reading Nutrition Labels' (educational
    seminar subject) 95
recollection of taxes 136
recruiting facilitators 162
reducing cardiovascular risks 91
reducing stress 71, 285, 308
Reed, J. 117, 230
registered nurses 78, 98, 164
relational frame theory *see* RFT
relational frames 53–4
*relationship of civility and psychological
    safety to an important organizational
    outcome: Costs incurred from sick leave
    usage* **161**
Renaudin, Nathalie 129
repetitive strain injuries 224
reports 85, 173
    advancement 275
    aggregated 272
    Aviva Family Finances Report 139
    case 113–16, 301
    Eurest Lunchtime Report 135
Republic of Mexico 131
research groups 77, 103, 284, 307
research projects 171, 243–4, 248–9, 252–5,
    284, 288
    first partnership-based intervention 241

funded 288
including intervention 5
rigorous 250
research teams 72, 97, 180, 240–41, 255
respondents 7, 73–4, 78, 81, 190, 271–2,
    300, 303, 310, 312, 314–16
Reynolds, S. 3, 281, 321
RFT 53
Rheinberg, F. 40
Rhodes, S. 300
Richard, L. 249
Richardson, K.M. 52, 61–2, 281
Richerson, P.J. 37, 39
Rinfret, N. 270
RIPOST researchers 241n3, 241
risk assessment 8, 170, 177–8, 240, 242,
    246, 250, 262, 264, 270–71, 276,
    299–300, 305
    annual 288–9
    phases of 252, 274, 277
    results 271
    surveys 301
    team-based 271
    traditional 305
risk factors 13, 96, 186, 188, 261, 264–5,
    271
    cardiovascular 92, 100, 102
    occupational 31–2
    organizational 244
    potential 7
risk prevention 14, 111, 305
risks 9, 14, 56, 59, 96, 98, 110, 130, 135,
    152, 264–5, 269, 282–3, 300, 316
    associated health-related 2
    chemical 265
    chronic disease 91
    higher 8
    identifying of 268, 275
    increased 132, 307
    interpersonal 160
    potential 305
    reduced 112
Risky Business 1–2
Robinson, Brandi 91
Roblin, Douglas W. 91, 101
Robson, L.S. 240–41
Rodin, J. 230

Roemer, L. 52
Rogers, S.J. 68
Rojek, C. 43
Rokeach, M. 38
role stressors 6
Romania 133
Rooney, J. 210
Rosenberg, E.L. 227
Ross, L. 151
Rossi, P.H. 98, 272
Rothstein, H.R. 52, 61–2, 281
Rowland, Jeanette 156
'Rudeness Rationales Scale' 176
Ruiz, F.J. 52
Russell, Margaret A. 156
Russia 6
Ryan, R. 40, 185

safety 8, 24–5, 27, 29, 31–2, 132, 151, 163,
    210, 215, 240, 264–5, 283–4, 287–91,
    294
  behaviours 210
  committees 28
  compliance 210
  consciousness 210
  employee 166
  food 96
  inspections 283, 289
  occupational 132, 282
  provisions 29
  psychological 152, 160–62
  workers 207
Sainsbury Centre for Mental Health 261
Saksvik, P.O. 12, 184, 240, 251, 263, 265,
    275, 309–11, 313–14
Salanova, M. 40
Saltzstein, A.L. 196
*sample size and effect estimates for achieving
    statistical significance* **100**
San Francisco Bay Area Rapid Transit
    District *see* BART
Saravanan, B. 41
Sashkin, M. 4
Sauter, S.L. 68, 70
Scandinavia 5–6
Schabracq, M.J. 2
Scharf, T. 169

Scharmer, C.O. 174
Schat, A. 206
Schaufeli, W. 43
Schein, E.H. 4, 152
Scheinder, F. 136
Schier, M.F. 230
Schlegel, R.J. 39
Schmidt, J. 95
Schön, D.A. 176
Schultz, A.B. 112
Schwartz, T. 2, 38
SDs 79–81, 100, 190–92
Seaverson, E.L. 95
Sebelius, K. 92
Second World War 130, 148
Seki, Keiko Sakakibara 183
Seligman, Martin 4, 150, 224–5, 228, 231,
    322
senior management 7, 10, 13, 172, 175–6,
    245–6, 248, 256, 265, 269, 272,
    275–6, 310
  commitment of 273, 275–6
  involvement 7
  participating organization's 172
  QWL 273
service providers 131, 138
shared management 240, 246, 249, 256
sickness absence 109, 112–22, 289 *see also*
    absence
  days 116, 119
  decreasing of 113, 116, 119–20
  frequency of 119
  levels 119
  measures of 115
  short-term 114, 119
Siegrist's work stress models 240
Siham Ali 135
SMI *see* stress management interventions
smoking 1, 30
  behaviours 231
  and cancers 30
  cessation of 26, 164, 300
social benefits 134, 142
social support 60, 68, 71, 148, 230–31,
    242, 244, 266, 321
Society for Occupational Health Psychology
    5, 235

standard deviations *see* SDs

standardized regression coefficients of work factors on fatigue, depression and work motivation (N=625) 192

Stanford presenteeism scale 99

Stang, D.J. 227

Stansfeld, S.A. 242

Steelman, L.A. 113

Steger, M.F. 39

Stenfors-Hayes, Terese 281

Stephenson, J. 109

steps of an organizational stress intervention **268**

Sternfeld, B. 101, 111

Stevens, Sonya 67

Stewart, W.F. 30, 112

Stockholm University 113

Stone, M.H. 306

Straightforward Incivility Scale 176

strain 69–71, 73–4, 79, 81, 206–7, 321–2

    decreasing of 72, 76

    increasing of 68

    occupational 58

    outcomes 68–9

    reduction methods 77

    reduction techniques 71

strategies 4–5, 24, 38, 41, 54, 77, 92–3, 103, 110, 148, 263, 266–7, 285, 304, 309

    active health promotion 314, 316

    adaptive 45

    avoidance 54

    catalogued using various qualitative research 241

    core 232

    cutting-edge 148

    delivery 76

    job-crafting 295

    lean 285

    long-term 230

    marketing 155

    positive reinforcement 304

    prevention 265

    recruitment 132

    short-term 77

    survival 37

    sustainability 25

traffic mitigation 141

    wellness 267

    working environment 285

Straus, S.G. 195

Strecher, V.J. 300

strengths 134, 150, 195, 223, 229–33, 304

    character 228–30

    consistent 225

    developing new 230

    innate 229

    personal 229

    perspective 229

    well-honed 229

stress 1–2, 4–6, 12–13, 30–31, 52, 55–7, 59–62, 67–72, 74–5, 230–31, 261–3, 265–7, 301–4, 314, 321

    categorizing of 69

    components 69

    creating of 322

    evaluating 11

    exacerbating 67

    experience of 8, 12, 52, 231

    increase of 69–70, 205, 207

    job-related 186, 188

    lessening of 293

    levels of 9, 43

    long-term 303

    mental 26

    monitoring of 308

    negative consequences of 71, 84

    normal 39

    occupational 69, 184, 261–3, 322

    organization's 308

    perceived 69, 71, 76, 80–82, 84

    preventing of 129, 324

    psychosocial 269

    reducing 70–72, 267

    short-term 314

    work-related 30–3, 61, 283, 285, 321

stress interventions 12, 75, 321, 323–4

    organizational 71, 261–2, 268

    post-traumatic 8

    programmes 75

    projects 265, 277

    studies 264

stress management 59, 61, 68–70, 301–4

    active 308

interventions 52, 58–9, 62
problem-focused 58
programmes 8, 52, 69, 71, 92
training 8, 59, 71
stress prevention 70, 137, 262, 266
guidelines 265
models of occupational 261–2
traditional occupational 321
stress reactions 52, 70
stress reduction 4, 9, 69, 300, 323–4
stressed individuals 304
stressors 11, 52, 57, 68–72, 77, 194
hindrance 194
job 184
potential 70
reducing 69, 71
role 6
work 8, 58
work-related 52
workplace 8, 12, 26
Stroedbeck, F. 38
Ström, L. 312
Stubbe, J.H. 120
sub-interventions 295
substance abuse counsellors 59 see also
counsellors
substance abusers 59
Sui, X. 109
Sullivan, S.E. 3
Sultan-Taïeb, H. 239
supervisors 8, 71, 77, 151, 155, 157, 162–3,
183–5, 191, 193–4, 196–7, 242, 244,
264
abusive 6, 206
group's 157
immediate 175
management 190, 194
supportive 69
work 153
support 5–8, 14, 26, 29, 112–13, 130–31,
148–9, 162–3, 174–5, 244–5, 247–9,
251–6, 273–6, 287–8, 313–15
continuous 152
emotional 180
explicit 175
hands-on 313
logistic 172

managerial 248, 252
organizational 152, 314
requested 284
services 139, 165
strong 262
supervisor 148
unanimous 252
surveys 102, 187, 302–4, 307, 312–14
All Employee Survey 153, 178, 194
Behavioral Risk Factor Surveillance
Survey 99–101
Buffett National Wellness Survey 7
ECQOTESST survey 240
National Survey of Health Conditions of
Working People 183
National Worksite Health Promotion
Survey 92
PRIMA-EF European-wide stakeholder
survey 32
Svensson, L. 288
Sweden 14, 30, 110, 113, 130, 140, 185,
281, 283
Swedish companies 283
Swedish hospitals 296
Swedish National Institute of Public Health
110
Swedish tax authorities 142
Swedish work environment legislation 283
Switzerland 143
systems 29, 92, 129, 131, 133, 137–8,
140–41, 143, 152, 231–2, 282, 284,
287, 291, 303
adaptive 304
bargaining 249
compensatory 119
comprehensive 148
cultural 45
early warning 305
educational 38
health delivery 6, 92–3
healthy cardiovascular 164
immune 228
institutionalized normative 38
integrated 282–3, 290
legal 38
meal vouchers 132
multiple 284

organizational   281–2
performance evaluation   194
physician-hospital   92
production   282, 294
quality improvement   283–4, 287, 294–5
redesigning of   148, 164
social   324
structured   176
sympathetic nervous   231
welfare   30–31

Takahashi, M.   195
Takala, J.   30
Taos Institute   232
tax exemptions   131, 141–3
tax-free parking benefits   141
tax incentives   136
tax revenue   131, 143
taxes   139–41, 143
    complementary   137
    ensuring a recollection of   136
taxonomies, development of   229
Taylor, Fredrick   71, 223, 230–31
teams   14, 150, 156, 177, 197, 225, 227,
        267–8, 270–77, 290, 323
    cohesive top leadership   4
    coordinating/steering   11
    cross-functional   195
    employer's leadership   94–5, 97, 100–102
    executive   162, 175
    high-functioning   156
    highest performing   227
    research   72, 97, 180, 240–41, 255
    steering/coordinating   11
technologists   78
Teed, M.   210
Tepper, B.J.   2, 206
Tepperman, L.   78
Terra, N.   71
Terry, P.E.   93, 101
Tesluk, P.E.   185
Tetrick, L.   4–6
thank you note from the second example   165
Theorell, T.   29, 71, 240, 242, 266
theoretical perspectives   37, 322
therapies
    acceptance and commitment   52

acceptance and commitment therapy
    52–3
Thomas, S.   69
Thomason, J.A.   71
time management   70–72, 303–4
Tims, M.   15
Ting-Toomey, S.   70
Titre-services providers   140
Titze, S.   111
TJC   154, 166
Tompa, E.   112, 115
Tomporowski, P.   119
total quality management see TQM
Totterdell, P.   68
Towers Watson (HR firm)   7
toxic organizational environments   3
Toyota Production Systems   284
TQM   15
training interventions   212–14
transformational leadership   205, 208
    behaviours   205, 208–16
    concepts of   213
    development of   213
    supports a sense of workplace spirituality
        210
    training   205, 211–14, 216, 218
    use of   209
Trapnell, P.D.   70
treatment for leadership   212
Trosten-Bloom, Amanda   232
Trudel, L.   240
Tsutsumi, A.   184, 240–41
Tudorel, A.   136
Turkey   137
turnover   1, 4, 7, 67, 163, 248, 251, 254
    employee   30, 206, 239
    high   247, 250, 253–4
    lower   184
    natural   251
Turpin, R.S.   99
Tvedt, S.D.   12
Tyler, P.   71
types of organizational leadership
    styles   208

Uegaki, K.   112–13, 115
Union Gas, Canada   5

United Nations 33
United Nations Human Rights Council 33
United States 2, 92, 130, 135, 141, 147, 165, 184–5
  adults 101
  Airforce 166
  and China 184
  Department of Veterans Affairs 147–8
  government 141
  government agencies 166
  healthcare agencies 166
  and HSE 265
  population 101
  Veterans Health Administration 147
Universal Employment Services Vouchers 129
universal interventions 8
Uruguay 130
UWES 178, 180

van der Doef, M. 242
van der Klink, J.J. 184
van Dierendonck, D. 210
van Dongen, J.M. 112
van Gelder, M.M.H.J. 300
van Woerkom, M. 322
Varela, F. 174
VAT 131, 136
VCRs 95
Vella-Brodrick, D. 41
Venkatesh, H. 29
Veterans Health Administration *see* VHA
Vézina, Michel 239–40, 244, 264, 270
VHA 147–9, 152–3, 161, 164–5, 169, 171–2, 175, 178, 180
  annual 153
  eight-item civility scale 176
  employee satisfaction 149
  employees 149, 162
  facilities 149
  healthcare services result 148
  leaders 148–9
  medical 154, 163
  organizations 149
  processes 176
  systems 176
  work environment 149

workforce 148–9
Villaume, Karin 299
Violanti, J.M. 71
viraaga, concept of 43
Virtanen, M. 29
Viswesvaran, C. 68
Vivona-Vaughan, E. 282
von Thiele Schwarz 116
voucher systems 130, 133, 136–7, 142–3
vouchers 129–33, 135–6, 139, 141–3
  childcare 129, 138–9
  commuting 141
  green 130, 141–2
  holiday 130, 142–3
  meal 130–31, 133–4, 142
  service 138, 140, 143
  transport 130, 141–2

Wager, N. 68
Wagner, G.R. 99
Walley, P. 284–5
Wanjek, Christopher 131–2, 134–5
Wantland, D.J. 300
Warr, P.B. 70
Warren, N. 153, 322
Waterman, R.H. 4
Watson, Irene 7, 74, 78, 156
Weaver, S.L. 231
web-based, surveys 300
web-based programmes 300, 305, 313, 316
web-based risk assessment surveys 299
web-based surveys 300
web-based tools 300–302, 308, 312, 315–16
Weber, T. 285
websites 7, 60, 77, 312
Weick, K.E. 310
Weiner, B.J. 93, 101
well-being 4–6, 24–5, 39–40, 42–5, 82–5, 129–32, 184, 205–6, 230, 265–7, 277, 302–6, 308, 312–16, 321–2
  decreasing 2, 68
  emotional 109
  human 227
  improving 265
  individual 3–4, 41, 206, 301
  interventions 11
  organizational level 300

perceived   41
personal   215
psychological   188, 210, 230–31, 267
self-rated   74
spiritual   210
systemic   148
wellness programmes   7, 29, 92, 95
comprehensive   29, 103
effective   7
endorsed worksite   92
germane to worksite health promotion
92
implementation of worksite   92–3
individually-tailored worksite   96
multifactorial worksite   92, 100
pilot-worksite   93–4, 103
post-worksite   100
six-month worksite   103
wellness strategies   267
Westerlund, H.   29
Weyman, A.   14
Whatmore, L.C.   12
Whelton, S.P.   101
white-collar workers   40, 111, 186
Whitney, Diana   232
WHO   23, 39, 130, 135, 184, 283
Global Plan of Action on Workers' Health
23
healthy workplace model and framework
28
member states   23–4, 30, 137, 140
Whysall, Z.   240–41, 248–50, 253, 255
Wilder, D.A.   304
Wilkinson, B.   282, 285
Williams, J.H.   68, 70, 101
Wilson, K.G.   92, 111
Wittenberg-Cox, A.   3
Wolever, R.Q.   71
Womack, J.P.   284
Work and Stress   5
work control   188–93, 198
work engagement   15, 169–70, 178–80, 304
work group leaders   303, 310, 314–15
work groups   290–91, 300–303, 305, 307–9,
312–13, 315 *see also* workplaces
hospital's   284
monitoring of   300

work motivation   183, 185, 187, 189,
191–4, 196, 271
work-related stress   30–33, 61, 283, 285, 321
mental health impact of   30
and psychosocial workplace hazards   23
work reorganization intervention   58
work stress   2, 52, 70, 267
and conflict interventions   69
work stressors   8, 58
worker compensation claims   74, 85
Worker Food Programme   133 *see also*
Brazilian World Food Programme
workers   23–32, 43, 52, 59–61, 130–32,
134–5, 137, 140–41, 143, 240, 243–4,
247–52, 254–6, 266, 321–3
assembly-line   40
canteen   253
comprehensive   11
food service   164
health of   23, 25, 32–3, 68
individual   45
low income   141
low-skilled   138
medically-approved   135
participation of   251, 253
social   59, 61
stigmatizing of   322
temporary   113
well-being of   25, 27
white-collar   40, 111, 186
workgroups   149–50, 152–5, 157–8, 162–3,
165, 169, 172–6, 180
conflicts   153
local   153
meetings   152
members   151–3, 157–8, 169
motivated   155
participating   157–8, 160, 162, 178
selection guidelines   154–5, 158
working environments   29, 160–61, 264,
285
physical   23, 29
present respective   271
stressful   186
*work–life balance and childcare: the vicious
circle* **138**
work–life conflicts   67–8, 73–4

workplace health 25, 27, 120
   interventions 111
   promotion initiatives 109
workplace accidents 135–6
workplace civility 3, 151, 174, 176–7, 180
   high 163
   improving 149
   increasing 3
   rating of 157
workplace culture 25, 173
workplace, health promotion 25
workplace injuries 1, 205, 207
workplace interventions 9, 12, 14, 69, 84, 240
   analysis of 170
   preventive 248
workplace leaders 163, 284, 299, 313, 315
workplace levels 115–16, 118, 284
workplace promotion of physical activity on health, effects of 111–12
workplace stress 2, 14, 52
   identified 6
   reduction programme, based 71
workplace stressors 8, 12, 26
workplaces, healthy 4, 24–5, 28–9, 33, 51–2, 57, 61–2, 130, 183–6, 193, 196, 223, 225, 231, 321
workshop intervention programme 72
workshop intervention results 74
workshops 59–60, 62, 68, 72, 75–6, 84, 96, 113, 288
   ACT 58–9
   classroom 212
   half-day 60
   health-related 75
   interactive 72

   introductory 290
   nine-hour 58
   one-day 212
   short 61
worksite health interventions 112–13
worksite interventions 52
worksite stress, management interventions 58–9
worksite wellness programme content topics 96
worksite wellness programme for reducing cardiovascular risks 91
worksite wellness programmes 91–8, 100–104
worksites 58, 91–2, 111
World Health Assembly *see* WHO
Worrall, L. 30, 185
Wrencher, Cliff 155
Wu, C. 44–5

Xavier University 166

Yamazaki, Y. 189, 194
Yancey, A. 110–11
Yarker, S. 266, 270–71
Ylipaavalniemi, J. 186, 188, 193
Youngstedt, S.D. 119

Zapf, D. 111
Zero Hunger programme 133
Zettle, R. 52
Zimmerman, M.A. 193
Zlobina, A. 41
Zwerling, C. 240
Zwetsloot, G.I.J.M. 282

If you have found this book useful you may be interested in other titles from Gower

Corporate Reputation
Managing Opportunities and Threats
Edited by Ronald J. Burke, Graeme Martin and Cary L. Cooper
Hardback: 978-0-566-09205-3
e-book PDF: 978-1-4094-2327-0
e-book ePUB: 978-1-4094-6039-8

Occupational Health and Safety
Edited by Ronald J. Burke, Sharon Clarke and Cary L. Cooper
Hardback: 978-0-566-08983-1
e-book PDF: 978-1-4094-3207-4
e-book ePUB: 978-1-4094-8663-3

The Fulfilling Workplace
The Organization's Role in Achieving
Individual and Organizational Health
Edited by Ronald J. Burke and Cary L. Cooper
Hardback: 978-1-4094-2776-6
e-book PDF: 978-1-4094-2777-3
e-book ePUB: 978-1-4094-6045-9

Human Frailties
Wrong Choices on the Drive to Success
Edited by Ronald J. Burke, Suzy Fox and Cary L. Cooper
Hardback: 978-1-4094-4585-2
e-book PDF: 978-1-4094-4586-9
e-book ePUB: 978-1-4724-0242-4

GOWER